Coin-Turning, Random Walks
and
Inhomogeneous Markov Chains

Coin-Turning, Random Walks
and
Inhomogeneous Markov Chains

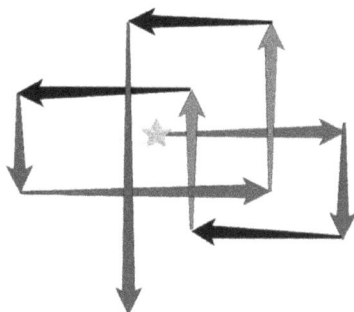

János Engländer
University of Colorado Boulder, USA

Stanislav Volkov
Lund University, Sweden

World Scientific

NEW JERSEY · LONDON · SINGAPORE · BEIJING · SHANGHAI · HONG KONG · TAIPEI · CHENNAI

Published by

World Scientific Publishing Co. Pte. Ltd.

5 Toh Tuck Link, Singapore 596224

USA office: 27 Warren Street, Suite 401-402, Hackensack, NJ 07601

UK office: 57 Shelton Street, Covent Garden, London WC2H 9HE

Library of Congress Control Number: 2024043048

British Library Cataloguing-in-Publication Data
A catalogue record for this book is available from the British Library.

ISBN 978-981-12-9312-2 (hardcover)
ISBN 978-981-12-9313-9 (ebook for institutions)
ISBN 978-981-12-9314-6 (ebook for individuals)

For any available supplementary material, please visit
https://www.worldscientific.com/worldscibooks/10.1142/13837#t=suppl

Desk Editors: Nambirajan Karuppiah/Angeline Husni

Typeset by Stallion Press
Email: enquiries@stallionpress.com

This book is dedicated to our families

Preface

We were very pleased and honored to receive Word Scientific's invitation to write a monograph on inhomogeneous Markov chains and related topics.

The idea of Markov chains goes back to the early 20th century, when Russian mathematician A. A. Markov invented them, in part, for statistical text analysis. As the saying goes, "there are full libraries written" by now about the theoretical and applied aspects of Markov chains. It is thus surprising to observe the scarcity of literature on chains that are *inhomogeneous* in time.

Applications of inhomogeneous Markov chains are of course numerous. David Aldous brought to our attention the following "real-life application" of inhomogeneous chains that was relevant to his *Probability in the Real World* course in Berkeley. "One can make a toy model of a sports match, to be decided by who gains the most points. At each time there is a point difference and a remaining time. In choosing a strategy, the obvious idea is

> Play conservatively when ahead, play boldly when behind,

and in particular one must play boldly when behind with little time remaining. So, the consequence of optimal strategy is that the point difference (in very simple models) will behave as an inhomogeneous Markov chain." The reader can find a numerical example at Aldous's web page: https://www.stat.berkeley.edu/~aldous/Real_World/RW.html.

In this volume, we present some results about inhomogeneous Markov chains and associated random walks. Before doing so, we review some material from the theory of homogeneous chains and classical random walks. We are painfully aware of the fact that we cannot even approximate a "theory" of inhomogeneous chains; all we do instead is present some natural questions and provide answers to them.

Even though there is almost no general theory on inhomogeneous chains, there were studies in the 20th century on weak ergodicity by such people as John Hajnal and Eugene Seneta; see, e.g., Seneta (2014). The most recent systematic studies we are aware of were done by Laurent Saloff-Coste and his student Jessica Zúñiga; see, e.g., Saloff-Coste and Zúñiga (2011). We say more about these in Chapter 4.

This is also the place to stress that we focus exclusively on the discrete-time framework. There exists some theory in continuous time, for instance, as pointed out to us by Sir John Kingman, in the work of Gerald Goodman; see, e.g., Goodman (1974) (cf. with our remark on diffusion processes at the beginning of Chapter 4.) In addition, the reader is warned that we use the term "chain" to distinguish discrete time from "process" in continuous time. However, some authors like Kai Lai Chung use "chain" to refer to countable rather than more general state spaces. So, the terminology is not completely unified in that respect.

We assume that the reader has a graduate-level background in probability (comparable to a typical US university course), which covers the major limit theorems and introduces the Markov property and stopping times. Measure theory is not essential for most of this book, as we deal with discrete models. The only exceptions are sections on scaling limits, where we use some measure-theoretic tools.

This book consists of 10 chapters, which we summarize as follows:

Chapter 1 reviews some basic facts and tools. Chapter 2 provides a concise (and necessarily incomplete) overview of the classical theory of time-homogeneous Markov chains. Here, we closely follow some parts of the excellent monograph Levin and Peres (2017).

Chapter 3 discusses the approach based on electrical networks, as well as some results on random walks in random environments and in random labyrinths.

In Chapter 4, *time-inhomogeneous* Markov chains are considered. Unfortunately, as pointed out above, it cannot be said that their "theory" is presented, simply because no such theory exists as of today, only a tool set that works for various particular cases. This is true regardless of whether one looks at the more probability-focused work of, e.g., Laurent Saloff-Coste and Jessica Zúñiga which we review in the chapter or the more matrix-focused studies by, e.g., Eugene Seneta (Seneta, 2006). Consequently, one has to be content here with having a somewhat random collection of existing results picked from the literature.

As George Pólya once said,

> There are many questions which fools can ask that wise men cannot answer.

Chapter 5 introduces what we call the "coin turning process." As every student knows, the simplest possible probabilistic scenario is coin tossing, when the consecutive tosses are independent, and one always tosses the same coin; thus mathematically, one deals with independent Bernoulli variables of the same parameter. But what happens when, instead of tossing the coin, one "turns" it over with a certain probability? And what if that probability changes in time? Mathematically, we are then handling a Markov chain with only two states. Even though this sounds like a pretty trivial model, it is not, and this is because, in general, we have a *time-inhomogeneous* chain. It may be surprising to learn how rich and nontrivial the set of questions raised by this model is. This chapter discusses several regimes that emerge according to the behavior of the sequence of "turning probabilities." In particular, the reader observes the breakdown of the law of large numbers and the central limit theorem, depending on the extent to which that sequence deviates from a sequence of constants.

In Chapter 6, we look at the *walk* induced by the coin-turning process. That is, the steps of the walk are Markovian (+1 or −1 according to what the "turning" coin shows), but the walk itself is not. Here we revisit the classical questions concerning transience/recurrence and scaling limits, for which the answers are known for simple random walks, and again, see different regimes. In particular, the scaling limit is not necessarily a Brownian motion or one of its relatives, but it may be a "zigzag process," that is, a piecewise linear one that turns at random times up and down.

An interesting observation, made first by Márton Balázs, is that the coin turning model can be related to "urn models," the most classical of which being of course the so-called *Pólya urn*. This connection is presented briefly in Chapter 7.

In Chapter 8, we explore another walk related to the coin turning walk of Chapter 6, which we dubbed *Rademacher walk*. If the turning probabilities p_n are very small but not summable, then the coin-turning walk will change its direction very rarely, although infinitely often. As a result, the walk can be decomposed into long stretches moving to the left and some other long ones moving to the right. Those lengths are of course random. In the Rademacher walk, the lengths will be deterministic, prescribed by a given (non-random) sequence of numbers. For some such sequences of numbers, this walk will be recurrent, and for some other ones, it will be transient. It turns out that this process is a very useful tool for understanding the coin-turning walk while being an intriguing model itself.

In Chapter 9, we generalize the coin-turning walk to higher dimensions. The steps are defined as follows. Let $Y_n \in \{\pm e_1, \ldots, \pm e_d\}$, where e_i are the $2d$ unit vectors in $\mathbb{R}^d$, and let Y_1 be chosen uniformly from these vectors. Let $p_2, p_3, \ldots$ be a given sequence of probabilities and let the vectors $Y_1, Y_2, \ldots$ form an inhomogeneous Markov chain with the transition matrix between times $n-1$ and n given by

$$(1 - p_n)\mathbf{I}_{2d} + \frac{p_n}{2d}A_{2d}, \ n \geq 2,$$

where $\mathbf{I}_{2d}$ is the $2d \times 2d$ identity matrix and

$$A_{2d} := \begin{pmatrix} 1 & 1 & \cdots & 1 \\ 1 & 1 & \cdots & 1 \\ \vdots & \vdots & \ddots & \vdots \\ 1 & 1 & \cdots & 1 \end{pmatrix}$$

is the $2d \times 2d$ matrix of ones. Now, our "conservative random walk" on $\mathbb{Z}^d$ is defined by having steps Y_i. The walk "updates" its direction with probability p_n at time n, and, for instance, when p_n

decays in time, it does it more and more reluctantly, hence the name[1] "conservative." This is a straightforward extension of the one-dimensional coin-turning walk, except that an update with probability p_n means that the likelihood of actually changing the direction is $p_n/2$, which is bounded above by $1/2$. Hence, the coin-turning walk yields a richer set of models than a conservative walk would in one dimension, as it includes not just "cooling" but also "heating" dynamics. Still, using update times instead of turning times seems to be the natural choice in higher dimensions and it dovetails better into existing similar models.

Finally, the concluding tenth chapter is in some sense a continuation of Chapter 7: we revisit the connection between random walks and certain urn models. The original motivation came from an open problem related to *Friedman urns*, introduced in the 1960s, and the topic is also related to *Lamperti walks*, initiated in the same decade. In this chapter, we study the nearest neighbor random walk on a line with a drift that depends both on the location of the walker and time. We obtain sufficient conditions for the transience or recurrence of such a walk, in terms of the parameters of its drift.

The first six chapters have several exercises appended as well — they are usually not difficult to solve.

Sadly, of course, those chapters might contain typos or errors. So, what to do if you encounter any of those? Well, the authors suggest that you recall George Pólya's famous saying:

> The traditional professor writes a, says b, means c; but it should be d.

One of the authors of this book heard another relevant story from Sir John Kingman. When Sir John was a student in Cambridge, one of his classmates discovered a possible error in J. L. Doob's book "Stochastic Processes." After some hesitation, the classmate finally wrote to Doob to inquire if it was an actual mistake or a misunderstanding on his/her part. The reply from Doob was as follows:

> If you got so far in the book without finding a mistake, you were not giving the book the attention it deserves.

---

[1]Note that in time-homogeneous settings, such a walk is sometimes called "persistent" or "Newtonian."

In summary, errors happen and we welcome your feedback on this book. If you spot any typos, errors, inconsistencies or perhaps have solutions to the open problems, please write to us. We intend to update this book with your suggestions/corrections on a web page or in a future edition.

We wish you happy reading!

Boulder, USA, 2024   *János Engländer*
                              `janos.englander@colorado.edu`
Lund, Sweden, 2024   *Stanislav Volkov*
                            `stanislav.volkov@matstat.lu.se`

Acknowledgment

Several discussions on these models and collaborations in various projects are gratefully acknowledged. The authors are thus indebted to the following colleagues (in alphabetical order): David Aldous, Márton Balázs, Codina Cotar, Edward Crane, Dmitry Dolgopyat, Sir John Kingman and Bálint Tóth. We are especially grateful to Dmitry, who thoroughly reviewed the entire manuscript! Finally, we apologize for any unintentional omissions of names that should be included on this list.

Our special thanks go to Rochelle Kronzek Miller, Nijia Liu and Nambirajan Karuppaiah at World Scientific for their very professional handling of the manuscript and patience during the process. The publisher's understanding concerning the delays caused by the global COVID-19 pandemic is greatly appreciated.

Grant support from the Simons Foundation (J.E.) and from the Swedish Research Council, Crafoordska Stiftelsen, and Stiftelsen GS Magnusons fond (S.V.), as well as the hospitality of the University of Colorado and of Lund University during the authors' visits are gratefully acknowledged.

Last but not least, we are grateful to our families for their patience and support during our visits to Lund and Boulder.

About the Authors

Janos Englander received his Diploma in Mathematics at Eötvös Loránd University, Budapest, and his Ph.D. in Mathematics from the Technion, Haifa.

He worked as a postdoctoral fellow at the Weierstrass Institute (Berlin) and at EURANDOM (Eindhoven) before joining the University of California and later the University of Colorado, where he is currently a professor at the Department of Mathematics. He is the author of a research monograph on spatial branching processes and his research interests include discrete and continuous probability theory and partial differential equations.

Stanislav Volkov received his Diploma in Mathematics and Ph.D. in Mathematics from Moscow State University, and his M.A. and M.Sc. in Economics from New Economics School and the University of Wisconsin-Madison.

He worked as a postdoctoral fellow at the Fields Institute (Canada) and EURANDOM (Eindhoven) before moving to a permanent position at the University of Bristol (UK). Since 2012, he has been a Professor of Mathematical Statistics at Lund University, Sweden.

His research interests include discrete probability, history-dependent random walks, and percolation.

Contents

Chapter 1

Preliminaries

This book discusses aspects of time-inhomogeneous Markov chains and related random walks. In this brief chapter, we review notation, as well as some technical tools that are needed in the rest of the chapters. Some of these are related to the Borel–Cantelli lemma, some are anti-concentration inequalities, some are basic results on martingales, while the last two parts concern moment problems, random measures and point processes. The first-time reader may want to skip straight to Chapter 2 and come back here when needed.

We start with frequently used notation.

1.1   Notation and Terminology

The following notations/terminologies are used.

(1) **Numbers and sets:**

- $\mathbb{N}$ denotes the non-negative integers and $\mathbb{N}_+ = \mathbb{Z}_+$ denote the positive integers.
- $\mathbb{R}$ denotes the reals and $\mathbb{R}_+$ denotes the positive reals.
- $\mathbb{Z}^n$ denotes the n-dimensional integer lattice and $\mathbb{R}^n$ denotes the n-dimensional Euclidean space for $n \in \mathbb{N}_+$.
- The symbol $\lfloor z \rfloor$ denotes the lower integer part (or floor) of $z \in \mathbb{R}$, that is, $\lfloor z \rfloor := \max\{n \in \mathbb{Z} \mid n \leq z\}$. Similarly, $\lceil z \rceil$ denotes the upper integer part (or ceiling) of z, that is, $\lceil z \rceil := \min\{n \in \mathbb{Z} \mid n \geq z\}$.

- $A \triangle B$ denotes the symmetric difference of sets A and B, that is, $A \triangle B := (A \setminus B) \cup (B \setminus A)$.
- $n!!$ is the semi-factorial of $n \geq 1$. That is, $n!! := 1 \cdot 3 \ldots n$ for n odd and $n!! := 2 \cdot 4 \ldots n$ for n even.
- The cardinality of the set A is denoted by $|A|$ or by $\text{card}(A)$.

(2) **Topology and measures:**

- By a *bounded rational rectangle* we mean a set $B \subset \mathbb{R}^d$ of the form $B = I_1 \times I_2 \times \cdots \times I_d$, where I_i is a bounded interval with rational endpoints for each $1 \leq i \leq d$. The family of all bounded rational rectangles is denoted by $\mathcal{R}$.
- The symbol δ_x denotes the Dirac measure (point measure) concentrated on x.
- The symbols "$\overset{w}{\Rightarrow}$" and "$\overset{v}{\Rightarrow}$" denote convergence in the weak topology and in the vague topology, respectively.
- Given a metric space, by the "Borels" or "Borel sets" of that space, we mean the σ-algebra generated by the open sets.

(3) **Functions:**

- $\delta_{i,j}$ is the Kronecker delta: $\delta_{i,j} = 1$ when $i = j$ and $\delta_{i,j} = 0$ otherwise.
- For $A \subset (0, \infty)$ and functions $f, g : A \to (0, \infty)$, the notation $f(x) = \mathcal{O}(g(x))$ means that $f(x) \leq Cg(x)$ if $x > x_0$ with some $x_0 > 0, C > 0$, while $f \sim g$ means that f/g tends to 1 given that the argument tends to an appropriate limit. The notation $f(x) = \Theta(g(x))$ means that $c \leq f(x)/g(x) \leq C \; \forall n$, with some $c, C > 0$.
- We use the notation $\mathbb{1}_B$ to denote the indicator function (characteristic function) of the event (set) B.
- I_α and K_α denote the Bessel-I and Bessel-K functions, respectively.

(4) **Probability:**

- For "infinitely often" and "almost surely," we use the standard abbreviations: i.o. and a.s.
- We write $\mathsf{Law}(X)$ for the distribution of the random variable X, and commonly used distributions are denoted as $\mathsf{Beta}(\alpha, \beta), \mathsf{Uniform}([a, b])$, etc.

- We write Corr and Cov for correlation and covariance.
- The state space of a Markov chain is usually denoted by S.
- The total variation distance between probability measures μ and ν is denoted by $\|\mu - \nu\|_{TV}$.
- (S,W)LLN is abbreviated as the (Strong,Weak) Law of Large Numbers.
- IID is abbreviated as "independent and identically distributed."
- Fidi's is abbreviated as "finite dimensional distributions" for stochastic processes.
- Finally, even though stochastic processes *should* be denoted by the letters X, Y, Z, etc., the value of X at time t by X_t and a "generic path" by X, we are going to be pretty inconsistent in the notation, and so we apologize to the reader.

(5) **Random measures:**

- The Poisson point process is abbreviated as PPP.
- For random measures, $\xrightarrow{vd}$ denotes convergence in law with respect to the vague topology.

(6) **Matrices:**

- The symbol $\mathbf{I}_d$ denotes the d-dimensional unit matrix, and $r(\mathbf{A})$ denotes the rank of a matrix $\mathbf{A}$.
- The transposed matrix of $\mathbf{A}$ will be denoted $\mathbf{A}^T$.
- The diagonal matrix with elements $a_{11}, a_{22}, \ldots, a_{nn}$ is denoted by $\text{diag}(a_{11}, a_{22}, \ldots, a_{nn})$.

1.2   Some Technical Tools

This brief preparatory subsection collects some technical tools that we frequently use in this book, providing an easy reference.

1.2.1   *An application of Lévy's Borel–Cantelli lemma*

Paul Lévy's version of the Borel–Cantelli lemma (also called the "extended/conditional Borel–Cantelli lemma") is a well-known and frequently used tool. For completeness, we recall it here (see, e.g., Corollary 5.29 in Breiman (1992)).

Theorem 1.1 (Lévy's Borel–Cantelli lemma). *Let $Y_1, Y_2, \ldots$ be a sequence of random variables and $A_n \in \sigma(Y_1, \ldots, Y_n)$. If*

$$A := \{\omega : \omega \in A_n \ i.o.\}; \quad B := \left\{\omega : \sum_1^\infty P(A_{n+1} \mid Y_n, \ldots, Y_1) = \infty\right\},$$

then $P(A \triangle B) = 0$.

In particular, Theorem 1.1 implies the following statement. Often, we refer to Proposition 1.1 implicitly when we use the phrase "by (Lévy's/conditional) Borel–Cantelli lemma."

Proposition 1.1 (Equivalent events). *Let $(\tau_k)_{k \geq 1}$ be a sequence of non-decreasing stopping times, taking values in $\mathbb{N} \cup \{\infty\}$ on a filtered probability space $(\Omega, \mathcal{F}, P, \{\mathcal{F}_n\}, n \geq 0)$, where $\mathcal{F} = \mathcal{F}_\infty := \sigma(\cup_n \mathcal{F}_n)$. Let $D(m) \subset C(m)$ be $\mathcal{F}_m$-measurable events and $E(m) := C(\tau_m)$; $\widehat{E}(m) := D(\tau_m)$ for $m \geq 1$. Let*

$$A := \{\omega : E(k) \ i.o.\} = \left\{\omega : \sum_{k=1}^\infty \mathbb{1}_{E(k)} = \infty\right\},$$

$$B := \{\omega : \widehat{E}(k) \ i.o.\} = \left\{\omega : \sum_{k=1}^\infty \mathbb{1}_{\widehat{E}(k)} = \infty\right\}.$$

If for some $\epsilon > 0$

$$P(D(\tau_k) \mid C(\tau_k), \ \mathcal{F}_{\tau_{k-1}}) \geq \epsilon, \ a.s. \ \forall k \geq 1, \tag{1.1}$$

then $P(A \triangle B) = 0$.

Proof. Define $\mathcal{G}_k := \mathcal{F}_{\tau_k}$. By Theorem 1.1,

$$P\left(A \triangle \left\{\sum_1^\infty P(C(k) \mid \mathcal{G}_{k-1}) = \infty\right\}\right) = 0,$$

while also,

$$P\left(B \triangle \left\{\sum_1^\infty P(D(k) \mid \mathcal{G}_{k-1}) = \infty\right\}\right) = 0.$$

On the other hand, by (1.1),

$$P\left(\left\{\sum_1^\infty P(C(k) \mid \mathcal{G}_{k-1}) = \infty\right\} \triangle \right.$$
$$\left.\left\{\sum_1^\infty P(D(k) \mid \mathcal{G}_{k-1}) = \infty\right\}\right) = 0,$$

hence $P(A \triangle B) = 0$. □

1.2.2 A slight extension of Lévy's Borel–Cantelli lemma

We continue the discussion about the conditional Borel–Cantelli lemma by presenting a slightly extended version in the following.

Lemma 1.1 (LBC for pairs of events). *Suppose that we have an increasing sequence of σ-algebras $\mathcal{G}_m, m \geq 1$ and a sequence of pairs of $\mathcal{G}_m$-measurable events $(A_m, E_m), m \geq 1$ such that for all $m \geq 1$,*

$$\mathbb{P}(A_m \mid E_{m-1}, \mathcal{G}_{m-1}) \geq \alpha_m \quad a.s. \ and \quad \mathbb{P}(E_m^c) \leq \varepsilon_m,$$

where the non-negative α_n and ε_n satisfy

$$\sum_{m \geq 1} \alpha_m = \infty, \quad \sum_{m \geq 1} \varepsilon_m < \infty. \tag{1.2}$$

Then $\mathbb{P}(A_m \ i.o.) = 1$.

Proof. Let $m > \ell \geq 1$ and $B_{\ell,m} = \bigcap_{i=\ell}^m A_i^c$. We need to show that for any $\ell \geq 1$

$$\mathbb{P}(B_{\ell,\infty}) = \mathbb{P}\left(A_\ell^c \cap A_{\ell+1}^c \cap A_{\ell+2}^c \cap \cdots\right) = 0. \tag{1.3}$$

We have for $m \geq \ell + 1$

$$\mathbb{P}(B_{\ell,m}) = \mathbb{P}\left(A_m^c \cap B_{\ell,m-1}\right) \leq \mathbb{P}\left(A_m^c \cap B_{\ell,m-1} \cap E_{m-1}\right) + \mathbb{P}\left(E_{m-1}^c\right)$$
$$= \mathbb{P}\left(A_m^c \mid B_{\ell,m-1} \cap E_{m-1}\right) \mathbb{P}\left(B_{\ell,n-1} \cap E_{m-1}\right) + \mathbb{P}\left(E_{m-1}^c\right)$$
$$\leq \mathbb{P}\left(A_m^c \mid B_{\ell,m-1} \cap E_{m-1}\right) \mathbb{P}\left(B_{\ell,m-1}\right) + \varepsilon_{m-1}$$
$$\leq (1 - \alpha_m)\mathbb{P}\left(B_{\ell,m-1}\right) + \varepsilon_{m-1}. \tag{1.4}$$

By induction over $m, m-1, m-2, \ldots, \ell+1$ in (1.4), we get that

$$
\begin{aligned}
\mathbb{P}(B_{\ell,m}) &\leq \varepsilon_{m-1} + (1-\alpha_m)[(1-\alpha_{m-1})\mathbb{P}\left(B_{\ell,m-2} \mid \mathcal{G}_m\right) \\
&\quad + \varepsilon_{m-2}] \leq \cdots \\
&\leq \varepsilon_{m-1} + (1-\alpha_m)\varepsilon_{m-2} + (1-\alpha_m)(1-\alpha_{m-1})\varepsilon_{m-3} + \cdots \\
&\quad + (1-\alpha_m)(1-\alpha_{m-1})\cdots(1-\alpha_{\ell+2})\varepsilon_\ell \\
&\quad + (1-\alpha_m)(1-\alpha_{m-1})\cdots(1-\alpha_{\ell+1}).
\end{aligned}
$$

Hence, for any integer $M \in (\ell, m)$,

$$
\begin{aligned}
\mathbb{P}(B_{\ell,m}) &\leq \varepsilon_{m-1} + \varepsilon_{m-2} + \cdots + \varepsilon_M \\
&\quad + (1-\alpha_m)(1-\alpha_{m-1})\cdots(1-\alpha_{M+1})\varepsilon_{M-1} \\
&\quad + \cdots + (1-a_m)(1-\alpha_{m-1})\cdots(1-\alpha_{\ell+2})\varepsilon_\ell \\
&\quad + (1-\alpha_m)(1-\alpha_{m-1})\cdots(1-\alpha_{\ell+1}) \\
&\leq [\varepsilon_{m-1} + \varepsilon_{m-2} + \cdots + \varepsilon_M] \\
&\quad + (1-\alpha_m)(1-\alpha_{m-1})\cdots(1-\alpha_{M+1}) \\
&\quad \times [\varepsilon_{M-1} + \varepsilon_{M-2} + \cdots + \varepsilon_\ell + 1].
\end{aligned}
$$

Fix any $\delta > 0$. By (1.2), we can find an M be so large that $\sum_{i=M}^{\infty} \varepsilon_i < \delta/2$. Then, again by (1.2), there exists an $m_0 > M$ such that

$$
\prod_{i=M+1}^{m_0} (1-\alpha_i) < \frac{\delta}{2\left(1 + \sum_{i=\ell}^{M-1} \varepsilon_i\right)}.
$$

Hence, for all $m \geq m_0$, we have $\mathbb{P}(B_{\ell,m}) \leq \delta/2 + \delta/2 = \delta$. Since $\delta > 0$ is arbitrary, and $B_{\ell,m}$ is a decreasing sequence of events in m, we conclude that $\mathbb{P}(B_{\ell,\infty}) = 0$, as required. □

1.2.3 The anti-concentration inequalities of Kolmogorov, Rogozin and Kesten

The next result is a handy anti-concentration inequality due to Kolmogorov and Rogozin[1] (see, e.g., Rogozin (1961)).

---
[1] It is also called the Döblin–Lévy–Kolmogorov–Rogozin inequality.

Definition 1.1. For a random variable Y define the *concentration function* as

$$a \mapsto Q(Y;a) = \sup_{x \in \mathbb{R}} \mathbb{P}(Y \in [x, x+a]), \ a \geq 0.$$

Let $S_n = X_1 + \ldots + X_n$, where X_i are independent random variables, which do not necessarily have the same distribution. The following result is an estimate on how concentrated the law of S_n can be.

Lemma 1.2 (Kolmogorov–Rogozin inequality). *There exists a universal constant $C > 0$ such that if $L > 0$ and $0 < a_1, a_2, \ldots, a_n \leq 2L$, then*

$$Q(S_n; L) \leq \frac{CL}{\sqrt{\sum_{i=1}^n a_i^2[1 - Q(X_i; a_i)]}}.$$

As a particular case, picking $L = 1/2 - \epsilon$ for a small $\epsilon > 0$ for a random walk S on $\mathbb{Z}$ with independent steps, we have the following bound. (Random walks are introduced more formally in Chapter 2.)

Corollary 1.1 (Application for random walks). *For some universal constant $C > 0$,*

$$\sup_{y \in \mathbb{Z}} \mathbb{P}(S_n = y) \leq \frac{C}{\sqrt{\sum_{i=1}^n [1 - Q(X_i; 1)]}}.$$

An interesting improvement, when $\max_{k \leq n} Q(X_k; L)$ is small, of the Kolmogorov–Rogozin inequality is due to Kesten.

Lemma 1.3 (Kesten's anti-concentration bound; Kesten (1969)). *There is a universal constant $C' > 0$ such that for any $L > 0$ and real numbers $0 < a_1, \ldots, a_n \leq 2L$, one has*

$$Q(S_n; L) \leq \frac{C'L \cdot \max_{k \leq n} Q(X_k; L)}{\sqrt{\sum_{i=1}^n a_i^2(1 - Q(X_i; a_i))}}.$$

In fact, $C' = 4\sqrt{2}(1 + 9C)$.

We use Corollary 1.1 in Chapter 7.4.

1.2.4   *Some basic facts about martingales*

Warning to the reader: In what follows, the time parameter t can be either discrete ($t = 0, 1, 2, \ldots$) or continuous ($t \in [0, \infty)$).

Definition 1.2 (Submartingale). Given the filtered probability space $(\Omega, \mathcal{F}, (\mathcal{F}_t)_{t \geq 0}, P)$, a stochastic process X is called a *submartingale* if

(1) X is adapted,[2]
(2) $E|X_t| < \infty$ for $t \geq 0$,
(3) $E(X_t \mid \mathcal{F}_s) \geq X_s$ (*P*-a.s.) for $t > s \geq 0$.

The process X is called a *supermartingale* if $-X$ is a submartingale. Finally, if X is a submartingale and a supermartingale at the same time, then X is called a *martingale*.

It is easy to check that if one replaces the filtration by the canonical filtration generated by X (i.e., one chooses $\mathcal{F}_t := \sigma(\bigcup_0^t \sigma(X_s))$), then the (sub)martingale property still holds. Hence, when the filtration is not specified, it is understood that the filtration is the canonical one.

Next, we recall the two most often cited results in martingale theory; they are both due to Doob. The first one is his famous "optional stopping" theorem.[3]

Theorem 1.2 (Doob's optional stopping theorem). *Let M be a martingale (with right continuous paths in the continuous case) and $\tau : \Omega \to [0, \infty]$ a stopping time with respect to the filtered probability space $(\Omega, \mathcal{F}, (\mathcal{F}_t)_{t \geq 0}, P)$. Then the process η defined by $\eta_t := M_{\min\{t, \tau\}}$ is also a martingale with respect to the same $(\Omega, \mathcal{F}, (\mathcal{F}_t)_{t \geq 0}, P)$.*

Replacing the word "martingale" by "submartingale" in both sentences produces a true statement too.

---

[2]by which we mean that $\sigma(X_t) \subset \mathcal{F}_t$ for $t \geq 0$.

[3]A.k.a. "Doob's optional sampling theorem," although in Doob's own terminology, the latter name referred to a more general result. Another, closely related version of optional stopping concerns two stopping times $S \leq T$ and whether the defining inequality of submartingales still holds at these times. In that version though, unlike here, the martingale must be "closable" by a last element $(M_\infty, \mathcal{F}_\infty)$.

The second one is an improvement on Markov inequality for sub-martingales.

Theorem 1.3 (Doob's inequality). *Let M be a submartingale (with right continuous paths in the continuous case) and $\lambda > 0$. Then, for $t > 0$,*

$$P\left(\sup_{0 \leq s \leq t} M_s \geq \lambda\right) \leq \frac{EM_t^+}{\lambda},$$

where $x^+ := \max\{x, 0\}$.

A well-known inequality for conditional expectations closely related to martingales is as follows.

Theorem 1.4 (Conditional Jensen's inequality). *Let X be a random variable on $(\Omega, \mathcal{F}, P)$ and $\mathcal{G} \subset \mathcal{F}$ be a σ-algebra. If f is a convex[4] function, then*

$$E(f(X) \mid \mathcal{G}) \geq f(E(X \mid \mathcal{G})).$$

(If the left-hand side is $+\infty$, the inequality is taken as true.)

Remark 1.1. (a) When $\mathcal{G} = \{\emptyset, \Omega\}$, one obtains the (unconditional) Jensen's inequality.

(b) The fact that a convex (concave) function of a martingale is a submartingale (supermartingale), provided it is integrable, is a simple consequence of Theorem 1.4.

A fundamental convergence theorem is as follows:

Theorem 1.5 (Submartingale convergence theorem). *Let M be a submartingale (with right continuous paths in the continuous case) with respect to the filtered probability space $(\Omega, \mathcal{F}, (\mathcal{F}_t)_{t \geq 0}, P)$. Assume that $\sup_{t \geq 0} E(X_t^+) < \infty$. Then X_t has a P-almost sure limit, X_∞ as $t \to \infty$, and $E|X_\infty| < \infty$.*

Here $X^+ := \max\{0, X\}$. Letting $Y := -X$, one gets the corresponding result for supermartingales.

One often would like to know when a martingale limit holds true in L^1 as well.

---

[4]By "convex" we mean convex from above, like $f(x) = |x|$.

Theorem 1.6 (L^1-convergence theorem). *Let M be a martingale (with right continuous paths in the continuous case) with respect to the filtered probability space $(\Omega, \mathcal{F}, (\mathcal{F}_t)_{t \geq 0}, P)$. The following conditions are equivalent:*

(1) *$\{M_t\}_{t \geq 0}$ is a uniformly integrable family.*
(2) *M_t converges in L^1 as $t \to \infty$.*
(3) *M_t converges in L^1 as $t \to \infty$ to a random variable $M_\infty \in L^1(P)$ such that M_t is a martingale on $[0, \infty]$ with respect to $(\Omega, \mathcal{F}, (\mathcal{F}_t)_{t \in [0,\infty]}, P)$. (Here $\mathcal{F}_\infty := \sigma(\bigcup_{t \geq 0} \sigma(M_t))$, and M_∞ is the "last element" of this martingale.)*
(4) *There exists a random variable $Y \in L^1(P)$ such that $M_t = E(Y \mid \mathcal{F}_t)$ holds P-a.s. for all $t \geq 0$.*

In fact, the relationship between M_∞ and Y of the last two conditions is that $M_\infty = E(Y \mid \mathcal{F}_\infty)$.

For an interesting read on the history of martingale theory, see Mazliak and Shafer (2022).

1.2.5   *Moment problems*

As we often carry out computations with moments, it will be useful to recall the *Hausdorff moment problem*, which deals with necessary and sufficient conditions for a given sequence to be the sequence of moments of some probability law (more generally, a Borel measure), supported on a closed bounded interval $[a, b]$.

The essential difference between this and other well-known moment problems is that the Hausdorff moment problem deals only with a *bounded* interval, while in the Stieltjes and Hamburger moment problems, one considers a half-line $[0, \infty)$ and the whole real line, respectively. These last two problems may have infinitely many solutions, whereas the Hausdorff moment problem, if it is solvable, always has a unique solution. See more on moment problems in Chapter 1 of Shohat and Tamarkin (1943).

For the relationship between the convergence of moments and convergence in law, see Chapter 5, Section 8.4 in Gut (2005). In particular, we have the following result.

Theorem 1.7 (Theorem 8.6. in Gut (2005)). *Let X and $X_1, X_2, \ldots$ be random variables with finite moments of all orders,* *and suppose that $E|X_n|^k \to E|X|^k$ as $n \to \infty$, for $k = 1, 2, \ldots$. If the moments of X determine the distribution of X uniquely, then $X_n \to X$ in law as $n \to \infty$.*

1.2.6 *Point processes and random measures*

Let ν be a locally finite measure on $\mathbb{R}^d$. A d-dimensional Poisson point process (PPP) with intensity measure ν is a random collection of points on $\mathbb{R}^d$, satisfying that if $X(B)$ is the number of points in a Borel set $B \subset \mathbb{R}^d$, then

(1) $X(B_1), X(B_2), \ldots, X(B_k)$ are independent if $B_i \cap B_j = \emptyset$ for $1 \le i \ne j \le k$, for $k \ge 1$,
(2) $X(B) \sim \mathsf{Poisson}(\nu(B))$.

In particular, using a slight abuse of notation, if $\nu(B) := \int_B \nu(x) \mathrm{d}x$ for some non-negative, measurable and locally integrable function ν on $\mathbb{R}^d$, then the PPP can be determined by the density function ν as well.

Remark 1.2. The reason Poisson point processes arise naturally is the Poisson approximation of the Binomial distribution. The reader can easily see this for the very simple case of constant density $\lambda > 0$ on the unit interval, using the following heuristic argument. Divide each interval into n subintervals, put into each one of them a point with probability λ/n, independently of each other (say, in the middle). As $n \to \infty$, the distribution of the number of points in a given interval is then going to tend to the Poisson distribution with parameter λ times the length of the interval. The independence of the number of points on disjoint intervals is also clear, and so is the extension to the real line, by performing this procedure independently on every unit interval with integer endpoints. The argument is similar for $\mathbb{R}^d$.

For a general measure, the only difference is that one has to use the more general version of the Poisson approximation theorem, where the sums of different probabilities converge. ◇

Next, we invoke some background on random measures that we utilize in the proof of Proposition 6.4. More material on random measures can be found in, e.g., Kallenberg (2017).

Assume that we are given a complete separable metric space S.

Definition 1.3 (Dissecting subsets). Denote by $\widehat{S}$ the set of all bounded Borel sets of S. A subset $\mathcal{I} \subset \widehat{S}$ is called *dissecting* if

(a) every open set $G \subset S$ is a countable union of sets in $\mathcal{I}$,
(b) every set $B \in \widehat{S}$ is covered by finitely many sets in $\mathcal{I}$.

The following lemma is a useful result concerning the weak convergence of random measures. (The measures are equipped with the vague topology; in the following, $\xrightarrow{vd}$ means convergence in law with respect to the vague topology.)

Lemma 1.4 (Theorem 4.11 in Kallenberg (2017)). *Let $\xi, (\xi_n)_n$ be random measures on S and let E denote the expectation for ξ. Furthermore, let*

(1) $\widehat{C}_s$ *be the set of all continuous compactly supported functions on S,*
(2) $\widehat{S}_{E\xi}$ *be the class of all bounded sets $A \subset S$ with $E\xi(\partial A) = 0$,*
(3) $\widehat{\mathcal{I}}_+$ *be the set of all non-negative simple $\mathcal{I}-$measurable functions for a fix dissecting semi-ring $\mathcal{I} \subset \widehat{S}_{E\xi}$.*

Then, $\xi_n \xrightarrow{vd} \xi$ as $n \to \infty$, if and only if $\xi_n(f) \xrightarrow{d} \xi(f)$ for all $f \in \widehat{C}_s$. Furthermore, in this condition, the class $\widehat{C}_s$ may be replaced by $\widehat{\mathcal{I}}_+$.

1.3   Exercises

Problem 1. Formulate Lemmas 1.2 and 1.3 for the case when the independent random variables have the same distribution.
Problem 2. Assuming that the X_i have finite variance, can you relate Lemmas 1.2 and 1.3 to the central limit theorem?
Problem 3. Let $d \geq 3$ and Z defined by $Z_n = (Z_n^{(1)}, \dots, Z_n^{(d)})$, $n \in \mathbb{N}$, a random process on $\mathbb{Z}^d$ such that the $Z^{(k)}$ are independent one-dimensional random walks on $\mathbb{Z}$. Using Corollary 1.1, give a sufficient condition for the transience of Z when, for each k, the steps of $Z^{(k)}$ are bounded below by 1 and independent but not necessarily identically distributed. (Transience means that with probability one, every bounded subset of $\mathbb{Z}^d$ is visited finitely often.)

Chapter 2

Time-Homogeneous Markov Chains

The idea of Markov chains was born in 1906 in Russia, when P. Chebyshev's student A. A. Markov (1856–1922, Figure 2.1) invented them, as an "extension of the law of large numbers to dependent quantities" as well as a tool for statistical text analysis, which he later used when analyzing Pushkin's Eugene (Yevgeniy) Onegin. It was around the same time that the theory of finite matrices with non-negative entries was developed, mainly by O. Perron and G. Frobenius in Germany.

The "fundamental equation" for Markov chains was derived independently, around 1930, by S. Chapman (1888–1970) in the U.K. and by A. N. Kolmogorov (1903–1987) in the Soviet Union, who called Markov chains "stochastically determined processes." The name "Markov chain" was suggested by A. Y. Khintchine (1894–1959) in 1934, while the first systematic treatment of Markov chains was carried out by R. M. Frechet (1878–1973) in the late 1930s. It is fair to say that today there is no branch of natural or social sciences that would be able to function without the use of Markov chains.

We start with a review of the theory of homogeneous Markov chains, concentrating on countable (and mostly finite) state spaces.

2.1 Transition Matrix and Chapman–Kolmogorov Equation

Let S be a countable set. We call it the **state space**. (Note, however, that the theory of Markov chains exists for uncountable state spaces as well.)

Fig. 2.1.   Andrey Andreyevich Markov (1856–1922). [Courtesy of Wikipedia]

Assumption: In fact, in this chapter, we always assume that S is finite, unless stated otherwise. When it is infinite, the matrices in the following definition are infinite ones, requiring extra care.

The following definition is fundamental.

Definition 2.1 (Markov chain). Consider a sequence of S-valued random variables $X_0, X_1, \ldots$:

(i) These variables form a **Markov chain** with state space S if for all $x, y \in S$, $n \geq 1$ and

$$H_{n-1} := \bigcap_{i=0}^{n-1} \{X_i = x_i\}$$

such that $\mathbb{P}(H_{n-1} \cap \{X_n = x_n\}) > 0$, the following holds:

$$\mathbb{P}(X_{n+1} = y \mid H_{n-1} \cap \{X_n = x\}) = \mathbb{P}(X_{n+1} = y \mid X_n = x). \tag{2.1}$$

(ii) If furthermore $\mathbb{P}(X_{n+1} = y \mid X_n = x)$ is the same for all n, then the Markov chain is called **time homogeneous** (or just **homogeneous**), in which case this conditional probability is denoted by P_{xy}.

(iii) The entries P_{xy} form the **transition matrix** P.

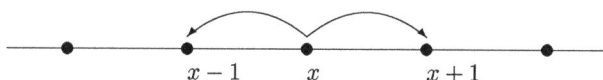

Fig. 2.2.   Random walk illustration.

In words, (2.1) means that "given the present, the future is independent of the past:" this is the **Markov property**, sometimes referred to as being "memoryless."

Remark 2.1. In the time-homogeneous case, $\sum_{y \in \mathsf{S}} P_{xy} = 1$ for all $x \in \mathsf{S}$, that is, P is a **stochastic matrix**. In this chapter, we always assume that the Markov chain is time homogeneous. Hence, it can be completely described by S and P.

Example 2.1 (Random walk). The prototypical example for an *infinite state* Markov chain is when $Y_1, Y_2, \ldots$ are IID (independent and identically distributed) random vectors in $\mathbb{Z}^d$, $d \geq 1$, with common law L, and the discrete-time stochastic process S is defined by $S_0 := 0$ and $S_n := Y_1 + \cdots + Y_n, n \geq 1$. Since $S_{n+1} = S_n + Y_{n+1}$ and Y_{n+1} is independent of the previous random variables, S is a Markov chain with the infinite but countable state space $\mathsf{S} = \mathbb{Z}^d$. This chain is called a d-dimensional **random walk** with step distribution L.

Often L is supported on the unit basis vectors and their negatives ($2d$ vectors), and S then in fact always moves on $\mathbb{Z}^d$ to a nearest neighbor. Such a random walk is called a **simple random walk**. When furthermore all the steps have the same probability $1/(2d)$, then it is a **symmetric simple random walk**. See Figure 2.2.

Definition 2.2. A **distribution** π (on S) is a row vector of length $|\mathsf{S}|$, whose entries are non-negative and sum to 1. When it possesses the invariance property $\pi = \pi P$, it is called a **stationary distribution or an invariant distribution**.

Now, suppose for simplicity that $\mathsf{S} = \{a, b\}$. Define

$$\mu_n := (\mathbb{P}(X_n = a \mid X_0 = a), \mathbb{P}(X_n = b \mid X_0 = a)).$$

Then $\mu_0 = (1, 0)$ and $\mu_1 = \mu_0 P$. Hence, $\mu_2 = \mu_1 P = \mu_0 P^2$, and similarly, $\mu_n = \mu_0 P^n$ and $\mu_{n+1} = \mu_n P$.

A natural question then is whether the limit $\lim_n \mu_n$ exists. Assume that P can be written as $\begin{pmatrix} 1-p & p \\ q & 1-q \end{pmatrix}$ with $p, q \in (0,1)$, and observe that any candidate π for a limit must be stationary, that is, $\pi = \pi P$ must hold, which, after a bit of algebra, leads to $\pi_1 = \frac{q}{p+q} = \pi(a)$ and $\pi_2 = \frac{p}{p+q} = \pi(b)$.

To check that π is indeed the limit, define $\Delta_n := \mu_n(a) - \frac{q}{p+q}$ and check that the recursion $\Delta_{n+1} = (1-p-q)\Delta_n$ holds. Now, if $p, q \in (0,1)$, then $|1-p-q| < 1$ and so $\Delta_n \to 0$ yielding that $\lim_n \mu_n = \pi$.

In fact, we even know the rate of convergence as $\Delta_n = \lambda^n \Delta_0$, where $\lambda := 1 - p - q$. Note that λ is an eigenvalue of P. Of course, 1 is an eigenvalue (with eigenvector π, having positive coordinates) too, but $\lambda < 1$. If we solve $(r, s)P = \lambda(r, s)$, then we get the eigenvector $(r, s) = (1/2, -1/2)$. That it has both positive and negative coordinates is in accordance with the Perron–Frobenius theory for matrices and so is the fact that 1 is a simple eigenvalue and π has positive entries. (In fact, 1 is the leading eigenvalue, meaning it has the maximal absolute value.) See Chapter 8 in Meyer (2023) for more on the Perron–Frobenius theory.

Now, what if we have more than two states? It's useful to define $\mathbb{P}_\mu$ and $\mathbb{E}_\mu$ when starting the chain randomly according to the probability distribution μ on S. With a slight abuse of notation, we write $\mathbb{P}_x$ and $\mathbb{E}_x$ instead of $\mathbb{P}_{\delta_x}$ and $\mathbb{E}_{\delta_x}$. If $x, y, z \in \mathsf{S}$ and $\mu_n^x := \mathbb{P}_x(X_n = z)$, then one easily checks that

$$\mu_{n+1}^x(y) = \sum_{z \in \mathsf{S}} \mathbb{P}_x(X_n = z)P_{zy} = \sum_{z \in \mathsf{S}} \mu_n^x(z)P_{zy} \qquad (2.2)$$

(note that $\mathbb{P}_x(X_n = z) = (P^n)_{xz}$). Equation (2.2) is one formulation of the **Chapman–Kolmogorov equation**.

We also conclude that the relations $\mu_{n-1}^x = \mu_n^x P$ and $\mu_n^x = \mu_0^x P^n$ hold, where of course $\mu_0^x = (0, 0, \ldots, 1, 0, \ldots, 0)$, and the weight 1 is assigned to the state $x \in \mathsf{S}$.

With the same μ as before and with $A \subset \mathsf{S}$, we write $\mu(A) = \sum_{x \in A} \mu(x)$ (instead of using $\mu(\{x\})$) and we note that $\mathbb{P}_{x.} = \mathbb{P}(x, \cdot)$ is a probability distribution for a given $x \in \mathsf{S}$.

Finally, a natural question is as follows: what is the significance of multiplying with the transition matrix *from the left*? Let $\mathsf{S} = \{x_1, \ldots, x_N\}$ and let f be a function $f : \mathsf{S} \to \mathbb{R}$. We can consider f

as a *column vector* of length N:

$$f = \begin{pmatrix} f(x_1) \\ f(x_2) \\ \vdots \\ f(x_N) \end{pmatrix}. \tag{2.3}$$

Now, for $1 \leq i, j \leq N$, abbreviate $P_{ij} := P_{x_i x_j}$. Then for $1 \leq i \leq N$, the ith coordinate of the column vector Pf is

$$(Pf)_i = \sum_{1 \leq j \leq N} P_{ij} f(x_j) = \mathbb{E}_{x_i} f(X_1).$$

Therefore, the answer to the above question is that Pf is the expectation vector $[\mathbb{E}_{x_1} f(X_1), \ldots, \mathbb{E}_{x_N} f(X_1)]^T$.

2.2 Basic Classification: Irreducible and Aperiodic Chains

Definition 2.3 (Communication classes). We say that y is **accessible** from y (and write $x \to y$) if there is an $n \geq 0$ with $P_{xy}^n > 0$. When $y \to x$ too, we say that the states **communicate** (and write $x \leftrightarrow y$). This is an equivalence relation because $P_{xx}^0 > 0$, and the induced classes are the **communication classes** of the chain.

Chains with only one class, that is, satisfying $x \to y$ for all $x, y \in S$, have special significance.

Definition 2.4 (Irreducibility). The Markov chain is **irreducible** if for $x, y \in S$ there is an $n \geq 1$ (possibly depending on x, y) such that $P_{xy}^n > 0$.

Before the following definition consider the example when the states are $1, 2, \ldots, N$, and the transitions are such that $P_{xy} = 0$ when $x + y$ is even. Then S can be partitioned into two subsets, E (even) and O (odd) such that with probability one, the current state of the chain alternates between being in these two sets, yielding some "periodicity" for the steps.

Definition 2.5 (Period of state). For $x \in S$, consider the possible times of return: $\mathcal{N}(x) := \{n \geq 1 \mid P_{xx}^n > 0\}$. If $\mathcal{N}(x) \neq \emptyset$, then the greatest common divisor (gcd) for the elements in $\mathcal{N}(x)$ is the **period** of x. (Otherwise, the period is defined to be $+\infty$.)

For irreducible chains, we can actually define the period for the whole chain.

Proposition 2.1 (Period of chain). *For irreducible chains, the period of x is the same for all $x \in S$, and this common quantity for all states is the* **period of the Markov chain***.*

Proof. Due to irreducibility, the period of x is finite, and furthermore, by symmetry, it is enough to show that the period of y divides it. That is, if we show that the gcd of the elements in $\mathcal{N}(y)$ (the period of y) divides any $m \in \mathcal{N}(x)$, then we are done. By irreducibility, there are $r, \ell \geq 1$ satisfying $P_{xy}^r, P_{yx}^\ell > 0$ and thus $r + \ell \in \mathcal{N}(x) \cap \mathcal{N}(y)$. Hence, if $m \in \mathcal{N}(x)$, then $\ell + m + r \in \mathcal{N}(y)$. Since also $r + \ell \in \mathcal{N}(y)$, and $\ell + m + r - r - \ell = m$, the gcd of the elements in $\mathcal{N}(y)$ divides m.   □

Definition 2.6 (Periodic/aperiodic irreducible chain). If the period of an irreducible Markov chain is 1, the chain is called **aperiodic**, and if it is larger than 1, then it is called **periodic**.

Note: In an aperiodic chain, it is not necessarily true that $1 \in \mathcal{N}(x)$ in a given state.

Example 2.2 (N-cycle, see Figure 2.3). If $S = 1, 2, \ldots, N$ and the transitions are such that $P_{xy} = 0$ unless $y - x = \pm 1$ modulo N, then periodicity depends on the parity of N. If N is even, then the chain is periodic with period 2. If N is odd, then the chain is aperiodic. If $P_{xy} = 1/2$ whenever $y - x = \pm 1$ modulo N, then the chain is called the **simple random walk on the N-cycle**.

Aperiodic irreducible chains[1] have some pleasant properties.

Proposition 2.2. *For an aperiodic irreducible chain, there is an integer r_0 such that for all $r \geq r_0$ and all $x, y \in S$, $P_{xy}^r > 0$.*

For example, it is easy to check that the smallest such r_0 is $r_0 = 2$ ($r_0 = 4$) for the simple random walk on the 3-cycle (5-cycle), see the exercises at the end of this chapter.

---

[1] Sometimes they are called "ergodic" or "regular."

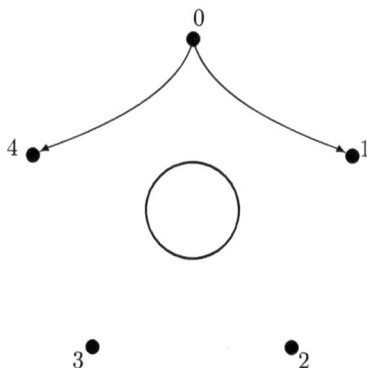

Fig. 2.3. Illustration of random walk on cycle.

Proof. The argument is based on a bit of number theory: we use the well-known[2] result of A. Schur that if $A \subset \mathbb{N}$ is closed under addition and the gcd of its elements is 1, then $|\mathbb{N} \setminus A| < \infty$. Applying this to $A = \mathcal{N}(x)$, it follows that for some $m(x)$, $\{m(x), m(x+1), \ldots\} \subset \mathcal{N}(x)$.

For $y \in S$, there is an $r(x,y)$ with $P_{xy}^{r(x,y)} > 0$ because of irreducibility, so for $m \geq m(x) + r(r,x)$, we have $P_{xy}^{m} \geq P_{xx}^{m-r(x,y)} P_{xy}^{r(x,y)} > 0$. Similarly, for any m such that $m \geq m^{*}(x) := m(x) + \max_{y \in S} r(x,y)$, we have $P_{xy}^{m} > 0$ simultaneously for all $y \in S$. Finally, for $m \geq \max_{x \in S} m^{*}(x) =: r_0$, we have $P_{xy}^{m} > 0$ for all $x, y \in S$. $\qquad \square$

Remark 2.2 (Lazy walk). Since the nearest neighbor random walk on the n-cycle is periodic (with period 2) when n is even, after an even (odd) number of steps, we are always in an "even state" ("odd state"). This means that we cannot have a limiting distribution for the walk.

It is easy, however, to modify the walk a bit so that the new walk does have a limit, by making the walk "lazy." Let the original transition matrix be P and the new one $Q := (I + P)/2$, where I is the identity matrix. That is, $Q_{ii} = 1/2$ while $Q_{i,i\pm1} = \frac{1}{2} P_{i,i\pm1}$. So, we first toss a fair coin, and only if it's heads, we apply the original

---

[2]See Lemma 1.30 in the Notes of Chapter 1 in Levin and Peres (2017).

transitions. Clearly, the new chain with matrix Q is aperiodic, and we see that such a chain possesses a limiting distribution.

2.3   Transience, (positive) Recurrence and Stationary Distribution

It is a well-known story how G. Pólya (1887–1985, Figure 2.4), after bumping into the same couple over and over again in a park, started to wonder if this *must* be so. Replacing the paths in the park by $\mathbb{Z}^2$ and considering a simple random walk on it, Pólya proved that this indeed is bound to happen.

More precisely, in 1924, exactly a century before this book was written, he proved the following.

Theorem 2.1 (Transience and recurrence in d-dimension). *Let $\tau_0^+ := \min\{n \geq 1 \mid X_n = \mathbf{0}\}$ denote the* **first return time** *to the origin for the simple random walk on $\mathbb{Z}^d$, $d \geq 1$. If $d = 1, 2$, then* $\mathbb{P}(\tau_0^+ < \infty) = 1$, *while* $\mathbb{P}(\tau_0^+ < \infty) < 1$ *holds[3] when $d \geq 3$.*

In fact, Theorem 2.1 answers the question about the couple in the park, as long as we replace the paths in the park with a perfect

Fig. 2.4.   G. Pólya. [Courtesy of Facts.net]

---

[3]In fact, the probability of return is roughly 0.34 in three dimensions and only around 0.19 in four dimensions.

two-dimensional lattice, as the distance vector between the couple and Pólya points to a random walk too, and this latter walk hits zero if and only if Pólya bumps into the couple. The fact that the latter walk takes place on a lattice which is rotated by 45 degrees compared to $\mathbb{Z}^2$ makes no difference. By S. Kakutani's famous quote,

> A drunken man will eventually find his way home, but a drunken bird may get lost forever.

Various proofs of Theorem 2.1 can be found in any number of textbooks, and we omit those here.

In general, for an irreducible Markov chain, for $x \in \mathsf{S}$, define

$$\tau_x^+ := \min\{n \geq 1 \mid X_n = x\}.$$

Definition 2.7 (Recurrence and transience). An irreducible Markov chain is called **recurrent** if $\mathbb{P}_x(\tau_y^+ < \infty) = 1$ for every $x, y \in \mathsf{S}$, and it is called **transient** otherwise.

Remark 2.3. It can be shown that for an irreducible Markov chain either $\mathbb{P}_x(\tau_y^+ < \infty) = 1$ for all $x, y \in \mathsf{S}$ or $\mathbb{P}_x(\tau_y^+ < \infty) < 1$ for all $x, y \in \mathsf{S}$.

For the simple random walk on $\mathbb{Z}^d$, $d \geq 1$, Pólya's theorem and translation invariance imply that, given $x \in \mathsf{S}$, if $d = 1, 2$, then $\mathbb{P}_x(\tau_x^+ < \infty) = 1$ (recurrence), while for $d \geq 3$, $\mathbb{P}_x(\tau_x^+ < \infty) < 1$ (transience).

However, an irreducible Markov chain with $|\mathsf{S}| < \infty$ behaves much more regularly.

Definition 2.8 (Positive and null recurrence). A recurrent Markov chain is called **positive recurrent** if $\mathbb{E}_x|\tau_y^+| < \infty$ for every $x, y \in \mathsf{S}$, and it is called **null recurrent** otherwise.

Proposition 2.3 (Finite state space). *Any irreducible finite state space Markov chain is positive recurrent.*

That is, the return time is not just finite, a.s., but its expected value is finite too. (For the simple random walk on $\mathbb{Z}^d$, $\mathbb{E}_x|\tau_x^+| = \infty$ for every $x \in \mathsf{S}$ in any dimension. Thus, even though the drunken man will find his way home, this, on average, will take an infinite amount of time, that is, null recurrence holds.)

Proof. We first claim that for some $m \geq 0$ and $\epsilon > 0$ the following holds: For any $z, w \in \mathsf{S}$, there is an $n \leq m$ with $P_{zw}^n > \epsilon$, hence $P_z(\tau_w^+ \leq m) > \epsilon$. To see why this is true, note that for given $z, w \in \mathsf{S}$, by irreducibility, $P_{zw}^\ell > 0$ for some $\ell = \ell(z, w)$. Thus, all we have to do is to pick $\epsilon := \min_{z, w \in \mathsf{S}} P_{zw}^{\ell(z,w)}$ and $m := \max_{z, w \in \mathsf{S}} \ell(z, w)$.

The previous claim implies that

$$\mathbb{P}_x(\tau_y^+ > km) \leq (1 - \epsilon)\mathbb{P}_x(\tau_y^+ > (k-1)m), \ k \geq 1,$$

hence $\mathbb{P}_x(\tau_y^+ > km) \leq (1 - \epsilon)^k, \ k \geq 1$. Since $a_i := \mathbb{P}(\tau_y^+ > i)$ is non-increasing, it follows from the above that

$$\mathbb{E}_x(\tau_y^+) = \sum_{i=0}^{\infty} \mathbb{P}(\tau_y^+ > i) \leq \sum_{k=0}^{\infty} m\mathbb{P}(\tau_y^+ > km) \leq \sum_{k \geq 0} m(1 - \epsilon)^k < \infty,$$

for each $x, y \in \mathsf{S}$. $\qquad\square$

Using Proposition 2.3, we now construct the stationary distribution for a (finite) irreducible chain.

Theorem 2.2 (Stationary distribution for an irreducible chain). *For an irreducible Markov chain, define the quantities* $\pi(x) := \frac{1}{\mathbb{E}_x(\tau_x^+)} \in (0, 1]$, $x \in \mathsf{S}$. *One has that* $\sum_{x \in \mathsf{S}} \pi(x) = 1$ *and* π *is the only distribution vector which is stationary for P, that is,* $\pi P = \pi$.

The intuition is that, on average, the chain visits state $x \in \mathsf{S}$ once every $\mathbb{E}_x(\tau_x^+)$ amount of time; thus it spends about $\pi(x)$ proportion of the time there, hence $\sum_{x \in \mathsf{S}} \pi(x) = 1$.

Proof. Let us fix $z \in \mathsf{S}$ and define $\tilde{\pi}(y) := \mathbb{E}_z V(y)$, where $V(y)$ is the number of visits to $y \in \mathsf{S}$ until returning to z; $\tilde{\pi}(z) = 1$. Then

$$\tilde{\pi}(y) = \mathbb{E}_z \sum_{n \geq 0} 1_{\{X_n = y\}} 1_{\{\tau_z^+ > n\}}$$

$$= \sum_{n \geq 0} \mathbb{P}_z\left(X_n = y, \tau_z^+ > n\right) \leq \mathbb{E}_z(\tau_z^+) < \infty.$$

Note that $\sum_{y \in \mathsf{S}} \tilde{\pi}(y) = \mathbb{E}_z(\tau_z^+)$.

We claim that $\tilde{\pi}P = \tilde{\pi}$. Indeed, on the one hand,

$$\sum_{x \in S} \tilde{\pi}(x) P_{xy}$$

$$= \sum_{x \in S} \sum_{n \geq 0} \mathbb{P}_z \left(X_n = x, \tau_z^+ > n \right) P_{xy}$$

$$= \sum_{n \geq 0} \left(\sum_{x \in S} \mathbb{P}_z \left(X_n = x, \tau_z^+ > n \right) P_{xy} \right)$$

$$= \sum_{n \geq 0} \mathbb{P}_z \left(X_{n+1} = y, \tau_z^+ > n \right) = \sum_{n \geq 1} \mathbb{P}_z \left(X_n = y, \tau_z^+ \geq n \right).$$

On the other hand, $\tilde{\pi}(y) = \sum_{n \geq 0} \mathbb{P}_z \left(X_n = y, \tau_z^+ > n \right)$. Taking the difference,

$$\tilde{\pi}P - \tilde{\pi} = \sum_{n \geq 1} \mathbb{P}_z \left(X_n = y, \tau_z^+ = n \right) - P \left(X_0 = y, \tau_z^+ > 0 \right)$$

$$= \mathbb{P}_z \left(X_{\tau_z^+} = y \right) - \mathbb{P}_z \left(X_0 = y, \tau_z^+ > 0 \right).$$

If $z \neq y$, then clearly the right-hand side equals $0 - 0 = 0$. If $z = y$, the right-hand side equals $1 - 1 = 0$.

Therefore, π defined by

$$\pi(y) := \frac{\tilde{\pi}(y)}{\sum_{y \in S} \tilde{\pi}(y)} = \frac{\tilde{\pi}(y)}{\mathbb{E}_z(\tau_z^+)}$$

is an invariant probability distribution (vector).

Since this is true for *any* $z, y \in S$, picking $z = y$ yields that

$$\pi(y) = \frac{\tilde{\pi}(y)}{\mathbb{E}_y(\tau_y^+)} = \frac{1}{\mathbb{E}_y(\tau_y^+)},$$

as $\tilde{\pi}(y) = \tilde{\pi}(z) = 1$ then.

Finally, uniqueness follows from the Perron–Frobenius theorem of linear algebra implying that the left-eigenvalue 1 of the matrix of an irreducible Markov chain is simple. (Alternatively, see Levin and Peres (2017, pp. 12–13) for a proof via harmonic functions.) □

2.4   Reversible Chains

Note that the definition of the Markov property is "symmetric" as far as past and future are concerned: given the present, the past and the future are independent. Hence, considering the process in "backward time" is a natural idea.

Let π be a probability distribution on S.

Definition 2.9. We say that π satisfies the **detailed balance equations** if $\pi(x)P_{xy} = \pi(y)P_{yx}$ for all $x, y \in S$.

Note that any such π is invariant because summing over $y \in S$ one obtains

$$\pi(x) = \sum_{y \in S} \pi(x)P_{xy} = \sum_{y \in S} \pi(y)P_{yx},$$

that is, $\pi P = \pi$.

If the detailed balance equations hold, then one can "reverse the chain" in the following sense.

Proposition 2.4. *Assume that π satisfies the detailed balance equations. Let $n \geq 1$ and $(x_0, x_1, \ldots, x_n) \in S^{n+1}$. If $Y_k := X_{n-k}$, then*

$$\mathbb{P}(Y_0 = x_0, \ldots, Y_n = x_n) = \mathbb{P}(X_0 = x_0, \ldots, X_n = x_n).$$

Proof. We have

$$\pi(x_0)P_{x_0 x_1} = \pi(x_1)P_{x_1 x_0},$$
$$\pi(x_1)P_{x_1 x_2} = \pi(x_2)P_{x_2 x_1},$$
$$\vdots$$
$$\pi(x_{n-1})P_{x_{n-1} x_n} = \pi(x_n)P_{x_n x_{n-1}}.$$

The statement follows by multiplying these equations and simplifying by $\pi(x_1)...\pi(x_{n-1})$. $\square$

Remark 2.4 (Self-adjointness). The more algebraically minded reader will note that one can define the inner product $\langle u, v \rangle_\pi := \sum_{x \in S} u(x)v(x)\pi(x)$ for vectors $u, v \in \mathbb{R}^{|S|}$ and then reversibility simply means self-adjointness: $\langle Pu, v \rangle_\pi = \langle u, Pv \rangle_\pi$.

More generally, we want to reverse time for any irreducible chain, thus obtaining a new chain. Then reversibility would mean that the new chain coincides with the old one when starting appropriately.

Definition 2.10 (Time reversal). Let P be an irreducible Markov chain and π the invariant distribution. The **time reversal** of P is $\widehat{P}$ defined by

$$\widehat{P}_{xy} := \frac{\pi(y)P_{yx}}{\pi(x)}.$$

Note that this definition makes sense as the entries of π are positive.

Note that $\widehat{P}$ is a stochastic matrix:

$$\sum_{y \in S} \widehat{P}_{xy} = \frac{1}{\pi(x)} \sum_{y \in S} \pi(y)P_{yx} = \pi(x)/\pi(x) = 1.$$

Of course, the chain with matrix P is reversible if and only if $P = \widehat{P}$.

Proposition 2.5. *The distribution π is also stationary (invariant) for $(\widehat{X}, \widehat{P})$ and*

$$\mathbb{P}_\pi \ (X_0 = x_0, X_1 = x_1, \ldots, X_n = x_n)$$
$$= \widehat{\mathbb{P}}_\pi(\widehat{X}_0 = x_n, \widehat{X}_1 = x_{n-1}, \ldots, \widehat{X}_n = x_0). \qquad (2.4)$$

The proof is straightforward computation and we omit it here. (See Levin and Peres (2017, p. 14).)

Example 2.3 (RW on the N-cycle). The reader is asked to verify at the end of this chapter that the nearest neighbor random walk on the N-cycle, $N \geq 2$, is not reversible when the probability of moving to the neighbor clockwise is $p \neq 1/2$, and in fact the time reversal is the modified walk when p is replaced by $q = 1 - p$.

Finally, for an irreducible and reversible Markov chain the transition matrix has a convenient property:

Proposition 2.6 (Diagonalizability). *If the matrix P corresponds to an irreducible and reversible chain, then it is similar to a symmetric matrix, hence all its eigenvalues are real and it is diagonalizable.*

Diagonalizability makes computations, like calculating powers, easy, and we see later that such a convenient situation is lost in general in the time-inhomogeneous setting. (See the discussion following Notation 4.1.)

Proof. By irreducibility, the vector π of the invariant distribution has positive entries. Let $Q := DPD^{-1}$, where

$$D := \mathrm{diag}(\sqrt{\pi(1)}, \ldots, \sqrt{\pi(n)}).$$

Clearly,

$$Q_{xy} = (DPD^{-1})_{xy} = \sqrt{\frac{\pi(x)}{\pi(y)}}\, P_{xy}.$$

Then the detailed balance equations

$$\pi(x)P_{xy} = \pi(y)P_{yx} \tag{2.5}$$

imply that Q is symmetric:

$$Q_{yx} = \sqrt{\pi(y)}P_{yx}\frac{1}{\sqrt{\pi(x)}} = \frac{\pi(y)P_{yx}}{\sqrt{\pi(y)}}\frac{1}{\sqrt{\pi(x)}} \stackrel{(2.5)}{=} \frac{\pi(x)P_{xy}}{\sqrt{\pi(y)}}\frac{1}{\sqrt{\pi(x)}}$$

$$= \sqrt{\pi(x)}P_{xy}\frac{1}{\sqrt{\pi(y)}} = Q_{xy}.$$

Hence, $P = D^{-1}QD$, with Q being a symmetric matrix, as stated.
□

2.5   Conditioning with Doob's h-Transform

A handy tool for Markov chains is Doob's h-transform.[4] The letter h refers to "harmonic."

Definition 2.11 (Harmonic function). A function $\mathsf{S} \to \mathbb{R}$ is called **harmonic** (with respect to the given transition matrix P) if $Ph = h$ holds.

---

[4]Named after J. L. Doob (1910–2004), the "founding father" of martingale theory.

If h is harmonic with respect to the transition matrix of the Markov chain X, then it is an easy exercise (left to the reader) to show that M defined by $M_n := h(X_n)$ is a martingale with respect to the canonical filtration of the chain.

Definition 2.12 (Doob's h-transform). Let $S \to \mathbb{R}$ be a positive harmonic function (with respect to the given transition matrix P). Define the transition matrix P^h by

$$P^h(x, y) := \frac{h(y)P_{xy}}{h(x)}.$$

We call P^h the **h-transformed transition matrix** and the corresponding Markov chain the **h-transformed chain**.

To see that P^h is indeed a transition matrix, write

$$\sum_{y \in S} P^h_{xy} = [h(x)]^{-1} \sum_{y \in S} h(y)P_{xy} = [h(x)]^{-1}Ph(x) = 1.$$

Remark 2.5. The condition that h is positive can be relaxed a bit. If $B \subset S$ is absorbing $(P(x, x) = 1, x \in B)$, then it is enough to require that $h > 0$ holds on $S \setminus B$. In this case, P^h_{xy} is defined as before when $x \notin B$ and for $x \in B$, $P^h_{x,x} := 1$.

If $a, b \in S$ are absorbing states, then define $\tau_a := \inf\{m \geq 0 \mid X_m = a\}$ and $\tau_b := \inf\{m \geq 0 \mid X_m = b\}$ and furthermore define $h(x) := P_x(\tau_b < \tau_a)$. This function is positive on $S \setminus \{a\}$. In fact, it is also harmonic because by the Markov property,

$$Ph(x) = \sum_{y \in S} h(y)P_{xy} = \sum_{y \in S} P_y(\tau_b < \tau_a)P_{xy} = P_x(\tau_b < \tau_a) = h(x).$$

Finally, it is clear that $h(a) = 0$ and $h(b) = 1$.

Claim 2.1 (Conditioning with h-transform). For the above h, the transition matrix P^h corresponds to the chain which is obtained from the original chain by conditioning on $\tau_b < \tau_a$.

Proof. For P^h, we obtain that $P^h_{a,a} = P^h_{b,b} = 1$, and for $x \neq a$,

$$P^h_{xy} = \frac{P_{xy}P_y(\tau_b < \tau_a)}{P_x(\tau_b < \tau_a)} = \frac{\mathbb{P}_x(X_1 = y, \tau_b < \tau_a)}{P_x(\tau_b < \tau_a)} = \mathbb{P}_x(X_1 = y \mid \tau_b < \tau_a),$$

as claimed. □

Perhaps the simplest example is that of a simple random walk on $\{0, 1, \ldots, n\}$, where 0 and n are absorbing states. Since, by an easy recursion, $h(k) := \mathbb{P}_k(\tau_n < \tau_0) = k/n$, the transition matrix for the chain conditioned to be absorbed at n is P^h given by $P^h_{xy} = \frac{y}{x} P_{xy}$.

2.6   Urn Models

A particularly useful Markov chain[5] is the **Pólya urn model,** introduced by Pólya and his student F. Eggenberger in 1923. In some sense, it is the opposite of the model of sampling without replacement, where every time a particular color is observed, it is less likely to be observed again. In a Pólya urn model, an observed color is *more* likely to be observed again. In the simplest case, we have an urn with two balls in it, one blue and one white. In each step, one chooses a ball at random from those already in the urn, returns it to the urn, *together* with another ball of the same color. This is a **reinforcement** mechanism, sometimes compared to the saying "the rich gets richer."

Let B_k (W_k) be the number of blue (white) balls in the urn after the addition of k balls, $k \geq 0$. The process $(B_k, W_k)_{k \geq 0}$ is a Markov chain with state space $\mathsf{S} := \{1, 2, \ldots\}^2$. A nice property is that B_k takes any possible value with the same probability.

Proposition 2.7 (Uniformity). *The distribution of B_k (or W_k) is uniform on the set $\{1, 2, \ldots, k+1\}$ for $k \geq 1$.*

Proof. The proof is by induction. The $k = 1$ case is obvious. For a given $1 \leq k \leq k+1$, one has the recursion

$$\mathbb{P}(B_k = j) = \left(\frac{j-1}{k+1}\right)\mathbb{P}(B_{k-1} = j - 1) + \left(\frac{k+1-j}{k+1}\right)\mathbb{P}(B_{k-1} = j).$$

Now, if the statement is true for $k-1$, then the right-hand side equals

$$\frac{1}{k}\left[\frac{j-1}{k+1} + \frac{k+1-j}{k+1}\right] = \frac{1}{k+1},$$

yielding the statement for k. □

---

[5]It is also a quintessential example of an exchangeable process.

There are many generalizations of the Pólya urn. For example, suppose that we return the ball along with c new balls of the same color. Usually one takes $c \geq 0$. Nevertheless, the model actually makes sense if $c < 0$ too. One then interprets this to mean that we remove the balls rather than add them; the process terminates when we run out of the balls. Clearly, $c = 0$ corresponds to sampling with replacement and $c = -1$ corresponds to sampling without replacement.

In the sequel, we assume that $c \geq 1$, while a is the starting number of blue balls and b is the starting number of white balls.

An intriguing property of the model is as follows. Let X_i be the indicator that we select a blue ball at time $i \geq 1$. It is not hard to see (left to the reader) that given a, b, c, n, the joint probability $\mathbb{P}(X_1 = x_1, \ldots, X_n = x_n)$ depends on $(x_1, x_2, \ldots, x_n)$ only through the number of blue balls $k = x_1 + \cdots + x_n$, namely, it equals

$$\frac{(a/c)^{(k)} (b/c)^{(n-k)}}{(a/c + b/c)^{(n)}}, \tag{2.6}$$

where $r^{(j)} := r(r+1)(r+2)\ldots(r+j-1)$. (Also, the dependence on a, b, c is only through a/c and b/c.) Thus, for any $n \geq 1$,

$$\mathbb{P}(X_1 = x_1, \ldots, X_n = x_n) = \mathbb{P}(X_{\sigma(1)} = x_1, \ldots, X_{\sigma(n)} = x_n)$$

whenever $\sigma(1), \ldots, \sigma(n)$ is a permutation of $1, 2, \ldots, n$. In such a case, we say that $X_1, X_2, \ldots$ is an **exchangeable** sequence of random variables. Exchangeable random variables have a beautiful theory, first developed systematically by B. de Finetti in the 1930s. The following classic theorem (see, e.g., Fristedt and Gray (1997)) is due to him.

Theorem 2.3 (de Finetti's theorem). *A sequence of indicator variables $X_1, X_2, \ldots$ is exchangeable if and only if there exists a probability measure Q on $[0, 1]$ such that for each $n \geq 1$ and $x_i \in \{0, 1\}$, $1 \leq i \leq n$,*

$$\mathbb{P}(X_1 = x_1, \ldots, X_n = x_n) = \int_0^1 \theta^{t_n} (1 - \theta)^{n - t_n} \, Q(d\theta),$$

where $t_n = \sum_1^n x_i$.

Furthermore, Q is uniquely identified as follows. The random variable $\lim_{n \to \infty} \frac{1}{n} \sum_1^n X_i =: \overline{X}_\infty$ exists almost surely and its distribution is Q.

Finally, $\mathbb{P}(X_1 = x_1, \ldots, X_n = x_n \mid \overline{X}_\infty = \theta) = \theta^{t_n}(1-\theta)^{n-t_n}$, for $\theta \in [0,1]$.

Thus, conditional on the value of the limiting frequency, the random variables are independent, and in law, we can obtain the sequence as a mixture of independent Bernoulli's.

Definition 2.13 (de Finetti measure). The probability measure Q appearing in de Finetti's theorem is called the **de Finetti measure**.

Remark 2.6. Since the 1930s, de Finetti's theorem has been generalized from indicator variables to general random variables and in several other ways to due to the work of Hewitt, Savage, Diaconis, Freedman and others.

Coming back to the Pólya urn, de Finetti's theorem implies a convergence result. We state it in a bit more general form. Instead of starting with one blue and one white ball, let us start it with $a \geq 1$ blue and $b \geq 1$ white balls. Let the proportion of blue balls selected in the first n trials be T_n and the proportion of blue balls in the urn be Z_n. Since $T_n = \frac{1}{n}\sum_1^n X_i$, where X_i is the indicator of the ith selection being a blue ball, and since the sequence $X_1, X_2, \ldots$ is exchangeable, de Finetti's theorem implies that the law of T_n converges to the de Finetti measure Q. One only has to identify Q.

Theorem 2.4 (Convergence for Pólya urn). *For the exchangeable sequence $X_1, X_2, \ldots$ defined by the Pólya urn, the de Finetti measure is $Q = \mathsf{Beta}(a/c, b/c)$. Equivalently, T_n has an almost sure limit as $n \to \infty$, distributed according to $\mathsf{Beta}(a/c, b/c)$.*

The proof can be found in Mahmoud (2009) (see Theorem 3.2. and its proof); alternatively, see Chapter 5, Example 5.9 in Rényi (1970).

Note: Since,

$$Z_n = \frac{a}{a+b+cn} + \frac{cn}{a+b+cn}T_n = o(1) + (1+o(1))T_n,$$

the a.s. limit of Z_n is the same.

Remark 2.7 (Proposition 2.7 through de Finetti). When $a = b = 1$, de Finetti's measure in Theorem 2.4 is uniform on the unit interval. Let W_k be as in Proposition 2.7. If $\mathcal{P}$ is uniformly distributed on $[0, 1]$, then conditionally on $\mathcal{P} = p$, the indicators of drawing white balls are i.i.d. Bernoulli(p); hence $W_k \sim \text{Bin}(k, p)$. Using a well-known identity for the beta function,

$$P(W_k = \ell) = \int_0^1 \binom{k}{\ell} p^\ell (1 - p)^{k-\ell} dp = \frac{1}{k+1}, \quad \ell = 0, 1, \ldots, k.$$

This re-proves Proposition 2.7. ◇

Another possible generalization is when there are d different colors. We start with d balls in the urn, each of a different color. The rule is the same as for the basic two-color model. A result analogous to Proposition 2.7 holds, Levin and Peres (2017, pp. 25–26).

2.7 Total Variation Distance and Convergence

If we want to quantify convergence results for Markov chains, then we need a notion of distance between probability distributions.

Definition 2.14 (Total variation distance). Let μ and ν be probability distributions on S. Their **total variation distance** is defined[6] as

$$\|\mu - \nu\|_{TV} := \max_{A \subseteq \mathsf{S}} |\mu(A) - \nu(A)|.$$

The reader is asked to check that

(i) this indeed defines a distance, that is, that the triangle inequality holds,
(ii) the following Definition 2.15 is in fact equivalent to Definition 2.14.

Definition 2.15 (Alternative definition of TVD). The total variation distance between μ and ν is defined as

$$\|\mu - \nu\|_{TV} := \frac{1}{2} \sum_{x \in \mathsf{S}} |\mu(x) - \nu(x)|.$$

---

[6]Sometimes a factor 2 is included.

Clearly, the second definition is easier to use as it is hard to maximize over all the subsets of S.

Remark 2.8. The total variation distance can be written slightly differently as

$$\|\mu - \nu\|_{TV} := \sum_{x \in S: \mu(x) \geq \nu(x)} |\mu(x) - \nu(x)| . \qquad (2.7)$$

The proof is left to the reader — see the exercises at the end of this chapter.

The following definition is extremely useful. We assume that $|S| < \infty$ although this is not essential.

Definition 2.16 (Coupling of distributions). Let μ and ν be probability distributions on S. Any joint variable $(X, Y) : \Omega \to S^2$ on some probability space Ω is a **coupling** between these distributions whenever the marginal distribution of X is μ and the marginal distribution of Y is ν, that is,

$$\mathbb{P}(X = x) = \mu(x), \mathbb{P}(Y = y) = \nu(y), (x, y) \in S^2.$$

Obviously, if (X, Y) is a coupling of μ and ν with $q(x, y) := \mathbb{P}(X = x, Y = y), x, y \in S$, then

$$\sum_{y \in S} q(x, y) = \mu(x), \quad \sum_{x \in S} q(x, y) = \nu(y).$$

Example 2.4. Consider tossing a fair coin. That is, let $S = \{0, 1\}$ and $\mu(0) = \mu(1)$. We want to couple two such coin tosses, that is, we take $\nu := \mu$. Then each one of these are couplings:

(1) Use independent tosses, that is, let X and Y be independent.
(2) Use the same toss, that is, $X = Y$.
(3) Turn the coin over after tossing it, that is, $Y := 1 - X$.

The total variation distance can in fact be expressed by using the notion of coupling.

Theorem 2.5 (TVD with coupling). *For two probability distributions μ and ν, one has*

$$\|\mu - \nu\|_{TV} = \inf_{(X,Y) \in \mathcal{C}} \mathbb{P}(X \neq Y),$$

where $\mathcal{C}$ denotes the collection of all possible couplings of μ and ν.

Remark 2.9.

(1) The intuition behind the result is as follows. Arranging the joint probability masses in an $|S| \times |S|$ table, our goal is to find a "close" coupling where the sum of the off-diagonal weights is small. The total variation distance is telling us how "close" such a coupling can be.

(2) When $\mu = \nu$, the coupling $X = Y$ shows that the right-hand side is indeed zero.

Theorem 2.6 (Infimum is minimum). *In fact*

$$\|\mu - \nu\|_{TV} = \min_{(X,Y)\in\mathcal{C}} \mathbb{P}(X \neq Y),$$

that is, the infimum is attained.

Definition 2.17 (Optimal coupling). If $\|\mu - \nu\|_{TV} = \mathbb{P}(X \neq Y)$, for some coupling (X,Y), then it is an **optimal coupling**.

Example 2.5. Let $S := \{a,b\}$, $\mu = (u, 1-u)$ and $\nu = (v, 1-v)$ with $0 \leq u \leq v \leq 1$. Then

$$\|\mu - \nu\|_{TV} = \frac{1}{2}[(v-u) + (1 - u - 1 + v)] = v - u.$$

In terms of couplings, if $p := \mathbb{P}(X = a, Y = a)$, then

$$\mathbb{P}(X \neq Y) = (v - p) + (u - p) = u + v - 2p.$$

Hence, we would like to find the largest possible p. The constraint is that

$$u + v - 1 \leq p \leq u.$$

We obtained that the optimal coupling is attained with $p := u = \min\{u, v\}$, and for that $\mathbb{P}(X \neq Y) = v - u$ holds. See Figure 2.5.

We now prove Theorem 2.5 together with Theorem 2.6.

Proof. First, $\|\mu - \nu\|_{TV} \leq \inf_{(X,Y)\in\mathcal{C}} \mathbb{P}(X \neq Y)$ holds because for any coupling (X,Y) and $A \subset S$ one has

$$\mu(A) - \nu(A) = \mathbb{P}(X \in A) - \mathbb{P}(Y \in A) \leq \mathbb{P}(X \in A, Y \notin A)$$
$$\leq \mathbb{P}(X \neq Y).$$

Hence, to complete the proof, it is enough to find an optimal coupling.

$\mathbb{P}(\cdot,\cdot)$	$X=a$	$X=b$
$Y=a$	p	$v-p$
$Y=b$	$u-p$	$1-u-v+p$

$\mathbb{P}(\cdot,\cdot)$	$X=a$	$X=b$
$Y=a$	u	$v-u$
$Y=b$	0	$1-v$

Fig. 2.5. The joint p.m.f. of X and Y and the optimal case with $p=u$ in case $u \leq v$.

Writing $a \wedge b$ for $\min\{a, b\}$, let

$$p := \sum_{x \in S}[\mu(x) \wedge \nu(x)] = \sum_{x \in S : \mu(x) \leq \nu(x)} \mu(x) + \sum_{x \in S : \mu(x) > \nu(x)} \nu(x) \in [0, 1].$$

Using (2.7),

$$p = 1 - \sum_{x \in S : \mu(x) > \nu(x)} [\mu(x) - \nu(x)] = 1 - \|\mu - \nu\|_{TV}. \tag{2.8}$$

It is useful to define two distributions α and β as follows:

$$\alpha(x) := (1-p)^{-1}[\mu(x) - \nu(x)], \text{ for } \mu(x) > \nu(x)$$

and $\alpha(x) := 0$ otherwise, and

$$\beta(x) := (1-p)^{-1}[\nu(x) - \mu(x)], \text{ for } \nu(x) > \mu(x)$$

and $\beta(x) := 0$ otherwise.

Now, toss a biased coin, for which the probability of heads equals p. According to the outcome, we choose the pair (X, Y). Since X and Y will depend on the outcome of the same experiment, they will form a coupling as long as X is distributed as μ and Y is distributed as Y.

The algorithm is as follows:

(a) If the coin lands on heads, then we set $X = Y = Z$, where Z takes the value $x \in S$ with probability $\gamma(x) := p^{-1}[\mu(x) \wedge \nu(x)]$. (If $p = 0$, then we define X and Y in an arbitrary manner.)

(b) Otherwise we pick X and Y independently, according to the distributions α and β, respectively.

Since $p\gamma + (1-p)\alpha = \mu$ and $p\gamma + (1-p)\beta = \nu$, X is distributed as μ and Y is distributed as Y. Furthermore, in case (b), $X \neq Y$ as α and β are singular measures (their supports are disjoint), that is, $X \neq Y$ if and only if case (b) holds and so $\mathbb{P}(X \neq Y) = 1 - p$. By (2.8), then $\mathbb{P}(X \neq Y) = \|\mu - \nu\|_{TV}$, as needed. $\square$

Using the notion of total variation distance, we now present the basic convergence result.

Theorem 2.7 (Convergence and speed estimate). *Let the matrix P correspond to an aperiodic irreducible chain with stationary distribution π. Then there exists an $\alpha \in (0,1)$ and a $C > 0$ such that*

$$\max_{x \in S} \|P_{x\cdot}^n - \pi\|_{TV} \leq C\alpha^n.$$

Proof. Let Π be the stochastic matrix which is such that each row is π. Then, for any stochastic matrix M, one has $M\Pi = \Pi$ because the entries in each row of M add up to one. It is also easy to check (left to the reader) that $\Pi M = \Pi$ holds for every matrix M such that $\pi M = \pi$.

By Proposition 2.2, we may and do pick an $r > 0$ such that all entries of P^r are positive. Since there are finitely many entries, there is a $\delta_0 > 0$ such that $P_{xy}^r \geq \delta_0 \pi(y)$, for all $x, y \in S$.

Make sure that $\delta_0 \in (0,1)$, fix a $\theta \in (1 - \delta_0, 1)$ and define the matrix Q via $P^r = (1-\theta)\Pi + \theta Q$, that is, $Q := \theta^{-1}(P^r - (1-\theta)\Pi)$. By the choice of θ, Q has non-negative entries, and a trivial computation shows that Q is stochastic too.

The key observation is that even if θ is close to one, for large k, the weight of Π in the decomposition on P^{rk} is very large:

$$P^{rk} = (1 - \theta^k)\Pi + \theta^k Q^k, \quad k \geq 1.$$

To see why this is true, note that for $k = 1$ the statement is just the definition of Q and if the statement holds for some $n \geq 1$, then

$$P^{r(n+1)} = P^{rn}P^r = [(1 - \theta^n)\Pi + \theta^n Q^n]P^r,$$

which, by the definition of Q, equals

$$[(1 - \theta^n)\Pi + \theta^n Q^n][(1 - \theta)\Pi + \theta Q].$$

Using that $\Pi P^r = \Pi, Q^n \Pi = \Pi, \Pi^2 = \Pi$ and $\Pi Q = \Pi$, the last expression indeed reduces to $(1 - \theta^{n+1})\Pi + \theta^{n+1}Q^{n+1}$, completing the induction.

Next, for $j = 0, 1, \ldots, r - 1$, we obtain

$$P^{rk+j} = (1 - \theta^k)\Pi P^j + \theta^k Q^k P^j.$$

But $\Pi P^j = \Pi$ because $\pi P^j = \pi$ by invariance. We thus obtain that

$$P^{rk+j} - \Pi = \theta^k (Q^k P^j - \Pi).$$

Take row x, sum the absolute values of the entries, and divide by two. Then the left-hand side becomes $\|P_{x\cdot}^{rk+j} - \pi\|_{TV}$, while the right-hand side becomes $\theta^k \|Q^k P^j - \Pi\|_{TV} \leq \theta^k$. We thus have $\|P_{x\cdot}^{rk+j} - \pi\|_{TV} \leq \theta^k$, for every $x \in S$ and $j = 0, 1, \ldots, r-1$, that is,

$$\|P_{x\cdot}^n - \pi\|_{TV} \leq \theta^{\frac{n-j}{r}},$$

and picking $j = r - 1$ yields the upper bound $\theta^{-\frac{r-1}{r}} (\theta^{1/r})^n =: C\alpha^n$, which works since $\alpha := \theta^{1/r} < 1$. $\square$

This discussion would not be complete without mentioning the work of W. Döblin[7] (1915–1940). His main result about Markov chains is as follows.

Theorem 2.8 (Döblin's theorem). *Let P be a transition matrix of a Markov chain with the property that for some state $j_0 \in S$ and $\epsilon > 0$, $P_{i,j_0} \geq \epsilon$ for all $i \in S$. Then there is a unique stationary vector (distribution) π and $\pi_{j_0} \geq \epsilon$. Furthermore, for all initial distributions μ, one has*

$$\|\mu P^n - \pi\|_{TV} \leq 2(1 - \epsilon)^n, \; n \geq 0. \tag{2.9}$$

Proof. If $\rho \in \mathbb{R}^S$ is any row vector, then it is easy to see that $\sum_{j \in S} (\rho P)_j = \sum_{i \in S} \rho_i$, and if also $\sum_{i \in S} \rho_i = 0$, then

$$\|\rho P^n\|_{TV} \leq (1 - \epsilon)^n \|\rho\|_{TV}, \; n \geq 1.$$

(The second part can be obtained by induction.) In this latter case,

$$|(\rho P)_j| = \left| \sum_{i \in S} \rho_i (P)_{i,j} \right|$$

$$= \left| \sum_{i \in S} \rho_i ((P)_{i,j} - \epsilon \delta_{j,j_0}) \right| \leq \sum_{i \in S} |\rho_i| ((P)_{i,j} - \epsilon \delta_{j,j_0}),$$

where $\delta_{i,j}$ is the Kronecker delta. Adding up for all j's, it follows that $\|\rho P\|_{TV} \leq (1 - \epsilon) \|\rho\|_{TV}$.

---

[7]His tragic story during WW2 is well known. His life was the subject of the 2007 movie "A Mathematician Rediscovered."

Next, if μ is a probability vector and $\mu_n := \mu P^n$, then $\mu_n = \mu_{n-m} P^m$ and $\sum_i ((\mu_{n-m})_i - \mu_i) = 1 - 1 = 0$. Thus,

$$\|\mu_n - \mu_m\|_{TV} \leq (1 - \epsilon)^m \|\mu_{n-m} - \mu\|_{TV} \leq 2(1 - \epsilon)^m,$$

proving the Cauchy property and the existence of a probability vector π to which the μ_i converge in total variation norm. The stationarity of π is trivial, and moreover,

$$(\pi)_{j_0} = \sum_{i \in S} \pi_i P_{i,j_0} \geq \epsilon \sum_{i \in S} \pi_i = \epsilon.$$

Then (2.9) and the uniqueness of the stationary vector both follow from the bound $\|\mu P^m - \pi\|_{TV} = \|(\mu - \pi) P^m\|_{TV} \leq 2(1 - \epsilon)^m.$ □

Remark 2.10. In fact, Döblin's theorem is valid even when $|S| = \aleph_0$. ◇

We conclude this chapter with noting that for time-homogeneous Markov chains, under appropriate conditions, the central limit theorem holds:

$$\mathsf{Law}\left(\frac{S_N - \mathbb{E}(S_N)}{\sqrt{N}}\right) \to \mathsf{Normal}\left(0, \sigma^2\right),$$

where $S_N := \sum_1^N X_i$ and σ^2 depends on the covariance structure. See Jones (2004) and also Example 5.1 of this book as an application.

2.8   Exercises

Problem 1. Let $P := \begin{pmatrix} 1 - p & p \\ q & 1 - q \end{pmatrix}$ with $p, q \in (0, 1)$. Verify that $\lambda := 1 - p - q$ is an eigenvalue of P with corresponding eigenvector $(r, s) = (1/2, -1/2)$.

Problem 2. An urn is given with a blue and b white balls in it, altogether $M = a + b$ balls. At each step, one ball is chosen randomly and is replaced by a ball of the *opposite* color. Is this chain reversible? Is it periodic? Prove that the invariant distribution is binomial with parameters $M, 1/2$.

Next, let n (the number of steps) converge to infinity. Does the chain have a limiting distribution? Can you establish some partial limits? (Hint: Prove that the distribution of the blue balls in the urn tends to some binomial distribution when $n \to \infty$ through even numbers and also when $n \to \infty$ through odd numbers. What can you say about the support of the limiting binomial distribution in those two cases?)

Problem 3. Verify that the nearest neighbor random walk on the N-cycle, $N \geq 2$, is not reversible when the probability of moving to the neighbor clockwise is $p \neq 1/2$, and in fact the time reversal is the modified walk when p is replaced by $q = 1 - p$.

Problem 4. You have k dollars, and at each step, you toss a fair coin, winning a dollar or losing one, according to the outcome. You stop this game the first time when you either have n dollars or you are bankrupt. Show that the expected duration of this game is $k(n - k)$ steps. Hint: The statement is clearly true for $k = 0, n$. Otherwise try to establish a connection between the expectations when starting at $k - 1, k$ and $k + 1$.

Problem 5. We have seen that for an aperiodic irreducible chain there is an integer r_0 such that for all $r \geq r_0$ and all $x, y \in \mathsf{S}$, $P_{xy}^r > 0$. Show that the smallest such r_0 is $r_0 = 2$ ($r_0 = 4$) for the simple random walk on the 3-cycle (5-cycle). Hint: For the 3-cycle check that $r = 1$ does not work but $r = 2$ and $r = 3$ do. Infer that $r = 2k + 3\ell = 2(k + \ell) + \ell$ works too if either $k \geq 1$ or $\ell \geq 1$, hence each $r \geq 2$ works. Argue similarly for the 5-cycle: $r = 1, 2, 3$ do not work but $r = 4, 5$ do. What is the general answer for the $2k + 1$-cycle?

Problem 6. Consider the simple random walk on $\mathbb{Z}$, starting at the origin. Show that the probability that the walker is at the origin after $n = 2k$ steps is asymptotical with $\frac{1}{\sqrt{\pi n}}(4p(1 - p))^n$ as $n \to \infty$ and that for $p \neq 1/2$, it is summable. Use the Borel–Cantelli lemma to conclude that the walk is transient when it is not symmetric.

Problem 7. Verify (2.6).

Problem 8. Prove that Definition 2.15 is equivalent to Definition 2.14. Then check the triangle inequality for the latter one.

Problem 9. Let $\mathsf{S} := \{a, b\}$ and $P := \begin{pmatrix} 1 - p & p \\ q & 1 - q \end{pmatrix}$ with $p, q \in (0, 1)$. Let $\lambda := 1 - p - q$. Let $\mu_0 := (1, 0)$ and let π denote the invariant distribution. We have already seen that

$$\pi = \left(\frac{q}{p+q}, \frac{p}{p+q} \right).$$

Let $\Delta_n := \mu_n(a) - \frac{q}{p+q}, n \geq 0$, where μ_n is defined after Definition 2.2. Prove by direct calculation that

$$\Delta_n = \lambda^n \Delta_0.$$

Problem 10. Verify (2.7).

Chapter 3

Absorption and Electric Networks

In this chapter, we discuss an important technique for the analysis of reversible Markov chains: the **method of electric networks**.

Consider a time-homogeneous Markov chain on a finite state space $S := \{0, 1, 2, \ldots, r\}$, $r \geq 1$, so that S has at least two states. Suppose that the state r is an absorbing state, that is, $P_{rr} = 1$ and $P_{rj} = 0$ for all $j \neq r$, and that it is the *only* absorbing state (or communication class), that is, r is accessible from any state. It is easy to see that in this case

$$\lim_{n \to \infty} \mathbb{P}_x(X_n = i) = \begin{cases} 1, & \text{if } i = r; \\ 0, & \text{if } i \in S \setminus \{r\}. \end{cases}$$

Let us define the time to absorption, τ, as

$$\tau = \min\{n \geq 0 : X_n = r\}.$$

This random variable is finite regardless of the value of X_0; in fact, it has a finite expectation. A simple explanation of this fact is as follows: since r is accessible from any x, and S is finite, there is an $\epsilon > 0$ such that $\mathbb{P}(\text{one of } X_{k+1}, X_{k+2}, \ldots, X_{k+r} \text{ equals } r \mid X_k = x) \geq \epsilon$ for all $x \neq r$. Indeed, r can be reached from x in a finite number of steps with positive probability, hence it can also be reached while never being in the same state twice, that is, at most in r steps (otherwise there are "loops" which may be "cut out" from the sequence of steps.)

Using the Markov property,

$$\mathbb{E}(\tau) = \sum_{n=0}^{\infty} \mathbb{P}(\tau > n) \le r \sum_{m=0}^{\infty} \mathbb{P}(\tau > rm) \le r \sum_{m=0}^{\infty} (1 - \epsilon)^m = r/\epsilon.$$

In fact, it is not very hard to compute $\mathbb{E}(\tau)$ explicitly. Let

$$P_0 = \begin{pmatrix} p_{00} & p_{01} & \cdots p_{0\,r-1} \\ p_{10} & p_{11} & \cdots p_{1\,r-1} \\ \vdots & \vdots & \ddots \vdots \\ p_{r-1\,0} & p_{r-1\,1} & \cdots p_{r-1\,r-1} \end{pmatrix}$$

be the $r \times r$ submatrix of the transition probability matrix P and $\mu = (\mu_0, \mu_1, \ldots, \mu_{r-1})$ be the $r \times 1$ column vector with $\mu_x = \mathbb{E}(\tau \mid X_0 = x)$. Then it is well known (see e.g., Taylor and Karlin (1998)) that

$$\mu = (\mathbf{I_r} - P_0)^{-1} \mathbf{1}, \tag{3.1}$$

where $\mathbf{I_r}$ is the $r \times r$ identity matrix and $\mathbf{1}$ is the $r \times 1$ column vector of ones.

The situation becomes more complicated if we have more than one absorbing state. In what follows, we treat only the situation when there are two absorbing states, say 0 and r, where we still assume that $S = \{0, 1, \ldots, r\}$. It is natural to ask now a more basic question: if the Markov chain starts from some state $x \in S$, what is the probability that it gets absorbed at r and not at 0? Such question is typical, for example, for the famous Gambler's ruin problem, a version of which is the following.

Two gamblers, A and B, play against each other, and at each round of the game they bet \$1. If player A wins, her capital increases by \$1 and the capital of player B decreases by \$1. If player A loses, the opposite occurs. Once one of the players has run out of money, the game stops. Assume that the chances of winning and losing are equal, i.e., $1/2$, and initially player A has $x \ge 1$ dollars while player B has $r - x \ge 1$ dollars. What is the probability that player B wins all the money?

One can model this game by a Markov chain S with $r + 1$ states where $X_n = x$ means that player A has x dollars; the transition probabilities are

$$P_{ij} = \begin{cases} 1/2, & \text{if } j = i - 1 \text{ and } i \in \mathsf{S} \setminus \{0, r\}; \\ 1/2, & \text{if } j = i + 1 \text{ and } i \in \mathsf{S} \setminus \{0, r\}; \\ 1, & \text{if } i = j = 0 \text{ or } i = j = r; \\ 0, & \text{otherwise.} \end{cases}$$

We can again define the stopping time

$$\tau = \min\{n \geq 0 : X_n = 0 \text{ or } X_n = r\}$$

and then the question we asked above is equivalent to finding $\mathbb{P}_x(X_\tau = r) = \mathbb{P}(X_\tau = r \mid X_0 = x)$ as a function of $x \in \mathsf{S}$; this quantity is well defined, as it is still easy to show that indeed $\tau < \infty$ a.s., similar to the case above.

Fortunately, there is a very neat technique to answer such questions whenever the underlying Markov chain is reversible, or at least can be made reversible by changing the transition probabilities for the absorbing states (0 and r in the example above); recall that Markov chain is called reversible if there is a probability distribution π on S which satisfies the detailed balance equations, that is, $\pi(x)P_{xy} = \pi(y)P_{yx}$ for all $x, y \in \mathsf{S}$; in our case, we require this equality for all $x, y \in \mathsf{S} \setminus \{0, r\}$, where 0 and r are the absorbing states. The method, introduced in Doyle and Snell (1984), consists in constructing an electric network corresponding to the Markov chain, and it goes as follows.

Consider an *irreducible, reversible* Markov chain on state space $\mathsf{S} = \{0, 1, \ldots, r\}$. Let $\mathcal{E}$ be an electric network with nodes $0, 1, \ldots, r$. Between each pair of nodes $i, j \in \mathsf{S}$, such that $P_{ij} > 0$, set a resistor with resistance $R_{ij} = 1/C_{ij}$ such that[1]

$$P_{ij} = \frac{C_{ij}}{\sum_{k \in \mathsf{S}, k \sim i} C_{ik}} = \frac{1/R_{ij}}{\sum_{k \in \mathsf{S}, k \sim i} 1/R_{ik}}, \tag{3.2}$$

where $k \sim i$ means that nodes k and i are connected in the model graph G corresponding to the chain, that is, $P_{ik} > 0$ (observe that the reversibility of the chain implies $P_{ki} > 0$ as well) and the set of the vertices of G is S.

---

[1]The quantity C_{ij} is called *conductance*.

It turns out that for reversible Markov chains (and only for them, in fact), such constructions are possible (moreover, the corresponding collection of resistors assigned to each edge is unique, up to multiplying each resistor by a fixed positive constant). Indeed, for each pair of nodes in S such that $P_{ij} > 0$, assign a resistor with the value

$$R_{ij} = (\pi(i)P_{ij})^{-1}, \tag{3.3}$$

where π is the measure appearing in the definition of reversibility. Note that it is unimportant how we choose indices i and j, since if we swap them, the value of the resistance will not change due to the equality $\pi(i)P_{ij} = \pi(j)P_{ji}$.

It is particularly easy to construct the electric network corresponding to a simple random walk on a non-oriented connected graph G with the set of nodes S. Recall that a simple random walk is defined as the Markov chain X_n taking values on the nodes of G such that from each node $x \in$ S the walk is equally likely to go to any of its immediate neighbors on G, that is,

$$\mathbb{P}(X_{n+1} = y \mid X_n = x) = \begin{cases} 1/N_x, & \text{if } x \text{ and } y \text{ are connected by} \\ & \quad \text{an edge,} \\ 0, & \text{otherwise,} \end{cases}$$

where N_x denotes the total number of edges coming out of x. Now to construct the corresponding electric network, simply set a resistor of value 1 to each edge of the graph G.

Next, apply a 1-volt battery between nodes 0 and 1, such that the voltage at 0 is 0, and the voltage at r is 1. It can be shown (see Doyle and Snell (1984)) that the voltage v_x at a particular node x has a simple probabilistic interpretation: it equals the ultimate hitting probability of r, i.e., $v_x = \mathbb{P}_x(X_\tau = r)$. The reason for this fact is as follows: the function h defined by $h(x) = \mathbb{P}_x(X_\tau = r)$ is a harmonic function for our Markov chain with the boundary values 0 and 1 at nodes 0 and r, respectively; from the physical properties (Ohm's law and Kirchhoff's current law[2]), voltage is also a harmonic function with the same boundary values. Since such a harmonic function is unique, they must coincide.

---

[2]Kirchhoff's current law requires that the total current, flowing into any node other than where the battery is attached, is 0.

One may wonder if the above representation with the help of electrical networks has some practical significance. The affirmative answer to this question is based on the fact that there are well-known techniques in electric networks which allow us to compute various characteristics using some "tricks." For example, when we have two resistors connected in parallel with resistances R_1 and R_2, respectively, then they can be replaced (see, e.g., Figures 3.3 and 3.4) by just one resistor with resistance $\left(R_1^{-1} + R_2^{-1}\right)^{-1}$, while all the voltages between the remaining nodes are unchanged. In the case of two resistors connected in series, we can replace them with one resistor with resistance $R_1 + R_2$.

For example, consider the Markov chain with the transition probabilities presented in Figure 3.1; the corresponding electric network is presented in Figure 3.2. Suppose that $\tau = \min\{n : X_n \in \{0, 4\}\}$ and we want to compute $\mathbb{P}(X_\tau = 4 \mid X_0 = 1)$. Then this probability equals the unknown voltage v at node 1, when voltages at 0 and 4 are equal to $v = 0$ and $v = 1$, respectively (see Figure 3.2). A similar statement holds for nodes 2 and 3 as well.

Next, we need a definition.

Definition 3.1 (Effective resistance). Given an electric network with the battery attached to two distinct nodes u and v, the **effective resistance** between them, denoted by $R_{\text{eff}} = R_{\text{eff}}(u, v)$, is the ratio between the difference in the potentials (voltage between the nodes) and the total current flowing through the network.

Loosely speaking, we may replace the entire network lying between u and v with just one resistor with the value R_{eff}.

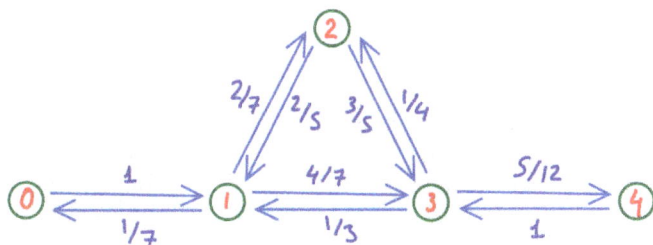

Fig. 3.1.   Model graph for a random walk on $\mathsf{S} = \{0, 1, 2, 3, 4\}$.

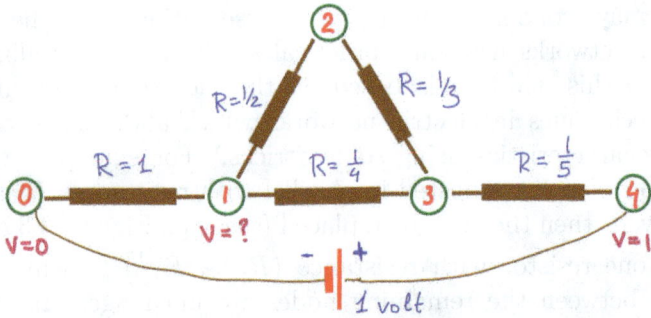

Fig. 3.2.   The electric network corresponding to the Markov chain.

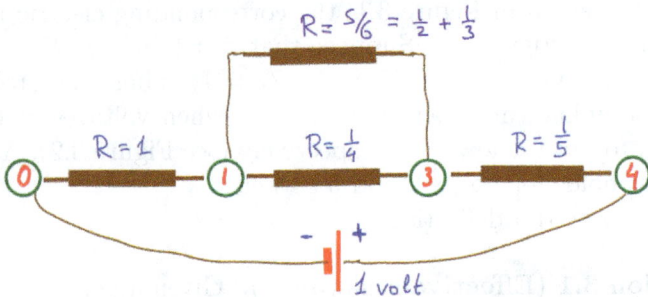

Fig. 3.3.   Merging resistors connected in series.

In order to perform the computation, we first need to compute the current going through the network, which equals $1/R_{\text{eff}}(0,4)$. To calculate $R_{\text{eff}}(0,4)$, we use the well-known rules for replacing sequences of resistors connected in series or in parallel.

First, we replace the two resistors $R_{12} = 1/2$ and $R_{23} = 1/3$ by the resistor equal to $5/6$, the sum of those, as they are connected in series, see Figure 3.3. Then we have two resistors connected in parallel, which we can replace by one with resistance $\left((5/6)^{-1} + 1/4)^{-1}\right)^{-1}$ (Figure 3.4). Finally, we have three resistors in series which we can replace by one (Figure 3.5). Hence the current going through the network equals $i = 1/R_{\text{eff}} = 130/181$. Now, by Ohm's law, the difference between the voltages at nodes 0 and 1 equals the resistance between them (which $= 1$) times the current emitting from the node 0. Consequently, the voltage at node 1 by Ohm's law equals

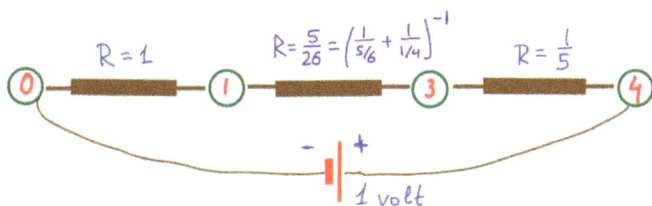

Fig. 3.4.   Merging resistors connected in parallel.

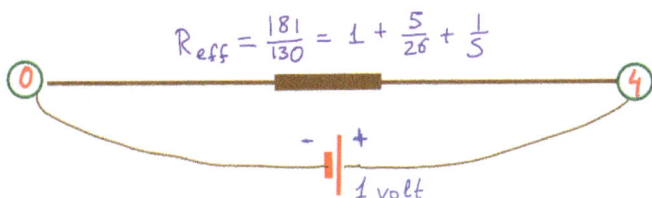

Fig. 3.5.   Merging resistors connected in series.

$v_1 = i \cdot 1 = 130/181$ and hence

$$\mathbb{P}(X_\tau = 4 \mid X_0 = 1) = \frac{130}{181}.$$

Another aspect arises if we want to *estimate* the probabilities rather than compute them exactly. In this case, we can use the so-called **Rayleigh's monotonicity law**, see Section 1.4 in Doyle and Snell (1984). This law states that if any of the resistances of the network are increased, then R_{eff} can only increase; if they are decreased, it can only decrease.

3.1   Transience and Effective Resistance

The method of electric networks is especially useful when we want to establish transience or recurrence of a given irreducible infinite-state space Markov chain. Throughout the rest of this chapter we assume that the model graph G corresponding to our Markov chain is *locally finite*, that is, the particle may jump to only a finitely many other sites. Recall that the chain is also assumed to be irreducible, and thus G is connected.

Indeed, when S is infinite, we can still construct the corresponding electric network by the method described above. Fix some vertex of the corresponding graph G, call it *the origin*, and denote by $\mathbf{0}$. For each node $x \in$ S, the distance to the origin, $\rho(x)$, is defined as the number of edges on a shortest path connecting $\mathbf{0}$ and x; we set $\rho(\mathbf{0}) = 0$. For $L \in \mathbb{N}$, we also define a "circle" of radius L as

$$D_L = \{x \in \mathsf{S} : \rho(x) = L\}.$$

Let $R^{(L)} > 0$ be the effective resistance between the origin and short-circuited D_L (thus we look only at the subnetwork $\mathcal{E}_L$ with the nodes in $\bigcup_{j=0}^{L} D_j$). The sequence $R^{(L)}$ is non-decreasing; indeed, for each $M > L$, we can bound the effective resistance between $\mathbf{0}$ and D_M below by using Rayleigh's monotonicity law for the effective resistance of a modified network, which is $\mathcal{E}_M$ with all the nodes in $D_L \cup D_{L+1} \cup \cdots \cup D_M$ being short-circuited. However, the effective resistance of the obtained network is exactly $R^{(L)}$.

Definition 3.2 (Effective resistance between 0 and ∞). The limit

$$R_\infty = \lim_{L \to \infty} R^{(L)} \in (0, +\infty]$$

exists by monotonicity and is called the **effective resistance between the origin and infinity**.

It turns out that this quantity is of great importance for the long-term behavior of the Markov chain.

Theorem 3.1 (Criterion for transience). *The Markov chain is transient if and only if $R_\infty < \infty$ for the corresponding electric network.*

Proof. Construct the electric network $\mathcal{E}$ corresponding to the chain, and consider the sequence of subnetworks $\mathcal{E}_L$, $L \in \mathbb{N}_+$, constructed by throwing away all the x with $\rho(x) > L$ and short-circuiting all vertices in D_L (see above). Let $x_1, \ldots, x_K \in D_1$ denote all the neighbors of the origin, and $v_L(x_1), \ldots v_L(x_k)$ be the voltages at those vertices when the origin is connected to one pole of the battery (with potential 0) and D_L is connected to the other pole (with potential 1).

Then,

$$v_L(x_k) = \mathbb{P}(X_n \text{ reaches } D_L \text{ before } \mathbf{0} \mid X_0 = x_k).$$

By the Markov property, we get that

$$v_L(x_k) = \mathbb{P}(X_n \text{ reaches } D_L \text{ before returning to } \mathbf{0} \mid X_0 = \mathbf{0})$$

$$= \sum_{k=1}^{K} P_{\mathbf{0}x_k} \mathbb{P}(X_n \text{ reaches } D_L \text{ before } \mathbf{0} \mid X_0 = x_k)$$

$$= \sum_{k=1}^{K} P_{\mathbf{0}x_k} v_L(x_k) = \sum_{k=1}^{K} \frac{1/R_{\mathbf{0}x_k}}{C_0} v_L(x_k)$$

$$= \frac{1}{C_0} \sum_{k=1}^{K} \frac{v_L(x_k)}{R_{\mathbf{0}x_k}} =: (*),$$

where $R_{\mathbf{0}x_k}$ is the value of the resistor connecting $\mathbf{0}$ and x_k, and $C_0 = \sum_{k=1}^{K} \frac{1}{R_{\mathbf{0}x_k}}$. Let $\imath_{0k} = \frac{v_L(x_k)}{R_{\mathbf{0}x_k}}$ denote the current from $\mathbf{0}$ to node x_k (by Ohm's law) and $\imath$ is the total current from the origin. Then,

$$(*) = \frac{1}{C_0} \sum_{k=1}^{K} \imath_{0k} = \frac{\imath}{C_0} = \frac{1}{C_0 R^{(L)}},$$

again by Ohm's law. Finally, since

$$\mathbb{P}_0(X_n \neq \mathbf{0} \text{ for all } n \in \mathbb{N}_+)$$

$$= \lim_{L \to \infty} \mathbb{P}_0 (X_n \text{ reaches } D_L \text{ before returning to } \mathbf{0})$$

$$= \lim_{L \to \infty} \frac{1}{C_0 R^{(L)}} = \frac{1}{C_0 R_\infty},$$

we obtain the statement of the theorem. $\qquad\square$

The reader may also consult Section 1.3.4 in Doyle and Snell (1984) for more details.

Consequently, in order to establish recurrence versus transience of random walks, we can estimate the value of R_∞. This is often done with the help of the so-called **cut and short-circuit method**, based on Rayleigh's monotonicity law. If we want to show that the random walk is transient, we need to show that R_∞ is finite. Instead of computing its value directly, which may turn out to be difficult or even impossible, we can get an upper bound for it by increasing the values of some resistors (sometimes even to infinity, which is equivalent to simply removing them from the network completely) and breaking out some connections. If we do it smartly, we might be able to estimate the effective resistance of the newly obtained network, and if it is finite, it would imply that the effective resistance of the original network was also finite.

Similarly, if we want to show that $R_\infty = \infty$, we may decrease the values of some resistors or even short-circuit some of the nodes (thus creating new connections). If we manage to show that the resistance of the obtained network is infinite, then so is the effective resistance of the original one.

For example, it is extremely easy to establish the recurrence of the simple symmetric random walk on $S = \mathbb{Z}^2$, i.e., the Markov chain which has equal probabilities to go up, left, down, or right. The electric network corresponding to this Markov chain is simply the network which has nodes at the points of $\mathbb{Z}_2$ and all the edges are replaced by unit resistors. Let $\mathbf{0} = (0,0)$ be the origin of this network.

Fix $L \in \mathbb{Z}_+$, and short-circuit all the nodes at distance L from the origin (see Figure 3.6), i.e., at

$$\tilde{D}_L = \{(x,y) \in \mathbb{Z}^2 : |x| + |y| = L\}.$$

By Rayleigh's monotonicity law, the effective resistance between $\mathbf{0}$ and $\tilde{D}_L$ is bounded below by the network in which, for each $i = 1, 2, \ldots, L-1$, we also short-circuit all the nodes at $\tilde{D}_i$ (separately for each i). Now it is very easy to compute the effective resistance of the new network. Let $\tilde{D}_0 = \mathbf{0}$. For each $i = 1, 2, \ldots, L$, between $\tilde{D}_i$ and $\tilde{D}_{i-1}$ we have exactly $4(2i - 1)$ resistors connected in parallel, hence the effective resistance between these "circles" equals $1/(4(2i - 1))$. As a result, we obtain L resistors connected now in series, with the total resistance

$$\sum_{i=1}^{L} \frac{1}{4(2i-1)} = \frac{1}{4}\left(1 + \frac{1}{3} + \frac{1}{5} + \frac{1}{7} + \cdots + \frac{1}{2L-1}\right).$$

Since the sum above diverges as $L \to \infty$, we conclude that $R_\infty = +\infty$ and hence the Markov chain is indeed recurrent.

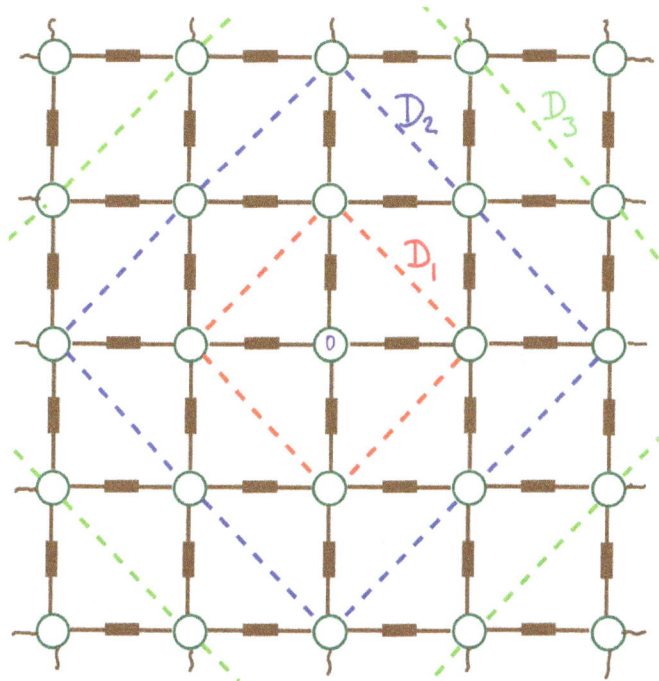

Fig. 3.6. Electric network corresponding to the simple symmetric random walk on $\mathbb{Z}^2$. All resistances are equal to 1.

3.2 Random Graphs and Random Environments

The method of electric networks is especially useful in the case when the underlying random walk takes place on a *random graph* or in a random environment (of course, only in the cases where we can construct a relevant network). We illustrate this in the following with two examples.

Random walk in random environment on $\mathbb{Z}_+$

Here we consider a version of the model introduced in Solomon (1975).

Let $\mathsf{S} = \mathbb{N}$ and X_n be a nearest neighbor random walk on S with random transition probabilities: namely, we assume that we are given a sequence of i.i.d. random variables $\xi_1, \xi_2, \ldots$ such that $\mathbb{P}(\xi_i \in (0,1)) = 1$. Once we have a realization of $\{\xi_i\}_{i=1}^\infty$, we fix it, and for this fixed environment, we define random walk on S by setting

$$p_{ij} = \mathbb{P}(X_{n+1} = j \mid X_n = i) = \begin{cases} \xi_i, & \text{if } j = i+1, \\ 1 - \xi_i, & \text{if } j = i-1, \end{cases}$$

for all $i \geq 1$, as well as, for definiteness, $X_{n+1} = 1$ when $X_n = 0$. We are interested to know if such a walk, called[3] (quenched) **random walk in random environment (RWRE)**, is recurrent for almost every environment.

It is fairly easy to construct the electric network corresponding to this walk. Let R_i, $i \in \mathbb{N}$, denote the value of the resistor between nodes i and $i+1$. Set $R_0 := 1$, then since

$$\frac{p_{i,i+1}}{p_{i,i-1}} = \frac{R_{i-1}}{R_i},$$

we have $R_i = \prod_{k=1}^{i} \frac{1-\xi_k}{\xi_k}$, $i \in \mathbb{N}_+$. It is easy to check that the Markov chain corresponding to these resistances is indeed the one above (i.e., with the given $p_{i,i+1}$s).

Let $\rho_i := \log \frac{1-\xi_i}{\xi_i}$. Then, since resistors R_1, R_2, etc. are connected in series, the effective resistance of the network from the origin 0 to infinity equals

$$R_\infty = \sum_{k=0}^{\infty} R_k = 1 + e^{\rho_1} + e^{\rho_1+\rho_2} + e^{\rho_1+\rho_2+\rho_3} + \cdots = 1 + \sum_{k=1}^{\infty} e^{S_k},$$

where $S_k := \rho_1 + \cdots + \rho_k$.

Theorem 3.2 (Recurrence/transience for RWRE). *Define $\mu := \mathbb{E}\rho_i$ and assume that $\mu < \infty$. If $\mu > 0$, then RWRE is recurrent a.s.; if $\mu < 0$, then RWRE is transient a.s.*

Remark 3.1. For simplicity, we do not deal here with the case $\mu = 0$ or when the walk is on $\mathbb{Z}$ rather than $\mathbb{Z}_+$; we refer interested reader to Sinaĭ (1982).   ◇

---

[3]The term "quenched" is coming from metallurgy; it refers to the fact that we first "freeze" the environment and consider the walk in this frozen environment; we are interested in properties which hold for almost every environment.

Proof. If $\mu > 0$, then by the Strong Law of Large Numbers, for a.e. ω, there exists an $N = N(\omega)$ such that $S_n/n \geq \mu/2$ for all $n \geq N$. Hence,

$$R_\infty \geq \sum_{n=N}^\infty e^{n\frac{S_n}{n}} \geq \sum_{n=N}^\infty e^{\frac{\mu n}{2}} = \infty,$$

and thus RWRE is recurrent by Theorem 3.1.

Similarly, if $\mu < 0$, then for a.e. ω, there exists an $N = N(\omega)$ such that $S_n/n \leq -|\mu|/2$ for all $n \geq N$. Hence,

$$R_\infty \leq R_0 + R_1 + \cdots + R_{N-1} + \sum_{n=N}^\infty e^{n\frac{S_n}{n}}$$

$$\leq R_0 + R_1 + \cdots + R_{N-1} + \sum_{n=N}^\infty e^{-\frac{|\mu|n}{2}} < \infty,$$

thus RWRE is transient. □

Random walk in random labyrinths

Another research area where the method of electric networks is especially useful is that of **random walks on random graphs**. We illustrate this here by presenting a model studied in Grimmett *et al.* (1996).

Suppose that we are given a probability distribution $\{p_O, p_{NW}, p_{NE}, p_T\}$ on the set of four elements $\tilde{S} = \{$O, NW, NE, T$\}$, such that $p_O > 0$. Each site of $\mathbb{Z}^2$ is either a "normal" site (O) or a two-sided North-West mirror (NW), or a two-sided North-East mirror (NE), or a tunnel (T), according to the distribution above, and independently for each site. We assume that the origin $\mathbf{0} = (0,0)$ is always a normal site. This collection of mirrors, tunnels and normal sites on $\mathbb{Z}^2$ is called a **random labyrinth**.

A particle performs a random walk $X_n \in \mathbb{Z}^2$ on the labyrinth according to the following rules. First, $X_0 := \mathbf{0}$. At normal sites, the particle behaves just like a simple symmetric random walker, that is, chooses each of the four directions equally likely, independent of the past. At any of the other three types of sites, the behavior of the particle is non-random. Namely, if the particle arrives to a tunnel site, it will continue moving in the same direction as before, that is, $X_{n+1} - X_n = X_n - X_{n-1}$. When the particle hits any of the mirrors, it gets reflected and changes its direction by 90 degrees using the following rules:

$$X_{n+1} - X_n = \begin{cases} (\pm 1, 0) & \text{if } X_n - X_{n-1} = (0, \pm 1); \\ (0, \pm 1) & \text{if } X_n - X_{n-1} = (\pm 1, 0) \end{cases}$$

(if X_n is a NE mirror),

$$X_{n+1} - X_n = \begin{cases} (\mp 1, 0) & \text{if } X_n - X_{n-1} = (0, \pm 1); \\ (0, \mp 1) & \text{if } X_n - X_{n-1} = (\pm 1, 0) \end{cases}$$

(if X_n is a NW mirror).

The effect of a NW or NE mirror is illustrated in Figure 3.7.

We would like to know whether such a random walk, referred to as a **random walk in a random labyrinth**, is recurrent (for almost every environment). Observe that recurrence can occur in two possible ways: (a) the set of all sites which are accessible from the origin is finite and thus the random walk occurs on a finite set of nodes, or (b) the set of accessible sites is infinite, but the random walk is still recurrent. While the probability of the event described in (a) is obviously positive (for example, if the sites $(0, \pm 1)$, $(\pm 1, 0)$ are NE and the sites $(-1, -1)$ and $(1, 1)$ are NW, see Figure 3.8), it is very difficult to discern if this probability is one or less; to the best of our knowledge, this question remains open. Note that if the graph does not have mirrors, i.e., $p_O + p_T = 1$, then the set of accessible sites is always infinite, and all the normal sites are accessible from the origin.

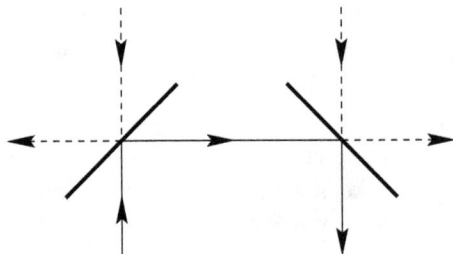

Fig. 3.7.   NE and NW mirrors reflect the particle as shown.

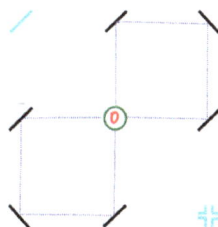

Fig. 3.8.   Random walk is localized.

However, with the help of the electric network method, we can at least show that the random walk in the labyrinth is a.s. recurrent, regardless of whether the situation (a) or (b) holds.

Suppose that the environment is fixed. Then the process X_n itself is not Markov, since the (non-random) value of $X_{n+1} - X_n$ depends on $X_n - X_{n-1}$ whenever X_n is not at a normal point. To construct the Markov chain corresponding to the (non-Markovian) process X, we construct the following graph G. The vertices of the graph G are all normal sites of $\mathbb{Z}^2$ which are accessible from the origin, and from now on we assume that G has more than one vertex (the other case is trivial). Now define the sequence of stopping times η_m as follows:

$$\eta_0 := 0,$$

$$\eta_{m+1} := \min\{n > \eta_m : X_n \text{ is at a normal site and } X_n \neq X_{\eta_m}\},$$

$$m \in \mathbb{N}.$$

Since whenever X_n is at a normal site, X_{n+1} equals one of the $(X_n + (1,0), X_n + (-1,0), X_n + (0,1), X_n + (0,-1))$ with equal probability independent of the past, the process Y defined by $Y_m = X_{\eta_m}$, $m \in \mathbb{N}$, is a time-homogeneous Markov chain on the vertices of G.

We say that two vertices u and w of G are *neighbors* on graph G if $\mathbb{P}(Y_{m+1} = w \mid Y_m = u) > 0$; note that $Y_{m+1} \neq Y_m$ by definition, hence G cannot have loops.

For almost every environment, each vertex of G has either 4 or 2 other vertices; the latter case occurs if there is a chain of mirrors and possibly tunnels such that whenever X leaves this vertex in a certain direction, it will come back to it without visiting other normal vertices; see Figures 3.9 and 3.10.

Hence, $\mathbb{P}(Y_{m+1} = w \mid Y_m = u) \in \{0, 1/4, 1/2\}$ for any pair of vertices of G; moreover, Y is a simple random walk on G.

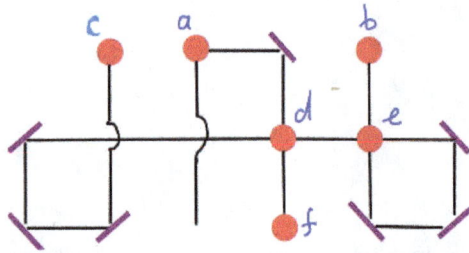

Fig. 3.9.   Node e has two neighbors: b and d (since we ignore the loop path resulted from the three mirrors located below and to the right from e). Node d has four neighbors: c, a, e and f.

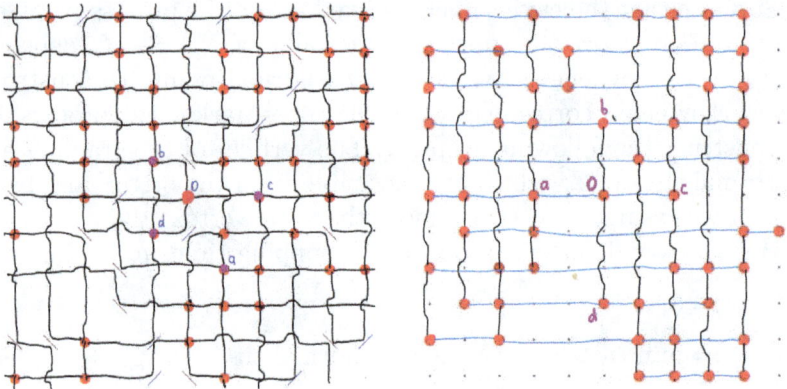

Fig. 3.10.   Node **0** is connected by edges of G to a, b, c and d. The graph on the left has mirrors as well as tunnels; the graph on the right has only normal points and tunnels.

Theorem 3.3 (A.s. recurrence).   *The Markov chain Y (and hence the process X too) is recurrent for almost all configurations.*

We can easily construct the electric network $\mathcal{E}$ corresponding to the Markov chain Y by replacing all the edges of G by unit resistors. Now, by Theorem 3.1, the statement of Theorem 3.3 follows immediately from the following.

Lemma 3.1.   *Either the graph G is finite or $R_\infty = \infty$ for the electric network $\mathcal{E}$.*

Proof. It suffices to show that $R_\infty = \infty$ a.s. on the event "G is infinite." We make use of the actual locations of the nodes of G in $\mathbb{Z}^2$.

First, recall the definitions of the distance from the origin $\rho(\cdot)$ and the circle D_L defined for the graph G, using the graph-theoretical distance. Let us also define the "geometric" circles $\tilde{D}_L$ as

$$\tilde{D}_L = \{(x,y) \in \mathbb{Z}^2 : |x| + |y| = L\}, \quad L \in \mathbb{N}.$$

Next, note that each edge of G corresponds to a unique path of edges of $\mathbb{Z}^2$, by which we mean a sequence of distinct $\mathbb{Z}^2$ edges such that two consecutive edges share a common vertex; that path starts at one vertex of G and ends at another vertex of G. Observe that this path may contain self-intersections, though no node can appear in the path more than twice,[4] and that all the interim nodes of this path are either mirrors or tunnels. By the *length of the path* we mean the total number of $\mathbb{Z}^2$ edges in it. For example, in Figure 3.9, the path from d to c has length 7 and intersects itself once at the tunnel node located just under the node c; the path from node d to a has length 2 and no self-intersections.

Now consider an arbitrary edge e of $\mathbb{Z}^2$ belonging to some such path. Let $m \in \mathbb{N}$. What is the probability that e belongs to a path of length at least $4m + 1$? Since this edge has a unit contribution to the path length, at least on one of the two sides of e, there should be a sequence of at least $2m$ distinct $\mathbb{Z}^2$ edges belonging to a path corresponding to an edge of G. However, each path of length $2m$ must pass through at least m nodes (as it can pass through the same node at most twice), and each of these nodes must *not* be a normal point. Hence,

$$\mathbb{P}(e \text{ belongs to a path of length } 4m + 1 \text{ or more}) \leq 2(1 - p_O)^m \tag{3.4}$$

(recall that $p_O > 0$ is the probability that the site is "normal"). If

$$C := \frac{3}{-\ln(1 - p_O)} > 0$$

---

[4]Note: "twice" happens when the path goes through a tunnel first time horizontally and the second time vertically, cf. the path from c to d in Figure 3.9; similar is the situation for mirrors.

and $m := C \ln(L)$, then (3.4) yields

$$\mathbb{P}(e \text{ belongs to a path of length } 4C \ln(L) + 1 \text{ or more})$$

$$\leq 2(1 - p_O)^{C \ln(L)} = \frac{2}{L^3}.$$

Now consider all $\mathbb{Z}^2$ edges with one endpoint in $\tilde{D}_L$ and the other in $\tilde{D}_{L-1}$. Since there are exactly $4(2L - 1)$ of those, we have

$$\mathbb{P}(B_L) \leq \frac{8(2L - 1)}{L^3} < \frac{16}{L^2},$$

where

$$B_L := \{\exists \text{ edge connecting } D_L \text{ and } D_{L-1}$$

$$\text{which belongs to a path of length } 4C \ln(L) + 1 \text{ or more}\}.$$

Let

$$L_0 := \max\{L \geq 1 : \ B_L \text{ occurs}\}. \tag{3.5}$$

Since $\sum_{L=1}^{\infty} \mathbb{P}(B_L) < \infty$, the Borel–Cantelli lemma yields that $L_0 < \infty$ a.s.

Next, let us modify the electrical network $\mathcal{E}$ corresponding to the random walk on G as follows. For each edge f of G corresponding to the path length $m \geq 1$, replace the unit resistor assigned to the edge f by m resistors of value $1/m$ connected in series. Also add $m - 1$ nodes to the electric network, corresponding to the nodes of $\mathbb{Z}^2$ the path goes through. Such a change will not affect the effective resistances between nodes of G, and hence instead of $\mathcal{E}$ we can study the newly obtained network $\mathcal{E}'$; see Figure 3.11.

Now, for each $L > L_0$, short-circuit all the nodes of $\tilde{D}_L$. Using obvious notation, by the monotonicity law, $R_\infty^{\mathcal{E}} \geq R_\infty^{\mathcal{E}'} \geq R_\infty^{\mathcal{E}''}$. Thus, it is enough to verify that $R_\infty^{\mathcal{E}''} = \infty$.

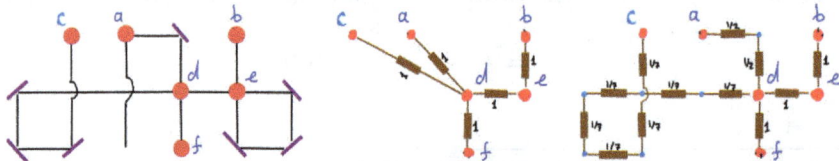

Fig. 3.11.   Random labyrinth, the electrical network $\mathcal{E}$ corresponding to it and the network $\mathcal{E}'$.

Since each $\mathbb{Z}^2$ edge e with the endpoints in D_L and D_{L-1} ($L \in \mathbb{N}_+$) can belong to a path of maximum length $4C \ln(L) + 1$ (recall the definition of L_0 in (3.5)), the resistor corresponding to e in $\mathcal{E}'$ has the value of at least $[4C \ln(L) + 1]^{-1}$, and all these $4(2L-1)$ resistors are connected in parallel. Consequently, the effective resistance between D_{L-1} and D_L is bounded from below by

$$\frac{[4C \ln(L) + 1]^{-1}}{4(2L-1)} = \left(\frac{1}{32C} + o(1)\right) \frac{1}{L \ln(L)}, \qquad (3.6)$$

where $o(1) \to 0$ as $L \to \infty$. Now we can replace all the resistors connecting D_{L-1} and D_L by one with the resistance bounded below by (3.6), and these resistors are connected in series. Since the quantities in (3.6) are not summable for $L = 1, 2, \ldots$, one has $R_\infty^{\mathcal{E}''} = \infty$, as required. $\square$

It turns out that if we generalize our random walk in the labyrinths to $\mathbb{Z}^d$, $d \geq 3$, then the resulting random walk will be a.s. transient on the event "the set of sites accessible from the origin is infinite." For the proof of this fact as well as further details, we refer the reader to the original article Grimmett *et al.* (1996).

3.3 Exercises

Problem 1. Prove formula (3.1) for the expectation of the time till absorption.

Problem 2. Prove that the resistances introduced in (3.3) will indeed result in the transition probabilities given by (3.2).

Problem 3. Prove that the resistances in Figure 3.2 will indeed result in the transition probabilities shown in Figure 3.1.

Problem 4. Try to prove the transience of the simple symmetric random walk in $\mathbb{Z}^3$ by showing that R_∞ is finite for the corresponding electric network.

Problem 5. Extend the result of Theorem 3.2 to the cases $\mu = -\infty$ and $\mu = +\infty$.

Problem 6. Consider a random walk in random environment (RWRE) on $\mathbb{Z}_+$ such that $\mathbb{P}_x(X_{n+1} = x+1) = 1 - \mathbb{P}_x(X_{n+1} = x - 1) = \xi_x$, where $\xi_x = a$ with probability q and $\xi_x = 1 - a$

with probability $1 - q$, and $a, q \in (0, 1)$, moreover the ξ_x are i.i.d. For which values of a and q is RWRE recurrent?

Problem 7. Prove that if a random labyrinth does not contain mirrors but only tunnels and normal points, then a.s. there is always a path connecting two normal sites, and hence the graph corresponding to Y_n is always infinite.

Problem 8. Consider the probability that the graph corresponding to the random walk in the labyrinth is infinite. For which set of parameters $(p_O, p_{NW}, p_{NE}, p_T)$, such that $p_O + p_T < 1$, can one be sure that this probability is positive?

Chapter 4

Time-Inhomogeneous Markov Chains

For a Markov chain, time inhomogeneity means that the transition probabilities $p_{ij}(k) := P(X_k = j \mid X_{k-1} = i)$ depend on k. Hence, we have a sequence of transition matrices $P_k := \{p_{ij}(k)\}$, $k \geq 1$, to deal with. We again assume that the state space S is finite, $|\mathsf{S}| = N \geq 1$.

4.1 The Difficulties

Time-inhomogeneous Markov chains are much less understood than their homogeneous counterparts. They exhibit a wider range of behavior. For example, they may reach a target distribution in finite time, as the following example shows. Suppose we have a deck of n cards and we want to reach the uniform distribution on permutations (order of cards). At step $i \geq 1$, we pick a card uniformly at random from the bottom $n - i + 1$ cards and insert it at position i. Then, after $n - 1$ steps, the deck is distributed uniformly, that is, each order has the same likelihood. Indeed, the top card has an equal chance of being any of the faces. The second one then has an equal chance of being any of the faces, except what the top card is, the two top cards thus have an equal chance of being any (ordered) pair of the faces. Continuing this argument, the top k cards have an equal chance of being any k-tuple of the faces, for $k = 3$, $k = 4, \ldots$ too, and finally also for $k = n$.

While an aperiodic irreducible homogeneous Markov chain on a finite state space admits a unique invariant probability measure, to which the time n distributions converge, the situation is much more complicated in the inhomogeneous case. In particular, the following type of inference is not at all clear to hold: assume that some given advantageous property holds for all the one-step transitions, does this then guarantee any nice property for the whole chain?

The largest contribution to the literature on time-inhomogeneous Markov processes seems to originate in research on diffusion processes with time-dependent coefficients, and on the corresponding second-order elliptic partial differential equations. Apparently, however, techniques developed in diffusion theory are hard to transfer to the qualitative analysis of finite state space general Markov chains.

Another research area motivating the analysis of inhomogeneous chains is the theory of time-inhomogeneous *random walks on groups*, providing many natural examples in the work of Saloff-Coste, Zúñiga and others (see Zúñiga (2008)).

Relatively little is known concerning the weak ergodicity (i.e., merging in total variation norm when starting with different distributions) and the asymptotic structure of inhomogeneous chains, as well as about the corresponding products of stochastic matrices. See Saloff-Coste and Zúñiga (2011) and the references therein.

Notation 4.1. The multistep transitions are now given by the forward product $P^{mn} := P_m P_{m+1} \cdots P_{n-1}$, $0 \leq n < m$. The distribution after n steps starting at $x \in \mathsf{S}$ is now denoted by $P^{0n}(x, \cdot)$ or by $P_{x, \cdot}^{0n}$.

To look at the situation from an algebraic perspective, recall that in linear algebra, a real square matrix A is called *diagonalizable* if it is similar to a diagonal matrix, i.e., if there exists an invertible matrix J and a diagonal matrix D such that $J^{-1} A J = D$, or equivalently $A = J D J^{-1}$. (Such J, D are not unique.) One can raise a diagonal matrix to a power by simply raising the diagonal entries to that power, and since $A^n = J D^n J^{-1}$, it is very easy to compute the powers of A once the decomposition with J and D is known. The diagonal entries of D are in fact the (not necessarily distinct) eigenvalues of the matrix A.

A set of matrices $\mathcal{A}$ is said to be *simultaneously diagonalizable* if there exists a single invertible matrix J such that $J^{-1} A J$ is a diagonal matrix for every $A \in \mathcal{A}$. It is a well-known fact in linear algebra that a set of diagonalizable matrices commutes if and only if the set is simultaneously diagonalizable. For example, the diagonizable matrices

$$A = \begin{pmatrix} 1 & 0 \\ 0 & 0 \end{pmatrix} \quad \text{and} \quad B = \begin{pmatrix} 1 & 1 \\ 0 & 0 \end{pmatrix}$$

are not simultaneously diagonizable as they do not commute.

Coming back to Markov chains and stochastic matrices, we have seen (Proposition 2.6) that a matrix corresponding to a reversible time-homogeneous chain is diagonalizable. Now, whenever P is a diagonizable stochastic matrix and $\mathbf{u}$ is a probability vector, $\lim_{n \to \infty} \mathbf{u} P^n$ can be computed by diagonalization. Indeed, there exists a basis formed by eigenvectors $\mathbf{v}_i$ of P and we can write $\mathbf{u} = \sum_{i=1}^{N} c_i \mathbf{v}_i$. By the Perron–Frobenius theorem, if $\mathbf{v}_i$ corresponds to the eigenvalue λ_i, then $\lambda_1 = 1$ and all other eigenvalues are less than one in absolute value. Hence, $\lim_{n \to \infty} \mathbf{u} P^n = \lim_{n \to \infty} \sum_{i=1}^{N} c_i \lambda_i^N \mathbf{v}_i = c_1 \mathbf{v}_1$. That is, the limit is a principal (Perron–Frobenius) eigenvector corresponding to the eigenvalue 1 with the additional constraint that the coordinates add up to one.

In the time-inhomogeneous case, if P_k, $1 \leq k \leq n$, happen to be simultaneously diagonalizable (because they commute and each one is diagonalizable) and so $P_k = J D_k J^{-1}$ for $1 \leq k \leq n$, then $\prod_1^n P_k = J \left(\prod_1^n D_k \right) J^{-1}$. Hence, if J and the D_k are known, then the forward product is easy to compute. Let $\lambda_{i,k}$, $i = 1, \dots, N$, be the eigenvalues of P_k with $\lambda_{1,k} = 1$. Then the previous method gives that $\mathbf{u} P_1 \cdots P_n = \sum_{i=1}^{N} c_i \Lambda_{i,n} \mathbf{v}_i$, where $\Lambda_{i,n} := \lambda_{i,1} \cdots \lambda_{i,n}$. The difference is that, unlike in the homogeneous case, now $\lim_{n \to \infty} \Lambda_{i,n} = 0$ is not guaranteed for any $i \geq 2$.

In most cases, however, simultaneous diagonalization is not possible even if each matrix is diagonalizable, as the matrices do not commute. This means that there is no easy way to analyze[1] the forward product for large n or, equivalently, the large time behavior of the time-inhomogeneous chain. It is possible to obtain certain estimates, however, using the notion of *singular values*.

---

[1]The classical matrix theory perspective on inhomogeneous Markov chains can be found in Seneta (2006).

4.2   Singular Values Bound

The importance of singular values lies in the following well-known decomposition result, which is an extension of the Hilbert–Schmidt theorem to the non-symmetric case. Let H and G be two (real or complex) Hilbert spaces of the same dimension, finite or countable. Let $A : H \to G$ be a compact[2] operator, with adjoint $A^* : G \to H$. Then there exist orthonormal bases (ϕ_i) of H and (ψ_i) of G and non-negative reals $\sigma_i = \sigma_i(H, G, A)$ such that $A(\psi_i) = \sigma_i \psi_i$ and $A^* \psi_i = \sigma_i \phi_i$. The non-negative numbers σ_i are called the singular values of A and are equal to the square root of the eigenvalues of the self-adjoint compact operator $A^* A : H \to H$ and also of $AA^* : G \to G$.

In particular, for an $N \times N$ real (or complex) matrix A, the singular values are the square roots of the eigenvalues of the symmetric (Hermitian) matrix $A^T A$, where A^T is the (conjugate) transposed, and these are non-negative real numbers, usually listed in decreasing order, as $\sigma_0(A) \geq \cdots \geq \sigma_{N-1}(A)$. When A is symmetric (Hermitian), the singular values coincide with the absolute values of the eigenvalues.

Before stating the result note that it is useful to write the total variation distance as

$$2d_{TV}(\mu, \nu) = \left(\sum_{y \in S} \left| \frac{\mu(y)}{\nu(y)} - 1 \right| \nu(y) \right)$$

because, by Jensen's inequality, it follows that for $p > 1$,

$$d_p(\mu, \nu) := \left(\sum_{y \in S} \left| \frac{\mu(y)}{\nu(y)} - 1 \right|^p \nu(y) \right)^{1/p}$$

controls the total variation distance as $2d_{TV}(\mu, \nu) \leq d_p(\mu, \nu)$. The following result is taken from Saloff-Coste and Zúñiga (2007), see Theorem 3.3 there.

Theorem 4.1 (Singular values bound; Saloff-Coste and Zúñiga). *Having a sequence of transition matrices* $(P_k)_{k \geq 1}$, *assume*

---

[2]A bounded linear operator is called compact if the image of the unit ball is pre-compact.

that they all admit the same probability distribution (vector) π as invariant distribution. Then

$$d_2(P^{0n}(x,\cdot),\pi) \leq \sqrt{\pi(x)^{-1}-1} \cdot \prod_1^n \sigma_1(P_j),$$

$$\sum_{x \in S} d_2(P^{0n}(x,\cdot),\pi(x)) \leq \sum_{i=1}^{N-1} \prod_{j=1}^n \sigma_i(P_j)^2.$$

For example, later on we study the case ("coin turning") when $S := \{-1,1\}$ and the transition matrices are symmetric:

$$P_k := \begin{pmatrix} 1-p_k & p_k \\ p_k & 1-p_k \end{pmatrix},$$

with some sequence $(p_k)_{k \geq 1}$ with $p_k \in [0,1]$. It is easy to see that $\sigma_0(P_k) = 1$ and $\sigma_1(P_k) = |1-2p_k|$ for all $k \geq 1$. Therefore,

$$d_2(P^{0n}(x,\cdot),\pi) \leq \prod_1^n |1-2p_j|$$

because in this case $\pi = (1/2,1/2)$. For instance, when $p_k = c/k$, $c > 0$, for large k, one obtains the bound const$\cdot n^{-c}$. We show later that in fact, whenever $\sum_i (p_i \wedge (1-p_i)) = \infty$, $P^{0n}(x,\cdot)$ converges in total variation norm to $\pi := (1/2,1/2)$ as $n \to \infty$.

Unfortunately, as the following example shows, singular value bounds are not always useful.

Example 4.1 (Adjacent transpositions). Consider a deck of $N \geq 2$ cards. On the state space $S = S_N$ of all permutations (so $|S| = N!$), let $P^{(j)} := \frac{1}{2}(I_N + K_j)$, $j = 1,\ldots,N-1$ where I_N is the identity matrix and K_j corresponds to the transposition of the jth and the $j+1$st cards. Note that $P^{(j)}$ is a symmetric matrix of size $N!$

We now define a time-inhomogeneous chain as follows. Let the transition matrix at time $i \geq 1$ be defined as $P_i = P^{(j)}$ if $i \equiv j$ (mod $N-1$). Direct computation shows that $\sigma_1(P^{(j)}) = 1$ for $j = 1,\ldots,N-1$; hence $\sigma_1(P_i) = 1$ for all $i \geq 1$. The singular value bound is consequently useless in this case, even though it is known (see Zúñiga (2008, p. 72)) that this shuffle (inhomogeneous chain) ensures convergence to the stationary distribution which is the uniform permutation of the cards (uniform distribution on S). Note that despite the inhomogeneity, we can turn the chain homogeneous if we consider the first $N - 1$ steps together as one "mega-step," etc.   ◇

4.3   Merging and Coupling

We need a definition for general time-inhomogeneous chains.

Definition 4.1 (Merging in TVN). We say that *merging occurs in total variation* if for any $x, y \in S$,

$$\lim_{n \to \infty} d_{TV} \left(P^{0n}(x, \cdot), P^{0n}(y, \cdot) \right) = 0.$$

Definition 4.1 is related to the technique of coupling Markov chains.

Definition 4.2 (Coupling of Markov chains). Let the (possibly inhomogeneous) chains X and Y both have the same state space S and transition matrices $(P_k)_{k \geq 1}$. Their *Markovian coupling* is a (two-dimensional) Markov chain Z with state space S^2 and transition matrices $(Q_k)_{k \geq 1}$, if the following holds. First, if X (resp. Y) starts at $x \in S$ (resp. $y \in S$), then Z starts at (x, y). Furthermore, for $n \geq 1$ and $u, v, i, j \in S$,

$$\sum_{j \in S} Q^n \left((u, v)(i, j) \right) = P^n(u, i),$$

$$\sum_{i \in S} Q^n \left((u, v)(i, j) \right) = P^n(v, j),$$

$$Q^n \left((u, u), (i, i) \right) = P^n(u, i).$$

The last equation means that once the two chains met, they stay together (merge). Sometimes this last condition is omitted in the definition, but one can always modify the coupling such that it actually holds.

For Markovian couplings, one has (see Theorem 5.4 in Levin and Peres (2017) or Saloff-Coste and Zúñiga (2009)) the fundamental bound

$$d_{TV}(P^{0n}(x, \cdot), P^{0n}(y, \cdot)) \leq P(T > n) \qquad (4.1)$$

if T is the coupling time of a coupling (X_n, Y_n) with starting points $X_0 = x$ and $Y_0 = y$. This easily follows (and left as an exercise) from

Theorem 2.6. By (4.1), merging in total variation (Definition 4.1) for the coupling of Markov chains follows when $T < \infty$ a.s.

Coupling remains a useful technique for time-inhomogeneous chains, but inhomogeneity results in very serious difficulties in the construction and analysis of couplings for specific examples (see Saloff-Coste and Zúñiga (2009)).

A little more generally[3] than in the coin turning model above, let us consider $S := \{-1, 1\}$ and the matrices

$$P_k := \begin{pmatrix} 1 - r_k & r_k \\ s_k & 1 - s_k \end{pmatrix}, k \geq 1$$

with some sequences $(r_k)_{k\geq1}$ and $(s_k)_{k\geq1}$ with $r_k, s_k \in [0, 1]$. The following result is (a slightly reformulated version of) Proposition 6.2 in Saloff-Coste and Zúñiga (2009).

Theorem 4.2 (No merging; Saloff-Coste and Zuñiga). *With r_k and s_k as above, define the set of indices $I := \{i \geq 1 \mid r_i + s_i > 1\}$ and $J := \{i \geq 1 \mid r_i + s_i < 1\}$. Merging in total variation fails if and only if the following three conditions are all satisfied:*

$$\begin{cases} \text{(i)} \sum_{i \in I}[(1 - r_i) + (1 - s_i)] < \infty; \\ \text{(ii)} \sum_{i \in J}(r_i + s_i) < \infty; \\ \text{(iii)} r_i + s_i \neq 1 \text{ for } i \geq 1, \text{ that is, } J = I^c. \end{cases} \quad (4.2)$$

In the symmetric $(s_i = r_i =: p_i)$ case, (4.2) is tantamount to

$$\begin{cases} \text{(a)} \sum_{i \in I}(1 - p_i) < \infty; \\ \text{(b)} \sum_{i \in J} p_i < \infty; \\ \text{(c)} p_i \neq 1/2 \text{ for } i \geq 1, \end{cases} \quad (4.3)$$

where now $I := \{i \geq 1 \mid p_i > 1/2\}$ and $J := \{i \geq 1 \mid p_i < 1/2\}$. Equivalently,

$$p_i \neq 1/2, \ i \geq 1 \quad \text{and} \quad \sum_i (p_i \wedge (1 - p_i)) < \infty.$$

---

[3]In the coin turning model, $r_k = s_k$.

In the symmetric case, we thus have that merging is equivalent to

$$\exists p_i = 1/2, \text{ or } \sum_i (p_i \wedge (1 - p_i)) = \infty. \qquad (4.4)$$

The reader may want to compare this with Proposition 6.1 and the whole Section 6.1. The proposition states that the non-summability is equivalent to "mixing," defined in Section 6.1; Theorem 6.1 states that (4.4) implies convergence to the uniform distribution. Having $p_i = 1/2$ of course means that we "forget" where we are and simply toss a fair coin. Clearly, if $p_i = 1/2$ for some $i \in \mathbb{N}$, then the process "gets symmetrized" from time i on.

Finally, we mention the recent monograph (Dolgopyat and Sarig, 2023). It focuses on local limit theorems for inhomogeneous Markov chains under the assumption that the chain is *uniformly elliptic*. We note already here that in the context of coin turning alluded to above and discussed in the following two chapters (Chapters 5 and 6), uniform ellipticity means that there is an $\varepsilon > 0$ for which $0 < \varepsilon < p_n < 1 - \varepsilon$ holds for all n. (Cf. Theorem 6.5.)

4.4   Exercises

Problem 1. We have seen that a matrix P corresponding to a (time-homogeneous) irreducible and reversible chain is diagonizable, as it is similar to a symmetric matrix. Recall that for a symmetric matrix, the eigenvectors corresponding to different eigenvalues are orthogonal. As far as P is concerned, in which sense (if any) can we claim that the eigenvectors, corresponding to different eigenvalues, are orthogonal? Try to define an inner product.

Problem 2. Show that $2d_{TV}(\mu, \nu) \le d_2(\mu, \nu)$.

Problem 3. Let $\mathsf{S} := \{-1, 1\}$ and

$$P_k := \begin{pmatrix} 1 - p_k & p_k \\ p_k & 1 - p_k \end{pmatrix},$$

with some sequence $(p_k)_{k \ge 1}$ such that $p_k \in [0, 1]$. Show that $\sigma_0(P_k) = 1$ and $\sigma_1(P_k) = |1 - 2p_k|$ for all $k \ge 1$. These are in fact absolute values of eigenvalues. (Why?) Show that the corresponding eigenvectors for $\sigma_1(P_k)$ are the same for all k (up to constant multiples).

Problem 4. Show that Theorem 2.6 implies bound (4.1).

Problem 5. Check that in Example 4.1, one indeed has $\sigma_1(P_i) = 1$ for all $i \geq 1$.

Chapter 5

Coin Turning

In this chapter, we study a specific time-inhomogeneous Markov chain. It is the simplest case in the sense that the state space has only two elements. Using a more intuitive language, we examine what happens if, instead of tossing a coin, we *turn it over* (from heads to tails and from tails to heads), with certain probabilities. So, now we are working with a Markov chain on the state space $S := \{T, H\}$ or $S := \{0, 1\}$.

To define the model more accurately, let p_n, $n = 2, 3, \ldots$, be a given deterministic sequence of numbers between 0 and 1. We define a time-dependent "coin turning process" X with $X_n \in \{0, 1\}$, $n \geq 1$, as follows. Let $X_1 = 1$ ("heads") or $= 0$ ("tails") with probability $1/2$. For $n \geq 2$, set recursively

$$X_n := \begin{cases} 1 - X_{n-1}, & \text{with probability } p_n; \\ X_{n-1}, & \text{otherwise,} \end{cases}$$

that is, we turn the coin over with probability p_n and do nothing with probability $1 - p_n$, independent of the sequence of the previous terms.

Remark 5.1. An alternative, equivalent definition would be that X_0 is deterministic and $p_1 := 1/2$. In the sequel, whenever we include p_1 in the formulas, **the convention will be that we use the latter**

definition, except in the few situations, when we explicitly point out that symmetry has been dropped.   ◇

Consider $\frac{1}{N}\sum_{n=1}^{N} X_n$, that is, the empirical frequency of 1's ("heads") in the sequence of X_n's. We are interested in the asymptotic behavior, in law, of this random variable as $N \to \infty$.

Since we are interested in limit theorems, we center the variable X_n; for convenience, we also multiply it by two, thus focusing on $Y_n := 2X_n - 1 \in \{-1, +1\}$ instead of X_n. We have

$$Y_n := \begin{cases} -Y_{n-1}, & \text{with probability } p_n; \\ Y_{n-1}, & \text{otherwise.} \end{cases}$$

Note that the sequence $\{Y_n\}$ can be defined equivalently as follows. Let

$$Y_n := (-1)^{\sum_{i=1}^{n} W_i}, \tag{5.1}$$

where $W_1, W_2, W_3, \ldots$ are independent Bernoulli variables with parameters $p_1, p_2, p_3, \ldots$, respectively, and $p_1 = 1/2$. The number of turns that occurred after the first toss up to n is $\sum_{i=2}^{n} W_i$. (Alternatively, one could start with X_0 being deterministic and then the number of turns would be $\sum_{i=1}^{n} W_i$.)

Remark 5.2 (Poisson binomial random variable). The number of turns that occurred after the first toss up to n, that is, $\sum_{i=2}^{n} W_i$, is a Poisson binomial random variable. See the review (Tang and Tang, 2023) on the properties of this distribution.   ◇

Representation (5.1) is important for the proofs, and following are some easy observations it implies. However, it is also important to consider the process Y as a non-homogeneous Markov chain with state space $\{-1, 1\}$, initial distribution $\mathbb{P}(Y_1 = 1) = \mathbb{P}(Y_1 = -1) = 1/2$ and doubly stochastic symmetric transition matrices $\begin{pmatrix} 1 - p_n & p_n \\ p_n & 1 - p_n \end{pmatrix}$, $n \geq 2$. Using the symmetry in the definition, Y_n ($n \geq 2$) has the same distribution as Y_1, namely, Bernoulli$(1/2)$. Hence, the limit theorems in this chapter involve particular cases of this two-state Markov chain, but the methods of our proofs tend to rely on the above representation for the random variables $\{Y_n\}$ rather than on Markovian techniques.

The following quantity plays an important role: for $1 \leq i < j \leq N$, let

$$e_{i,j} := \prod_{k=i+1}^{j} (1 - 2p_k). \tag{5.2}$$

Note that, as an easy computation reveals, the second eigenvalue of the matrix $M_n := \begin{pmatrix} 1 - p_n & p_n \\ p_n & 1 - p_n \end{pmatrix}$ is $\lambda_2(n) := 1 - 2p_n$, with the same eigenvector $\mathbf{v} := (1, -1)$ for all n, hence

$$e_{i,j} = \prod_{k=i+1}^{j} \lambda_2(k) =: \lambda_{i,j}, \tag{5.3}$$

where $\lambda_{i,j}$ is an eigenvalue of the product matrix $M_{i+1}M_{i+2} \cdots M_j$, with eigenvector $\mathbf{v}$.

Using representation (5.1) for the random variables $\{Y_n\}$, we have

$$Y_j = Y_i \cdot (-1)^{\sum_{k=i+1}^{j} W_k}, \quad \text{for } 1 \leq i \leq j,$$

(where $\sum_{i+1}^{i} := 0$) and hence if $i = 1$, then we get $\mathbb{E}(Y_j) = e_{1,j}\,\mathbb{E}(Y_1)$. In particular, since we assumed $p_1 = 1/2$, $\mathbb{E}(Y_1)$ as well as all consecutive $\mathbb{E}(Y_j)$ equal 0 for all $j \geq 2$. In fact, for arbitrary $\{p_n, \, n \geq 1\}$ satisfying $p_n \neq 1/2$, $n \geq 2$, the entire sequence $(Y_n)_{n\geq 1}$ is centered in expectation (equivalently, $\mathbb{E}(X_n) = 1/2, n \geq 1$) if and only if $p_1 = 1/2$.

Throughout the sequel, for $N \geq 1$ we set

$$T_N := X_1 + \cdots + X_N, \quad S_N := Y_1 + \cdots + Y_N.$$

Then $S_N = 2T_N - N$ and hence limit theorems we establish in the following for S_N/N easily imply analogous results for $T_N/N = S_N/(2N) + 1/2$. At a more elementary level, we first observe that S_N is symmetric in distribution about zero (hence its odd moments vanish), and as a result, T_N is symmetric about $N/2$. The symmetry of the law of S_N about zero follows from the symmetry in the definition of the model. For the characteristic function, one obtains, using symmetry, that

$$\varphi_{S_N}(t) = \mathbb{E}e^{itS_N}$$

$$= \mathbb{E}\cos\left(t\sum_{n=1}^{N} Y_1(-1)^{\sum_{k=2}^{n} W_k}\right) = \mathbb{E}\cos\left(t\sum_{n=1}^{N}(-1)^{\sum_{k=2}^{n} W_k}\right)$$

(where $\sum_{i+1}^{i} := 0$). Using Corr and Cov for correlation and covariance, respectively, one also has for $i < j$ that

$$\mathsf{Corr}(Y_i, Y_j) = \mathsf{Cov}(Y_i, Y_j) = \mathbb{E}(Y_i Y_j) = \mathbb{E}(-1)^{\sum_{k=i+1}^{j} W_k}$$

$$= \prod_{i+1}^{j} \mathbb{E}(-1)^{W_k} = \prod_{k=i+1}^{j} (1 - 2p_k) = e_{i,j}, \qquad (5.4)$$

$$\mathbb{E}(Y_j \mid Y_i) = Y_i \mathbb{E}(-1)^{\sum_{k=i+1}^{j} W_k} = e_{i,j} Y_i.$$

Corollary 5.1 (Correlation estimate). *Assume that* $\lim_{k \to \infty} p_k = 0$ *and let* $n_* \in \mathbb{N}$ *be such that* $p_k \leq 1/2$ *for* $k \geq n_*$. *For* $n_* \leq i < j$,

$$\exp\left(-2 \sum_{k=i+1}^{j} p_k\right) \cdot \prod_{i+1}^{j} (1 - r_k) \leq e_{i,j} \leq \exp\left(-2 \sum_{k=i+1}^{j} p_k\right),$$

where $r_k := 2p_k^2 e^{2p_k}$, *which is tending to zero rapidly.*

Furthermore, for any given $C > 1$, *there exists an* $n_* \in \mathbb{N}$ *such that for* $n_* \leq i < j$,

$$\exp\left(-2 \sum_{k=i+1}^{j} C p_k\right) \leq e_{i,j} \leq \exp\left(-2 \sum_{k=i+1}^{j} p_k\right).$$

Proof. Use the remainder theorem for Taylor series, yielding

$$0 \leq e^{-2p_k} - (1 - 2p_k) \leq 2p_k^2,$$

that is,

$$\exp\left(-2p_k\right) \cdot (1 - r_k) \leq 1 - 2p_k \leq \exp\left(-2p_k\right),$$

and multiply these inequalities to get the first statement.

For the second statement, use that for sufficiently small positive x,

$$e^{-Cx} \leq 1 - x \leq e^{-x}. \qquad \square$$

Similarly to (5.4), if $K = 2m$ is a positive even number, and $i_1 < i_2 < \cdots < i_K$ then, using the fact that[1]

$$\sum_{k=1}^{i_1} W_k + \sum_{k=1}^{i_2} W_k + \cdots + \sum_{k=1}^{i_K} W_k$$

$$= \sum_{k=i_1+1}^{i_2} W_k + \sum_{k=i_3+1}^{i_4} W_k + \cdots + \sum_{k=i_{K-1}+1}^{i_K} W_k \quad (\text{mod } 2),$$

we obtain that

$$\mathbb{E}(Y_{i_1} Y_{i_2} \cdots Y_{i_K}) = \mathbb{E}(-1)^{\sum_{k=1}^{i_1} W_k + \sum_{k=1}^{i_2} W_k + \cdots + \sum_{k=1}^{i_K} W_k}$$

$$= \mathbb{E}(-1)^{\sum_{k=i_1+1}^{i_2} W_k + \sum_{k=i_3+1}^{i_4} W_k + \cdots + \sum_{k=i_{K-1}+1}^{i_K} W_k}$$

$$= \mathbb{E}(-1)^{\sum_{k=i_1+1}^{i_2} W_k}$$

$$\cdot \ \mathbb{E}(-1)^{\sum_{k=i_3+1}^{i_4} W_k} \cdots \mathbb{E}(-1)^{\sum_{k=i_{K-1}}^{i_K} W_k}$$

$$= e_{i_1,i_2} e_{i_3,i_4} \cdots e_{i_{K-1},i_K}. \tag{$*$}$$

We close this section with introducing some frequently used notation.

Notation: In the sequel, Bessel I_α and Bessel K_α denote the modified Bessel function of the first kind (or Bessel-I function) and the modified Bessel function of the second kind (or Bessel-K function), respectively.

Writing out these functions explicitly, one has

$$\text{Bessel } \mathsf{I}_\alpha(x) = \sum_{m=0}^{\infty} \frac{1}{m! \Gamma(m+\alpha+1)} \left(\frac{x}{2}\right)^{2m+\alpha},$$

and

$$\text{Bessel } \mathsf{K}_\alpha(x) = \frac{\pi}{2} \frac{\text{Bessel } \mathsf{I}_{-\alpha}(x) - \text{Bessel } \mathsf{I}_\alpha(x)}{\sin(\alpha\pi)},$$

---

[1]We use the convention now that $p_1 := 1/2$.

if α is not an integer (otherwise it is defined through limits), where Γ is Euler's gamma function. See, e.g., in Abramowitz and Stegun (1964, Sections 9–10) and Bowman (1958, formula (6.8).

5.1   Supercritical Cases

First, if $\sum_n p_n < \infty$, then by the Borel–Cantelli lemma, only finitely many turns will occur a.s.; therefore, the X_j's will eventually become all ones or all zeros, and hence

$$\lim_{N\to\infty} \frac{T_N}{N} = \zeta \text{ a.s.,}$$

where $\zeta \in \{0,1\}$. By the symmetry of the definition with respect to heads and tails (or, by the bounded convergence theorem), ζ is a Bernoulli$(1/2)$ random variable.

5.2   The Critical Case

Fix $a > 0$ and let

$$p_n = \frac{a}{n}, \quad n \ge n_0$$

for some $n_0 \in \mathbb{N}$. Denote by $\mathsf{Beta}(a,a)$ the symmetric (about the point $1/2$) Beta distribution with $a > 0$, with density

$$f_{\mathsf{Beta}(a,a)}(x) = \frac{[x(1-x)]^{a-1}}{B(a,a)}$$

on the unit interval (the normalizing constant is $B(a,a) := \Gamma^2(a)/\Gamma(2a)$, using Euler's Gamma function), and moment generating function

$$M_{\mathsf{Beta}(a,a)}(t) = 1 + \sum_{k=1}^{\infty} \left(\prod_{l=1}^{\infty} \frac{a+l-1}{2a+l-1} \right) \frac{t^k}{k!}$$

$$= e^{t/2} \left(\frac{t}{4} \right)^{\frac{1}{2}-a} \Gamma\left(a + \frac{1}{2} \right) \mathsf{Bessel\ I}_{a-\frac{1}{2}}\left(\frac{t}{2} \right). \quad (5.5)$$

Theorem 5.1. *The law of $\frac{1}{N}\sum_{i=1}^{N} X_i$ converges to* Beta(a, a) *as* $N \to \infty$.

Remark 5.3. It turns out that the convergence in distribution cannot be strengthened to convergence in probability; see, e.g., the example in Section 2 of Gantert (1990).

Proof. We verify the statement by analyzing the moments of S_N. The odd moments are all zeros from the symmetry of S_N around 0; on the other hand, for even K, we can use the multinomial theorem:

$$\mathbb{E}(S_N^K) = I + K! \sum_{1 \le i_1 < i_2 < \cdots < i_K \le N} \mathbb{E}(Y_{i_1} Y_{i_2} \cdots Y_{i_K}),$$

where I stands for the sum of those products in which not all terms are different. Note that $|Y_i^l| = 1$ for any $l \ge 1$ (and $Y_i^l \equiv 1$ for l even). Therefore, $|I| \le m(N, K)$, where $m(N, K)$ is the number of such products. But $m(N, K) \le N \cdot N^{K-2} = N^{K-1}$ because each such product can be written (not uniquely) as $Y_{i_\ell}^2 \cdot Y_{i_1} Y_{i_2} \cdots Y_{i_{K-2}} = Y_{i_1} Y_{i_2} \cdots Y_{i_{K-2}}$, where the numbers $i_\ell, i_1, \ldots, i_{K-2}$ are between 1 and N and are not necessarily distinct. Hence, also using (5.5), we get

$$\mathbb{E}S_N^K = K! \sum_{1 \le i_1 < i_2 < \cdots < i_K \le N} \mathbb{E}(Y_{i_1} Y_{i_2} \cdots Y_{i_K}) + \mathcal{O}(N^{K-1})$$

$$= K! \sum_{1 \le i_1 < i_2 < \cdots < i_K \le N} e_{i_1, i_2} e_{i_3, i_4} \cdots e_{i_{K-1}, i_K} + \mathcal{O}(N^{K-1}).$$

$$(5.6)$$

Let us now analyze the elements in the sum above. From (5.4), for $j > i > \max\{2a, n_0\}$, we have

$$e_{i,j} = \exp\left\{ \sum_{n=i+1}^{j} \log\left(1 - \frac{2a}{n}\right) \right\} = \exp\left\{ \mathcal{O}\left(\frac{j-i}{i^2}\right) - 2a \sum_{n=i+1}^{j} \frac{1}{n} \right\}$$

$$= \exp\left\{ \mathcal{O}\left(\frac{j-i}{i^2}\right) - 2a \log\left(\frac{j}{i}\right) \right\} = \frac{i^{2a}}{j^{2a}} \cdot \left(1 + \mathcal{O}\left(\frac{j-i}{i^2}\right)\right).$$

$$(5.7)$$

Consequently, (5.6) can be approximated as

$$\mathbb{E}S_N^K = K! \sum_{1 \leq i_1 < i_2 < \cdots < i_K \leq N} \frac{i_1^{2a}}{i_2^{2a}} \cdot \frac{i_3^{2a}}{i_4^{2a}} \cdot \ldots \cdot \frac{i_{K-1}^{2a}}{i_K^{2a}} + \mathcal{O}(N^{K-1})$$

$$= \frac{K!\, N^K \left(1 + \mathcal{O}(N^{-1})\right)}{(1+2a) \cdot 2 \cdot (3+2a) \cdot 4 \cdots \cdots (K-1+2a) \cdot K} + \mathcal{O}(N^{K-1})$$

$$(5.8)$$

(the contribution from the terms where $i_1 \leq \max\{2a, n_0\}$ as well as the other remainder terms in the formula for $e_{i,j}$ is of order at most N^{K-1}), as follows from

Lemma 5.1. *Assume $K = 2k$ is an even positive integer. Then as $N \to \infty$*

$$f(K, N) := \sum_{1 \leq i_1 < i_2 < \cdots < i_K \leq N} \frac{i_1^{2a}}{i_2^{2a}} \cdot \frac{i_3^{2a}}{i_4^{2a}} \cdot \ldots \cdot \frac{i_{K-1}^{2a}}{i_K^{2a}}$$

$$= \frac{N^{2k}}{2^k\, k!\, (1+2a)(3+2a) \cdot (2k-1+2a)} + O(N^{2k-1}).$$

Proof. Throughout the proof, we will use the elementary fact that for any $\beta > 0$

$$\sum_{j=1}^{n-1} j^\beta = \frac{n^{\beta+1}}{\beta+1} - c_{n,\beta}\, n^\beta \qquad (5.9)$$

where $0 \leq c_{n,\beta} \leq 1$, which, in turn, follows from an observation that

$$\sum_{j=1}^{n-1} x^\beta \leq \int_0^n u^\beta \, du \leq \sum_{j=1}^n x^\beta.$$

Let us rewrite $f(K, N)$ as

$$f(K, N) = \sum_{3 \leq i_3 < i_4 < \cdots < i_K \leq N} \frac{i_3^{2a}}{i_4^{2a}} \cdot \ldots \cdot \frac{i_{K-1}^{2a}}{i_K^{2a}} \cdot \left[\sum_{\substack{i_1, i_2: \\ 1 \leq i_1 < i_2 < i_3}} \frac{i_1^{2a}}{i_2^{2a}} \right]$$

From (5.9), we get that the expression in the square brackets equals

$$\sum_{i_2=2}^{i_3-1} \left[\frac{1}{i_2^{2a}} \sum_{i_1=1}^{i_2-1} i_1^{2a} \right] = \sum_{i_2=2}^{i_3-1} \left[\frac{i_2}{1+2a} - c_{i_2,2a} \right] = \frac{i_3^2}{2(1+2a)} + O(i_3)$$

$$= \frac{i_3^2(1+o(1))}{2(1+2a)}$$

where $o(1)$ goes to zero as $i_3 \to \infty$. Hence

$$f(K,N) = \frac{1}{2(1+2a)} \sum_{3 \le i_3 < i_4 < \cdots < i_K \le N} \frac{i_3^{2a+2}(1+o(1))}{i_4^{2a}} \cdot \frac{i_5^{2a}}{i_6^{2a}} \cdots \cdots \frac{i_{K-1}^{2a}}{i_K^{2a}}$$

$$= \frac{1+o(1)}{2(1+2a)} \sum_{5 \le i_5 < i_6 < \cdots < i_K \le N} \frac{i_5^{2a}}{i_6^{2a}} \cdots \cdots \frac{i_{K-1}^{2a}}{i_K^{2a}}$$

$$\cdot \left[\sum_{\substack{i_3, i_4: \\ 3 \le i_3 < i_4 < i_5}} \frac{i_3^{2a+2}}{i_4^{2a}} \right],$$

where the very last $o(1)$ expression is meant in the sense that $N \to \infty$, however this requires some tedious but straightforward argument, left to the reader.

The expression in the square brackets again by (5.9) equals

$$\sum_{i_4=4}^{i_5-1} \left[\frac{1}{i_4^{2a}} \sum_{i_3=3}^{i_4-1} i_3^{2a+2} \right] = \sum_{i_4=4}^{i_5-1} \frac{i_4^3(1+o(1))}{3+2a} = \frac{i_5^4(1+o(1))}{4(3+2a)}$$

so

$$f(K,N) = \frac{1+o(1)}{2(1+2a)\,4(3+2a)} \sum_{5 \le i_5 < i_6 < \cdots < i_K \le N} \frac{i_5^{2a+4}(1+o(1))}{i_6^{2a}}$$

$$\cdot \frac{i_7^{2a}}{i_8^{2a}} \cdots \cdots \frac{i_{K-1}^{2a}}{i_K^{2a}}.$$

By repeating this argument, we eventually get

$$f(K,N) = \frac{1 + o(1)}{2(1+2a)\,4(3+2a)\cdots 2(k-2)(K-5+2a)} {2(k-1)(K-3+2a)}$$

$$\times \sum_{K-1 \le i_{K-1} < i_K \le N} \frac{i_{K-1}^{2a+K-2}}{i_K^{2a}}.$$

This final sum equals

$$\sum_{i_K=K}^{N} \left[\frac{1}{i_K^{2a}} \sum_{i_{K-1}=K-1}^{i_K-1} i_{K-1}^{2a+K-2} \right] = \sum_{i_K=K}^{N} \frac{i_K^{K-1}(1+o(1))}{2a+K-1}$$

$$= \frac{N^K}{K(K-1+2a)}(1+o(1)).$$

and hence we get the desired formula.   □

Since we are working on a compact interval, we may conclude (see, e.g., in Durrett (1996, Section 2, Exercise 3.27) that $S_N/N \to \xi_a$ in distribution, where ξ_a is distributed on $[-1,1]$ and has the following moments:

$$\mathbb{E}\left[\xi_a^K\right] = \begin{cases} 0, & K \text{ is odd;} \\ \dfrac{(2m)!}{2^m \cdot m! \cdot (2a+1)(2a+3)\cdots(2a+(2m-1))}, & \\ & K = 2m \text{ is even,} \end{cases}$$

which, for even moments, can be equivalently written as

$$\mathbb{E}\left[\xi_a^{2m}\right] = \frac{(2m)!\,\Gamma(a+1/2)}{2^{2m}\Gamma(m+a+1/2)}. \tag{5.10}$$

The moment generating function of ξ_a is

$$M_a(t) = \mathbb{E}e^{t\xi_a} = 1 + \sum_{m=1}^{\infty} \frac{t^{2m}}{2^m \cdot m! \cdot \prod_{i=1}^{m}(2a+2i-1)}$$

$$= \text{Bessel } I_{a-1/2}(t)\Gamma(a+1/2)(t/2)^{1/2-a}.$$

Let $\zeta_a := (\xi_a + 1)/2$. We know that $\frac{1}{N} \sum_{i=1}^{N} X_i \to \zeta_a$ in distribution, and using (5.5),

$$\mathbb{E}e^{t\zeta_a} = e^{t/2} M_a(t/2)$$
$$= e^{t/2} \text{Bessel } \mathsf{I}_{a-1/2}(t/2)\Gamma(a+1/2)(t/4)^{1/2-a} = M_{\mathsf{Beta}(a,a)}(t),$$

completing the proof. □

Remark 5.4 (Particular cases and densities). Note that in particular, for $a = 1, a = 1/2$ and $a = 3/2$, the limiting law $\mathsf{Beta}(a, a)$ of the relative frequencies in Theorem 5.1 is $\mathsf{Uniform}([0, 1])$, the Arcsine law and the transformed semicircle law[2] on $[0, 1]$, respectively.

Turning to S_N/N, the transformation $x \mapsto 2x - 1$ yields that $\lim_{N\to\infty} \mathsf{Law}(S_N/N)$ equals to $\mathsf{Uniform}([-1, 1])$, the transformed Arcsine law on $[-1, 1]$ and Wigner's semicircle law on $[-1, 1]$, respectively. Concerning the corresponding densities on $[-1, 1]$, we have the following explicit formulas.

- **Transformed Arcsine law:** Let $a = 1/2$. Then

$$\mathbb{E}e^{t\xi} = \sum_{m=0}^{\infty} \frac{t^{2m}}{(2^m \cdot m!)^2} = \text{Bessel } \mathsf{I}_0(t) = \frac{1}{\pi} \int_0^{\pi} e^{t\cos(\theta)} \, d\theta;$$

consequently (see, e.g., Abramowitz and Stegun (1964), formula 29.3.60) $\xi_{1/2}$ has the transformed Arcsine density

$$f_{\xi_{1/2}}(x) = \begin{cases} \dfrac{1}{\pi\sqrt{1 - x^2}}, & -1 < x < 1; \\ 0, & \text{otherwise.} \end{cases}$$

---

[2]The density is a semi-ellipse.

- **Wigner's semicircle law on** $[-1,1]$: Let $a = 3/2$. Then

$$\mathbb{E}e^{t\xi} = \frac{2\,\mathsf{Bessel}\,\mathsf{I}_1(t)}{t} = \frac{1}{\pi}\int_0^\pi e^{t\cos(\theta)}\cos(\theta)\,d\theta$$

$$= \mathsf{Bessel}\,\mathsf{I}_0(t) - \mathsf{Bessel}\,\mathsf{I}_2(t);$$

consequently (see Abramowitz and Stegun (1964), formula 9.6.19) $\xi_{3/2}$ has the Wigner semicircle density

$$f_{\xi_{3/2}}(x) = \begin{cases} \dfrac{2}{\pi}\sqrt{1-x^2}, & -1 < x < 1; \\ 0, & \text{otherwise.} \end{cases}$$

- **General case:** The density of ξ_a is given by

$$f_{\xi_a}(x) = \frac{\Gamma(a+1/2)}{\Gamma(a)\sqrt{\pi}}\left(1-x^2\right)^{a-1}$$

for $-1 < x < 1$. Indeed, for $m \in \mathbb{N}$, we have

$$\int_{-1}^1 x^{2m}\left(1-x^2\right)^{a-1}\,dx = \int_0^1 y^{m-1/2}\left(1-y\right)^{a-1}\,dy$$

$$= \mathsf{Beta}(m+1/2, a) = \frac{\Gamma(m+1/2)\Gamma(a)}{\Gamma(m+a+1/2)},$$

which is consistent with the moments $\mathbb{E}\xi^{2m}$ given by (5.10).

5.3   Subcritical Case

Now fix $\gamma, a > 0$ and let

$$p_n = \frac{a}{n^\gamma}, \quad n \geq n_0$$

for some n_0. Note that $\gamma > 1$ corresponds to the supercritical case studied in Section 5.1, so from now on assume $0 < \gamma < 1$.

Theorem 5.2 (Non-classical CLT). *The law of $S_N/N^{(1+\gamma)/2}$ converges to* $\mathsf{Normal}(0, \sigma^2)$, *where*

$$\sigma := \frac{1}{\sqrt{a(1+\gamma)}}. \tag{5.11}$$

Proof. Let $\eta_{a,\gamma,N} := S_N/N^{(1+\gamma)/2}$. Let $\eta_{a,\gamma}$ be normally distributed with variance σ^2, where σ is as in (5.11). We prove that $\lim_{N\to\infty} \eta_{a,\gamma,N} = \eta_{a,\gamma}$ in law.

Let $K \in \mathbb{Z}_+$. In the proof, we use that

$$\mathbb{E}\eta_{a,\gamma}^K = \begin{cases} 0, & \text{if } K \text{ is odd}; \\ \sigma^K(K-1)!!, & \text{if } K \text{ is even}, \end{cases} \tag{5.12}$$

where $(K-1)!! = (K-1)(K-3)\cdots 5\cdot 3\cdot 1$ (see, e.g., Durrett (1996), Chapter 2, Section 3.e).

Let A $(A > a)$ be a given constant. By Corollary 5.1, there exists an $n_* = n_*(a, A, \gamma)$ (w.l.o.g. assume that $n_* > n_0$) such that for $j > i > n_*$

$$\exp\left\{-2A \sum_{n=i+1}^{j} \frac{1}{n^\gamma}\right\} \le e_{i,j} \le \exp\left\{-2a \sum_{n=i+1}^{j} \frac{1}{n^\gamma}\right\}.$$

Using the fact that $x^{-\gamma}$ is decreasing and bounding the sum by the integral, we have

$$\frac{(j+1)^{1-\gamma} - (i+1)^{1-\gamma}}{1-\gamma} = \int_{i+1}^{j+1} \frac{dx}{x^\gamma} \le \sum_{n=i+1}^{j} \frac{1}{n^\gamma} \le \int_{i}^{j} \frac{dx}{x^\gamma}$$

$$= \frac{j^{1-\gamma} - i^{1-\gamma}}{1-\gamma},$$

yielding

$$\exp\left(-\frac{2A}{1-\gamma}\left[j^{1-\gamma} - i^{1-\gamma}\right]\right) \le e_{i,j};$$

$$e_{i,j} \le \exp\left(-\frac{2a}{1-\gamma}\left[(j+1)^{1-\gamma} - (i+1)^{1-\gamma}\right]\right), \tag{5.13}$$

that is, using the shorthand $c := 2a(1-\gamma)^{-1}$ and $d := 2A(1-\gamma)^{-1}$,

$$\exp\left(-d[j^{1-\gamma} - i^{1-\gamma}]\right) \le e_{i,j} \le \exp\left(-c[(j+1)^{1-\gamma} - (i+1)^{1-\gamma}]\right).$$

One can check that $\sup_N \mathbb{E}\eta_{a,\gamma,N}^2 < \infty$ (see the computation below with $m = 1$), and thus Chebyshev's inequality implies that

$(\eta_{a,\gamma,N})_{N \geq 1}$ is a tight sequence of random variables. Hence, it is enough to show that each subsequential limit is the same.

Assume that $(N_l)_{l \geq 1}$ is a subsequence and $\lim_{l \to \infty} \mathsf{Law}(\eta_{a,\gamma,N_l}) = \mathcal{L}$. Since

$$\left| \frac{Y_1 + \cdots + Y_{n_*-1}}{N_l^{\frac{1+\gamma}{2}}} \right| \leq \frac{n_* - 1}{N_l^{\frac{1+\gamma}{2}}},$$

one has $\mathcal{L} = \lim_{l \to \infty} \mathcal{L}_{N_l,A}$ too, where

$$\mathcal{L}_{N_l,A} := \mathsf{Law}\left(\frac{Y_{n_*} + \cdots + Y_{N_l}}{N_l^{\frac{1+\gamma}{2}}} \right),$$

and in fact, this limit must be the same for any $A > a$ (and corresponding $n_* = n_*(a, A, \gamma)$). Informally, this just means that we may "throw away a finite chunk of the sequence of Y_i's" (at the beginning) without affecting its limit.

Let us denote the even moments of $\mathcal{L}$ by $M_{2m} \in [0, \infty]$, $m \geq 1$, while we note again that the odd moments must be zero by symmetry. Also, $M_{N_l,A,K}$ denotes the Kth moment under $\mathcal{L}_{N_l,A}$.

We show in the following that for a fixed $A > a$ and $K = 2m$, $m \geq 1$,

$$\frac{(2m-1)!!}{[A(1+\gamma)]^m} \leq \liminf_{l \to \infty} M_{N_l,A,K} = \liminf_{l \to \infty} \mathbb{E}\left[\frac{Y_{n_*} + \cdots + Y_{N_l}}{N_l^{\frac{1+\gamma}{2}}} \right]^K$$

$$\leq \limsup_{l \to \infty} M_{N_l,A,K} = \limsup_{l \to \infty} \mathbb{E}\left[\frac{Y_{n_*} + \cdots + Y_{N_l}}{N_l^{\frac{1+\gamma}{2}}} \right]^K$$

$$\leq \frac{(2m-1)!!}{[a(1+\gamma)]^m}. \tag{5.14}$$

Once (5.14) is shown, it will follow from the upper estimate and from the relation $\mathcal{L} = \lim_{l \to \infty} \mathcal{L}_{N_l,A}$ for all $A > a$ that

$$\lim_{l \to \infty} M_{N_l,A,K} = M_K \tag{5.15}$$

for all $K \geq 1$ and all $A > a$. Since (5.14) holds for any $A > a$, letting $A \downarrow a$ and using (5.14) and (5.15), one has that in fact

$$M_K = \frac{(2m-1)!!}{[a(1+\gamma)]^m}.$$

In summary, we obtain that for any fixed $A > a$,

$$\lim_{l \to \infty} M_{N_l,A,K} = \frac{(2m-1)!!}{[a(1+\gamma)]^m}. \tag{5.16}$$

At the same time, we recall that the normal distribution is uniquely determined by its moments, and therefore the convergence toward a normal law is implied by the convergence of all the moments (see, e.g., Durrett (1996), Section 2.3.e). In our case, (5.16) along with (5.12) imply $\mathcal{L} = \lim_{l \to \infty} \mathcal{L}_{N_l,A} = \mathsf{Normal}(0, \sigma^2)$. Therefore, it only remains to prove (5.14).

Let us start with the upper estimate in (5.14). For $K = 2m$, one has

$$\mathbb{E}\left[Y_{n_*} + \cdots + Y_N\right]^K = I + K! \sum_{n_* \leq i_1 < i_2 < \cdots < i_K \leq N} \mathbb{E}(Y_{i_1} Y_{i_2} \cdots Y_{i_K}),$$

where I are lower order terms, as it is shown in the following. Using $(*)$ along with (5.13), we may continue with

$$\leq I + K! \sum_{n_*+1 \leq i_1 < i_2 < \cdots < i_K \leq N+1} \exp\left(c U_{i_1,\dots,i_k}\right), \tag{5.17}$$

where

$$U_{i_1,\dots,i_k} := i_1^{1-\gamma} - i_2^{1-\gamma} + i_3^{1-\gamma} - i_4^{1-\gamma} + \cdots + i_{K-1}^{1-\gamma} - i_K^{1-\gamma}.$$

By the calculation in the Appendix, the right-hand side of (5.17) is

$$I + K! \times \frac{N^{K(1+\gamma)/2}}{c^m(1-\gamma^2)^m \, m!} \cdot (1 + o(1)).$$

By the same token,

$$\mathbb{E}\left[Y_{n_*} + \cdots + Y_N\right]^K \geq I + K! \sum_{1 \leq i_1 < i_2 < \cdots < i_K \leq N} \exp\left(d U_{i_1,\dots,i_k}\right)$$

$$= I + \frac{K! \, N^{K(1+\gamma)/2}}{d^m(1-\gamma^2)^m \, m!} \cdot (1 + o(1)).$$

The reason the remaining terms, collected in I, are of lower order is as follows. Apart from the already estimated term, in the expansion for $\mathbb{E}(Y_{n_*} + \cdots + Y_N)^K$ for $r = 1, 2, \ldots, K - 1$, we also have to sum up the terms of the type

$$\mathbb{E}(Y_{i_1}^{p_1} Y_{i_2}^{p_2} \cdots Y_{i_r}^{p_r}), \quad \text{where } n_* \le i_1 < \cdots < i_r \le N,$$

$$\text{all } p_j \ge 1, \text{ and } p_1 + p_2 + \cdots + p_r = K.$$

Since $Y_i = \pm 1$, and thus $Y_i^p = 1$ if p is even and $Y_i^p = Y_i$ if p is odd, it suffices to estimate only the sums

$$\mathcal{R}(r; \ell_1, \cdots, \ell_r; N; K; \gamma) := \sum \mathbb{E}(Y_{i_1} Y_{i_2} \cdots Y_{i_r}),$$

where the summation is taken over all sets $(i_1, \ldots, i_r)$ such that $i_{k+1} \ge i_k + \ell_k$, $1 \le \ell_k \le K$, for all k, $i_1 \ge 1$ and $i_r \le N$. However, since $r \le K - 1$, each of the sums $\mathcal{R}(r; \ell_1, \ldots, \ell_r; N; K; \gamma)$ is at most of order $N^{r(1+\gamma)/2} \le N^{(K-1)(1+\gamma)/2}$, precisely by the same arguments which were used to estimate the sum in (5.17). The number of those sums can be large, as it is the number of integer partitions of K, but it depends only on K and does not increase with N.

Consequently, for $m \ge 1$, we have

$$(\text{I}) \le \liminf_{l \to \infty} \mathbb{E} \left[\frac{Y_{n_*} + \cdots + Y_{N_l}}{N_l^{\frac{1+\gamma}{2}}} \right]^K$$

$$\le \limsup_{l \to \infty} \mathbb{E} \left[\frac{Y_{n_*} + \cdots + Y_{N_l}}{N_l^{\frac{1+\gamma}{2}}} \right]^K \le (\text{II}),$$

where

$$(\text{II}) := \frac{(2m)!}{[c(1-\gamma^2)]^m \, m!} = \frac{(2m)!}{[2a(1+\gamma)]^m \, m!} = \frac{(2m)!}{2^m \, m!} \cdot [a(1+\gamma)]^{-m}$$

$$= \frac{(2m-1)!!}{[a(1+\gamma)]^m},$$

and by similar computation,

$$(\text{I}) := \frac{(2m-1)!!}{[A(1+\gamma)]^m}.$$

The proof is complete.   □

5.4 When Does the Law of Large Numbers Hold for General Sequences $\{p_n\}$?

A natural question to ask is when does S_N obey the strong (weak) law of large numbers. The following result gives a partial answer.

For a positive even number K, introduce the shorthand

$$\mathcal{E}(N,K) := N^{-K} \sum_{1 \leq i_1 < i_2 < \cdots < i_K \leq N} e_{i_1,i_2} e_{i_3,i_4} \cdots e_{i_{K-1},i_K}, \qquad (5.18)$$

and note that

$$\mathbb{Var}\left(\frac{S_N}{N}\right) = \frac{1}{N} + 2\mathcal{E}(N,2), \qquad (5.19)$$

where

$$\mathcal{E}(N,2) := N^{-2} \sum_{1 \leq i < j \leq N} e_{i,j}. \qquad (5.20)$$

The first condition in the following theorem may look reminiscent of Kolmogorov's sufficient condition for the strong law of large numbers.

Theorem 5.3 (Laws of large numbers). *Recall* (5.18) *and* (5.20). *We have the following:*

(a) *(Strong Law) Assume that at least one of the following two conditions holds:*
(C1)

$$\sum_N \frac{\mathcal{E}(N,2)}{N} < \infty.$$

(C2) *For some positive even number K,*

$$\sum_N \mathcal{E}(N,K) < \infty. \qquad (5.21)$$

Then SLLN holds, that is, $S_N/N \to 0$ a.s.

(b) *(Weak Law) The WLLN holds if and only if for each positive even number K,*

$$\lim_{N \to \infty} \mathcal{E}(N,K) = 0.$$

(c) *(No LLN) If for each positive even number K,*

$$\exists \lim_{N \to \infty} \mathcal{E}(N, K) =: \mu_K > 0,$$

and

$$\sum_{K \in 2\mathbb{N}} \left(\frac{1}{\mu_K} \right)^{1/K} = \infty, \tag{5.22}$$

then the law of large numbers breaks down, and in fact, $\mathrm{Law}(S_N/N)$ *converges to a law which has zero odd moments and even moments* $\{\mu_K\}$.

Note that (5.22) is the so-called *Carleman condition*, guaranteeing that the μ_K's correspond to at most one probability law (see Theorem 3.11, Section 2, in Durrett (1996)).

Proof. We use the facts about the *method of moments* for weak convergence discussed in the proof of Theorem 5.2, along with the fact that from (5.6) it follows that

$$\mathbb{E} \left(\frac{S_N}{N} \right)^K = \frac{K!}{N^K} \sum_{1 \le i_1 < i_2 < \cdots < i_K \le N} \mathbb{E}(Y_{i_1} Y_{i_2} \cdots Y_{i_K}) + \mathcal{O}(1/N)$$

$$= K! \mathcal{E}(N, K) + \mathcal{O} \left(\frac{1}{N} \right). \tag{5.23}$$

(a) Let us consider the two assumptions separately:
   Under (C1), the statement follows from Theorem 1 in Lyons (1988), as $e_{i,j} = \mathrm{Cov}(Y_i, Y_j)$.
   Under (C2), along the lines of Theorem 6.5 in Section 1 of Durrett (1996), we note that for $\varepsilon > 0$, one has

$$\mathbb{P} \left(\left| \frac{S_N}{N} \right| > \varepsilon \right) \le \frac{\mathbb{E} S_N^K}{\varepsilon^K N^K}$$

by the Markov inequality (recall that K is even). Since, by (5.8) and (5.23), the expression on the left-hand side of (5.21) is the leading order term in $\mathbb{E} S_N^K$, by (5.21), we have $\sum_N \mathbb{P} (|S_N/N| > \varepsilon) < \infty$, and thus, by the Borel–Cantelli lemma, $\mathbb{P}(|S_N/N| > \varepsilon \text{ i.o.}) = 0$, which implies the statement.

(b) Since $|S_N/N| \leq 1$, we know that S_N/N converges to zero in law (i.e., in probability, since the limit is deterministic) if and only if all its moments converge to zero. (One possible direction is to realize that the kth moment is the same as $\mathbb{E}f^K(S_N/N)$, where $f(x) = x$ on the unit interval, $f(x) := 1$ for $x > 1$ and $f(x) = -1$ for $x < -1$; then f is bounded and continuous. The other direction is also known since the deterministically zero distribution is uniquely determined by its moments.) By symmetry, it is enough to check the even moments, for which we know (5.23). The statement then follows from the fact that the remainder term is $\mathcal{O}(1/N)$.

(c) Assume that the conditions in (c) hold. Since the moments of S_N/N converge (the odd moments are zero by symmetry), the corresponding laws are tight and, by the Carleman condition, all subsequential limits are the same. That is, as $N \to \infty$, $\mathrm{Law}(S_N/N)$ converges to a law with moments given by μ_K, and since $\mu_K > 0$, the limit cannot be deterministically zero. ☐

The following corollary proves a conjecture in Dietz and Sethuraman (2007) when $1/2 \leq \gamma < 1$.

Corollary 5.2. *When $p_n = a/n^\gamma$ with $0 < \gamma < 1, a > 0$ (sub-critical case), the Strong Law of Large Numbers holds: $S_N/N \to 0$, $\mathbb{P}$-a.s. (Observe that in view of Theorem 5.2, convergence in probability is immediate.)*

Proof. We have seen in the proof of Theorem 5.2 that all moments, and in particular the second moment of the ratio $S_N/N^{\frac{1+\gamma}{2}}$, converge as $N \to \infty$. Thus, $\mathbb{E}(S_N^2) \sim N^{1+\gamma}$ and

$$\mathcal{E}(N, 2) = \frac{1}{2}\left(\mathbb{E}(S_N^2/N^2) - 1/N\right) \sim N^{\gamma-1},$$

hence condition (C1) of Theorem 5.3 is satisfied. (Here $f_N \sim g_N$ means that $\lim_{N\to\infty} f_N/g_N$ exists and is positive.) ☐

Corollary 5.3 (Monotonicity). *Assume that $p_n \leq \hat{p}_n \leq 1/2$ for all n. If the weak law of large numbers holds for the sequence $\{p_n\}$, then it also holds for the sequence $\{\hat{p}_n\}$.*

Proof. This result follows from Theorem 5.3(b) along with the definition of $\mathcal{E}(N, K)$ in terms of the $e_{i,j}$ and the fact that $e_{i,j} = \prod_{k=i+1}^{j}(1 - 2p_k)$ is monotone decreasing in the p_k's for each given $1 \le i < j$. $\qquad\square$

5.5   Giving Up Symmetry

Now we show that, in the supercritical case as well as in the set-ups of Theorem 5.1 and of Theorem 5.2, the initial condition being symmetric (i.e., X_1 is equally likely to be 0 or 1) is actually not essential for the limiting distributions.

Thus, in this section, we assume, without the loss of generality, that $X_1 \equiv 0$ and thus $Y_1 \equiv 1$ and $Y_k = (-1)^{W_2 + \cdots + W_k}$, $k \ge 2$.

In the supercritical case, we have again $T_N/N \to \zeta \in \{0, 1\} \sim$ Bernoulli(q) a.s., but because of the lack of symmetry, we can no longer claim that $q = 1/2$. Our next statement gives the exact value of q for any sequence of $\{p_n\}$. In particular, if at least one of p_i's is $1/2$, then $q = 1/2$, which is already clear from the symmetry.

Proposition 5.1 (No symmetry, supercritical case). *Let us define* $e_{1,\infty} = \prod_{i=2}^{\infty}(1 - 2p_i)$, *consistently with our previous definition. Then*

$$q = \mathbb{P}(\zeta = 1) = \frac{1 + e_{1,\infty}}{2}.$$

Proof. Since we are in the supercritical regime, only finitely many turns occur a.s., and hence $Y_n = Y_\infty \in \{-1, 1\}$ for all large n; as a result, $Y_n \to Y_\infty$ a.s. Hence, using Cesáro mean, $S_N/N \to Y_\infty$ a.s. as well. By the bounded convergence theorem, $\mathbb{E}Y_\infty = \lim_{n \to \infty} \mathbb{E}Y_n = e_{1,\infty}$. Since $Y_\infty = 2\zeta - 1$ and $\mathbb{E}\zeta = q$, we have that $2q - 1 = e_{1,\infty}$. This completes the proof. $\qquad\square$

Let us focus next on the critical case. In this case, equation (5.4) still holds and so does (5.5) for even K, but for odd $K = 2m + 1$ we have

$$\mathbb{E}(Y_{i_1} Y_{i_2} \cdots Y_{i_K}) = e_{0,i_1} e_{i_2,i_3} \cdots e_{i_{K-1},i_K}. \tag{5.24}$$

Calculation (5.8) remains valid for even K, however, if K is odd, one cannot claim any more that $\mathbb{E}S_N^K = 0$. At the same time, a calculation similar to (5.8) immediately shows that if $K = 2m + 1$, then $\mathbb{E}S_N^K = \mathbb{E}S_N^{2m+1} = \mathcal{O}(N^{2m}) = o(N^K)$, and hence the rescaled odd moments tend to zero, while the even moments are the same as in the original model. Hence, the limiting distribution must be the same.

A similar argument holds for the subcritical case as well. Indeed, the even moments $\mathbb{E}S_N^K$ remain the same, while the odd moments for $K = 2m + 1$ will be $\mathbb{E}S_N^{2m+1} = \mathcal{O}\left(N^{m(1+\gamma)}\right) = o\left(N^{K(1+\gamma)/2}\right)$ due to (5.24) and the lemma presented in the Appendix.

5.6 Further Heuristic Arguments and a Conjecture

To avoid ambiguity, by the "classical CLT" we mean the situation where, after normalizing by the standard deviation, the limit has a standard normal distribution, *and* the standard deviation itself is of order $\sqrt{N}$; a "non-standard CLT" will mean that the standard deviation (and thus the fluctuation) is of a different order.

Consider a sum of $N \geq 1$ variables having the same law with finite variance. As is well known, the two "extreme cases" for a sum are the independent case, when the variance is linear, leading to the central limit theorem, and the one when all the variables are identical and the variance grows like N^2. (We assume non-negative correlation, otherwise the sum can be deterministically zero, for instance.) By analogy then (after recalling that in our model

$$\mathbb{V}\mathrm{ar}\,(S_N) = N + 2N^2 \mathcal{E}(N, 2)$$

holds), it seems that the first crucial question is whether

$$\mathcal{E}(N, 2) = \mathcal{O}(1/N), \quad N \to \infty \tag{5.25}$$

is still the case. If (5.25) is true, then $\mathbb{V}\mathrm{ar}\,(S_N)$ is of order N, and one can expect that the classical CLT holds. This happens when $p_n \equiv p \in (0, 1)$.

In a situation when (5.25) fails, one should know at least if

$$\mathcal{E}(N, 2) = o(1) \tag{5.26}$$

holds. Indeed, we know from Theorem 5.3(b) that the exact criterion for WLLN to hold is that

$$\mathcal{E}(N, K) = o(1), \text{ for all } K = 2m. \tag{5.27}$$

In light of this, we make the following conjecture.

Conjecture 5.1 (Classical vs. non-standard CLT). *Let $p_n \in [0, 1]$ for $n \geq 1$, and assume that (5.27) holds:*

(i) *If (5.25) holds for $\{p_n\}$, then the proportion of heads obeys classical CLT. (See Example 5.1.)*

(ii) *If (5.25) fails for $\{p_n\}$, then there is a non-standard CLT for the proportion. (See Example 5.2.)*

If (5.27) fails for $\{p_n\}$, then WLLN is no longer valid for the proportion, that is, the proportion is not concentrated about $1/2$ at all. (See Examples 5.3 and 5.4.)

Examples supporting the discussion and Conjecture 5.1

In the following examples, the deviations from the classical CLT are becoming more marked as we go from Example 5.2 to Example 5.3 to Example 5.4. Recall that $S_N := Y_1 + \cdots + Y_N$ and $T_N := X_1 + \cdots + X_N$ with $X_N = (Y_N + 1)/2$; the frequency of heads is T_N/N.

Example 5.1 (Markov chain CLT). Consider the case $p_n = c$ for all $n \geq 1$, where $0 < c < 1$. If $c = 1/2$, we get an i.i.d. sequence of $+1$'s and -1's and the classical central limit theorem applies.

Now assume $c \neq 1/2$. Then the outcomes are not independent. Indeed, denoting $\kappa := 1 - 2c \in (-1, 1)$, we have

$$N^2 \mathcal{E}(N, 2) = \frac{\kappa(N - 1)}{1 - \kappa} - \frac{\kappa^2 \left(1 - \kappa^{N-1}\right)}{(1 - \kappa)^2}.$$

Therefore, the variance is still of order N but the constant has changed. Recall that $\mathsf{Cov}(Y_i, Y_j) = e_{i,j} = \kappa^{j-i}$, $i < j$, and, following Jones (2004), define $\sigma_c^2 := 1 + 2\sum_{j=2}^{\infty} \mathsf{Cov}(Y_j, Y_1) = 1 + 2\sum_{j=2}^{\infty} \kappa^{j-1} = \frac{1-c}{c}$, assuming that $Y_1 \sim \mathsf{Bernoulli}(1/2)$. In this case, since we are dealing with a time-homogeneous Markov chain, it is well known (see Jones (2004)) that

$$\mathsf{Law}\left(\frac{S_N}{\sqrt{N}}\right) \to \mathsf{Normal}\left(0, \sigma_c^2\right),$$

$$\text{i.e., } \mathsf{Law}\left(\frac{T_N - N/2}{\sqrt{N}}\right) \to \mathsf{Normal}\left(0, \frac{\sigma_c^2}{4}\right).$$

Therefore, unless $c = 1/2$, the classical CLT is slightly changed, since $\mathbb{Var}(S_n) \sim \sigma_c^2 n$. It is also clear that the limiting normal variance can be arbitrarily large when c is sufficiently small and thus turns occur very rarely. On the other hand, it can be arbitrarily small if c is sufficiently close to 1 and thus turns occur very frequently. (If the turns were *certain*, then the limiting variance would vanish of course.)

Example 5.2 (Classical CLT breaks down). Consider the case $p_n := a/n^\gamma$ with $0 < \gamma < 1$. Then

$$N^2 \mathcal{E}(N, 2) = \Theta\left(\frac{N^{\gamma+1}}{2a(1+\gamma)}\right),$$

that is, $\mathbb{Var}(S_N)$ is of order $N^{\gamma+1}$, and the power is strictly between 1 and 2. Now, (5.27) is true, WLLN is still in force, and S_N/N is still around zero (the proportion of heads is around $1/2$). But (5.25) is false. The closer γ to 1, the more the situation differs from the classical CLT. We now have a non-standard CLT, with larger than classical fluctuations.

Example 5.3 (LLN breaks down). Consider the case when $p_n = 1/n$ for $n \geq 2$. Then $e_{i,j} = \frac{(i-1)i}{(j-1)j}$. Consequently, elementary calculation yields that

$$N^2 \mathcal{E}(N, 2) = \sum_{I(N)} \frac{(i-1)i}{(j-1)j} = \frac{1}{3}\binom{N-1}{2},$$

where $I(N) := \{i, j : 1 \leq i < j \leq N\}$, which is of order N^2, that is, (5.25) and even (5.27) are false, causing the law of large numbers to break down, and S_N/N is no longer around zero. This means that the correlation is as strong as in the case of identical variables, and the fluctuations are now of order N, destroying LLN. Similar is the situation when $p_k = \frac{a}{k}$ with $a > 0$.

In terms of the relative frequency of heads, instead of being around the $\delta_{1/2}$ distribution, now it tends to the $\mathsf{Beta}(a, a)$ distribution.

Example 5.4 (Extreme limit). Consider the case when $\sum_n p_n < \infty$. Then $\liminf_{N\to\infty} \mathcal{E}(N, 2) > 0$ holds (hence (5.25) and even (5.27) are false).

Indeed, as we know, the limit of S_N/N is "extreme": $\frac{1}{2}(\delta_{-1} + \delta_1)$, which is as far away from δ_0 as possible! (That is, for the frequencies of heads, one obtains $\mathsf{Beta}(0,0) = \frac{1}{2}(\delta_0 + \delta_1) \equiv \mathsf{Bernoulli}(1/2)$.)

We conclude this section with an open problem.

Problem 5.1 (Monotonicity for SLLN). Is it true that if SLLN holds for the sequence $\{p_n\}$ then it also holds for the sequence $\{\hat{p}_n\}$, whenever $1/2 \geq \hat{p}_n \geq p_n$ for all n?

Note that the corresponding statement for WLLN is true, as it is implied by Corollary 5.3.

5.7   Some Exact Calculations

It turns out that in some particular instances, in the critical case, we can compute the distribution of T_n explicitly. Assume that $X_1 \sim \mathsf{Bernoulli}(1/2)$ and define p_n, $n \geq 2$, as follows:

- Let $a \geq -1$. If $p_n = \frac{1}{n+a}$, then

$$
\mathbb{P}(T_n = k) = \begin{cases} \dfrac{1+a}{n+a}, & k = 0 \text{ or } k = n; \\[2ex] \dfrac{2}{n+a}, & k = 1, 2, \ldots, n-1. \end{cases}
$$

- Let $a \geq 0$. If $p_n = \frac{2}{n+a}$, then

$$
\mathbb{P}(T_n = k) = \begin{cases} \dfrac{a(a+1)}{2(n+a)(n+a-1)}, & k = 0 \text{ or } k = n; \\[2ex] \dfrac{2a^2 + (3n-4)a - 3n + 6k(n-k)}{(n+a)(n+a-1)(n+a-2)}, & k = 1, \ldots, n-1. \end{cases}
$$

- Let $a \geq 1$. If $p_n = \frac{3}{n+a}$, then

$$
\mathbb{P}(T_n = k) = \frac{Q_{n,k,a}}{(n+a-4)(n+a-3)(n+a-2)(n+a-1)(n+a)},
$$

where

$$Q_{n,k,a} = 3a^4 + (9n - 24)a^3 + (6n^2 + 24M - 54n + 57)a^2$$
$$+ (6M(5n - 16) - 24n^2 + 87n - 36)a$$
$$+ 30(M - n)(M - n + 1)$$

for $k = 1, 2, \ldots, n - 1$, where $M = k(n - k)$, and

$$Q_{n,k,a} = \frac{1}{2} \left[a^5 + ((n - 3) + (n - 4))a^2(a^2 - 1) \right]$$
$$+ \frac{1}{2} \left[((n - 3)(n - 4) - 1)a^3 - (n - 3)(n - 4)a \right]$$

for $k = 0$ and $k = n$.

To obtain the above formulas, one can use induction over n for quantities $\mathbb{P}(T_n = k, X_n = 1)$ and $\mathbb{P}(T_n = k, X_n = 0)$ in the same way as it is done further in this book in the proof of Claim 7.1. The algebra involved becomes quite complicated, compared to that in the proof of Claim 7.1, however it can easily be checked using, e.g., Mathematica^{TM} or Maple^{TM}.

5.8 Appendix

In this appendix, we estimate the quantity

$$Q(n_*, N) := \sum_{n_* \leq i_1 < i_2 < \cdots < i_K \leq N}$$
$$\times \exp\left(c \left[i_1^{1-\gamma} - i_2^{1-\gamma} + \cdots + i_{K-1}^{1-\gamma} - i_K^{1-\gamma} \right] \right),$$

for large N and fixed n_*, γ, c, with $K = 2m$, $m \geq 1$. As needed for equation (5.17), we show that it asymptotically equals $\frac{N^{K(1+\gamma)/2}}{c^m (1-\gamma^2)^m \, m!}$

as $N \to \infty$. The result immediately follows from the next statement, as $Q(n_*, N)$ (asymptotically) does not depend on n_* as $N \to \infty$.

Lemma 5.2. *For $l = 0, 1, \ldots, m - 1$, we have*

$$Q(n_*, N) = \sum \exp\left\{ c \left[i_{2l+1}^{1-\gamma} - i_{2l+2}^{1-\gamma} + \cdots + i_{2m-1}^{1-\gamma} - i_{2m}^{1-\gamma} \right] \right\} \times Z_l, \tag{5.28}$$

where the sum is taken over $i_{2l+1}, i_{2l+2}, \ldots, i_{2m}$ such that $n_* + 2l \le i_{2l+1} < i_{2l+2} < \cdots < i_{2m-1} < i_{2m} \le N$, and $Q(n_*, N) = Z_m$ with

$$Z_l := \rho_l \cdot \frac{i_{2l+1}^{(1+\gamma)l}}{l! c^l (1 - \gamma^2)^l} + o\left(N^{(1+\gamma)l} \right),$$

where $\rho_l \to 1$ as $N \to \infty$.

Proof. We are going to prove the statement by induction on l.

For $l = 0$, it is true. Now, assume that we have established (5.28) for some $l \ge 0$. Then $Q(n_*, N)$ equals

$$\sum_{n_* + 2l + 2 \le i_{2l+3} < i_{2l+4} < \cdots < i_{2m-1} < i_{2m} \le N}$$

$$\times \exp\left\{ c \left[i_{2l+3}^{1-\gamma} - i_{2l+4}^{1-\gamma} + \cdots + i_{2m-1}^{1-\gamma} - i_{2m}^{1-\gamma} \right] \right\} \sum_{n_* + 2l \le i_{2l+1} < i_{2l+2} < i_{2l+3}}$$

$$\times \exp\left\{ c [i_{2l+1}^{1-\gamma} - i_{2l+2}^{1-\gamma}] \right\} \left[\frac{\rho_l \, i_{2l+1}^{(1+\gamma)l}}{l! c^l (1 - \gamma^2)^l} + o\left(N^{l(1+\gamma)} \right) \right],$$

where the sum in the second line is taken over i_{2l+1} and i_{2l+2} only. We shall estimate the following sum.

First, note that

$$(*) = \sum_{n_* \le i_{2l+1} < i_{2l+2} < N} \exp\left\{ c [i_{2l+1}^{1-\gamma} - i_{2l+2}^{1-\gamma}] \right\},$$

where each expression is between 0 and 1, can be very well approximated by the corresponding integral, since, whenever $y \le x$, $|\tilde{x} - x| \le 1$ and $|\tilde{y} - y| \le 1$, the ratio

$$\frac{e^{c[\tilde{y}^{1-\gamma} - \tilde{x}^{1-\gamma}]}}{e^{c[y^{1-\gamma} - x^{1-\gamma}]}}$$

is bounded above by $e^{c_1[y^{-\gamma}+x^{-\gamma}]}$, where $c_1 > 0$ is some constant. Hence, outside of the area where x and y are both smaller than $\sqrt{N}$, the above ratio is very close to 1, while the double sum over that area can be at most N. Therefore, as $N \to \infty$,

$$\sum_{n_* \le i_{2l+1} < i_{2l+2} < N} \exp\left\{c[i_{2l+1}^{1-\gamma} - i_{2l+2}^{1-\gamma}]\right\}$$

$$= (1 + o(1)) \int_{n_*}^{N} \int_{y}^{N} e^{cy^{1-\gamma} - cx^{1-\gamma}} \, dx \, dy + \mathcal{O}(N).$$

To calculate the inner integral, observe

$$\int e^{-cx^{1-\gamma}} \cdot \left[1 - \frac{\gamma x^{\gamma-1}}{c(1-\gamma)}\right] dx = -\frac{x^\gamma e^{-cx^{1-\gamma}}}{c(1-\gamma)} + \text{const},$$

implying

$$R(y, N) \le \int_{y}^{N} e^{-cx^{1-\gamma}} \, dx \le (1 + \psi(y)) \times R(N), \text{ where}$$

$$R(y, N) := \frac{y^\gamma e^{-cy^{1-\gamma}}}{c(1-\gamma)} - \frac{N^\gamma e^{-cN^{1-\gamma}}}{c(1-\gamma)} \text{ and } \psi(y) = \left[1 - \frac{\gamma y^{\gamma-1}}{c(1-\gamma)}\right]^{-1}.$$

$$(5.29)$$

Note that $\psi(y) \downarrow 1$ as $y \to \infty$, hence, since $y \ge n_*$,

$$\int_{n_*}^{N} \left[\int_{y}^{N} e^{-cx^{1-\gamma}} \, dx\right] e^{cy^{1-\gamma}} \, dy \le \int_{n_*}^{N} (1 + \psi(n_*)) R(y, N) e^{cy^{1-\gamma}} \, dy$$

$$\le \frac{1 + \psi(n_*)}{c(1-\gamma)} \int_{0}^{N} y^\gamma \, dy = \frac{N^{1+\gamma}}{c(1-\gamma^2)}.$$

Consequently, $(*) = \mathcal{O}(N^{1+\gamma})$ and

$$\sum_{n_* + 2l \le i_{2l+1} < i_{2l+2} < i_{2l+3}} \exp\left\{c[i_{2l+1}^{1-\gamma} - i_{2l+2}^{1-\gamma}]\right\} \cdot o\left(N^{l(1+\gamma)}\right)$$

$$= o\left(N^{(l+1)(1+\gamma)}\right).$$

The next step is to compute

$$\sum_{n_*+2l \leq i_{2l+1} < i_{2l+2} < i_{2l+3}} i_{2l+1}^{(1+\gamma)l} \times \exp\left\{c\left[i_{2l+1}^{1-\gamma} - i_{2l+2}^{1-\gamma}\right]\right\}. \qquad (5.30)$$

W.l.o.g. assume that $N^{\gamma/2} > n_* + 2l$, then (5.30) can be split into two terms, depending whether i_{2l+1} is smaller or larger than $N^{\gamma/2}$.

In the case $i_{2l+1} \leq N^{\gamma/2}$, we have

$$\sum_{n_*+2l \leq i_{2l+1} < i_{2l+2} < i_{2l+3}, \ i_{2l+1} \leq N^{\gamma/2}} i_{2l+1}^{(1+\gamma)l} \times \exp\left\{c\left[i_{2l+1}^{1-\gamma} - i_{2l+2}^{1-\gamma}\right]\right\}$$

$$\leq N^{\gamma/2} \sum_{1 < i_{2l+2} < N} i_{2l+1}^{(1+\gamma)l} \leq N^{\gamma/2} \cdot N \cdot N^{(1+\gamma)l} = o\left(N^{(l+1)(1+\gamma)}\right).$$

For $i_{2l+1} > N^{\gamma/2}$, the sum in (5.30) is well approximated by the integral

$$\int_{N^{\gamma/2}}^{b} \int_{y}^{b} y^q e^{c[y^{1-\gamma} - x^{1-\gamma}]} \, \mathrm{d}x \, \mathrm{d}y = \int_{N^{\gamma/2}}^{b} y^q e^{cy^{1-\gamma}} \left[\int_{y}^{b} e^{-cx^{1-\gamma}} \, \mathrm{d}x\right] \mathrm{d}y,$$

where $q := (1+\gamma)l$ and $b := i_{2l+3}$. We are again allowed to do this as for $|\tilde{x} - x| \leq 1$ and $|\tilde{y} - x| \leq 1$ the ratio

$$\frac{\tilde{y}^q e^{c[\tilde{y}^{1-\gamma} - \tilde{x}^{1-\gamma}]}}{y^q e^{c[y^{1-\gamma} - x^{1-\gamma}]}}$$

is bounded above by $\left(1 + \frac{2l}{y}\right) e^{c_2(x^{-\gamma} + y^{-\gamma})}$, where $c_2 > 0$ is some constant. Observing that $x \geq y \geq N^{\gamma/2}$ yields that this ratio is very close to 1, and thus the double sum in (5.30) equals

$$(1 + o(1)) \int_{N^{\gamma/2}}^{b} y^q e^{cy^{1-\gamma}} \left[\int_{y}^{b} e^{-cx^{1-\gamma}} \, \mathrm{d}x\right] \mathrm{d}y + o\left(N^{(l+1)(1+\gamma)}\right). \tag{5.31}$$

From (5.29), since $y \geq N^{\gamma/2}$ and hence is large, we get that the inner integral equals $(1+o(1))\, R(y,b)$. Therefore, the main expression in (5.31), up to a factor $1 + o(1)$ and the remainder term, equals

$$\int_{N^{\gamma/2}}^{b} \frac{y^{q+\gamma}}{c(1-\gamma)} - \frac{y^q b^\gamma e^{cy^{1-\gamma}-cb^{1-\gamma}}}{c(1-\gamma)}\, dy = \frac{b^{q+1+\gamma} - N^{(q+1+\gamma)\gamma/2}}{(q+1+\gamma)c(1-\gamma)}$$

$$- \int_{N^{\gamma/2}}^{b} \frac{y^q b^\gamma e^{cy^{1-\gamma}-cb^{1-\gamma}}}{c(1-\gamma)}\, dy =: \frac{i_{2l+3}^{(l+1)(1+\gamma)} - o\left(N^{(l+1)(1+\gamma)}\right)}{(l+1)\cdot c(1-\gamma)^2} - (**).$$

Now, the only remaining step is to show that the integral (**) is of smaller order; then the induction step is finished. To this end, fix some $\gamma < \theta < 1$, and do the following estimation:

$$(**) \propto \int_{N^{\gamma/2}}^{b} y^q b^\gamma e^{cy^{1-\gamma}-cb^{1-\gamma}}\, dy$$

$$\leq \int_0^{b-b^\theta} y^q b^\gamma e^{cy^{1-\gamma}-cb^{1-\gamma}}\, dy + \int_{b-b^\theta}^{b} y^q b^\gamma e^{cy^{1-\gamma}-cb^{1-\gamma}}\, dy$$

$$\leq \int_0^{b-b^\theta} y^q b^\gamma e^{c[(b-b^\theta)^{1-\gamma}-b^{1-\gamma}]}\, dy + b^\theta \cdot b^{q+\gamma}$$

$$\leq b^{q+\gamma}\left[b \cdot e^{-c(1-\gamma+o(1))b^{\theta-\gamma}} + b^\theta\right]$$

$$= o\left(b^{q+1+\gamma}\right) = o\left(N^{(l+1)(1+\gamma)}\right),$$

since $b \leq N$.

Finally, the second equality in the statement of lemma follows from repeating the above steps verbatim with the sum

$$\sum_{n_*+2m<i_{2m-1}<i_{2m}<i_{2m+1}} \exp\left\{c\left[i_{2m-1}^{1-\gamma} - i_{2m}^{1-\gamma}\right]\right\} \times Z_{m-1},$$

by replacing i_{2m+1} by $N+1$. $\qquad\square$

5.9 Exercises

(1) Regarding Example 5.3, compute $N^2 \mathcal{E}(N,2)$ when $p_k = a/k$ with $a \in (0,1)$, and estimate it when $p_k = a/k$ for large k's, with $a > 0$.

(2) In the supercritical case, consider $\zeta := \lim_{N \to \infty} T_N/N \in \{0,1\} \sim$ Bernoulli(q). Assuming that $p_i \neq 1/2$ for all $i \geq 2$, let

$$N := \text{card}\{i \geq 2 : p_i > 1/2\}.$$

By supercriticality, $N < \infty$. Concerning the parameter of the Bernoulli distribution, show that $q < 1/2$ if and only if N is odd.

(3) Consider a sum of $N \geq 1$ variables having the same law with finite variance. Make the following statement rigorous and verify it:

> The two "extreme cases" (in terms of variance) for a sum are the independent case, when the variance is linear, and the one when all the variables are identical and the variance grows like N^2.

(Hint: Assume non-negative correlation and use the Cauchy–Schwarz inequality.)

5.10   Notes

The problem has a history, going back to at least the 1950s. In the following, we give a review of the relevant achievements in the past and compare them to the results presented in this chapter.

The case $p_n = 1/n$ was already introduced in R. Dobrushin's thesis[3] in the 1950s and it is attributed to Bernstein (see Dobrušin (1956)). Dobrushin did not seem to explicitly identify the limiting frequency of heads with the uniform distribution. However, in the paper Dietz and Sethuraman (2007), the authors proved that for the more general $p_n = a/n$ case $(a > 0)$, the limiting frequency is Beta(a, a) — see their Theorems 1.3 and 1.4. (They consider a state space consisting of $m \geq 2$ points and so they treat the more general Dirichlet distributions.) Therefore, in the $p_n = a/n$ case, our contribution is providing a different proof only. The proof in Dietz and Sethuraman (2007) is significantly longer and more complicated than ours, however we only consider $m = 2$.

---

[3]He considered his thesis work a continuation of the work of Markov, Bernstein, Sapogov and Linnik on time-inhomogeneous Markov chains.

The case $p_n = a/n^\gamma$ is also treated in Dietz and Sethuraman (2007), albeit only the weak law of large numbers (SLLN when $0 < \gamma < 1/2$). The authors note that "simulations suggest that actually a.s. convergence might hold also on the range $1/2 \leq \gamma < 1$"; in our case, we prove this statement in Corollary 5.2. Fluctuations about the mean are not considered in Dietz and Sethuraman (2007) though.

The situation with the central limit theorem is more interesting in this case. First, Dobrushin's central limit theorem for inhomogeneous Markov chains (Theorem 1.1 in Sethuraman and Varadhan (2005)) only provides the statement in the sense that, after centering and normalizing with the standard deviation, the limit is standard normal. We did not find, however, any result in the literature identifying the order of the standard deviation, which we do provide. (See the estimates on p. 411 and also Corollary 15 on Peligrad (2012, p. 421)).

Second, and more importantly, Dobrushin's theorem only applies to the case when $0 < \gamma < 1/3$. Although the condition given in that result (formula (1.3) in Sethuraman and Varadhan (2005)) is known to be optimal (see Section 2 in Sethuraman and Varadhan (2005)), this is only true in the very general setting in which the theorem stated. It is therefore interesting, we believe, that we also prove the CLT in the $1/3 \leq \gamma < 1$ case excluded in Dobrushin's result. (In Peligrad (2012), Dobrushin's condition was improved by Peligrad, but it is still not applicable in our case when $1/3 \leq \gamma < 1$: formula (7) in Peligrad (2012) is actually more stringent then the Dobrushin condition (9).)

To the best of our knowledge, Theorem 5.3 was first proven in Engländer and Volkov (2018).

Random walks related to "coin turning" in one dimension are treated in Engländer *et al.* (2020) and in higher dimensions in Engländer and Volkov (2022) (where they are called conservative walks).

Chapter 6

The Coin-Turning Walk and Its Scaling Limit

In Chapter 5, the coin-turning process $X_1, X_2, \ldots$ has been studied with a focus on $\overline{X}_N := \frac{1}{N} \sum_{n=1}^{N} X_n$, that is, the empirical frequency of 1's ("heads") in the sequence of X_n's, and the centered version $Y_n := 2X_n - 1 \in \{-1, +1\}$ was introduced too. Recall also Remark 5.1 and that the latter can be defined equivalently as

$$Y_n := (-1)^{\sum_{i=1}^{n} W_i}, \qquad (6.1)$$

where $W_1, W_2, W_3, \ldots$ are independent Bernoulli variables with parameters $p_1, p_2, p_3, \ldots$, respectively, and $p_1 = 1/2$, and that

$$Y_n := \begin{cases} -Y_{n-1}, & \text{with probability } p_n; \\ Y_{n-1}, & \text{otherwise.} \end{cases}$$

Define also $\mathcal{F}_n := \sigma(Y_1, Y_2, \ldots, Y_n)$, $n \geq 1$. We have $Y_j = Y_i(-1)^{\sum_{i+1}^{j} W_k}$, $j > i$, yielding the fundamental formulas (5.4). The quantities $e_{i,j}$ appearing in (5.4) play an important role throughout this chapter. (The reader may also recall the interpretation of these quantities as eigenvalues.)

Before we define our basic object of interest, we point out that, unlike for example in the previous chapter, we **collect some of the more involved proofs in a separate subsection** (Subsection 6.4), and so those do not appear immediately after the corresponding results.

Definition 6.1 (Coin-turning walk). The random walk[1] S on $\mathbb{Z}$ corresponding to the coin turning is called the *coin-turning walk*. Formally, $S_n := Y_1 + \cdots + Y_n$ for $n \geq 1$; we can additionally define $S_0 := 0$, so the first step is to the right or to the left with equal probabilities. As usual, we then can extend S to a *continuous time process*, by linear interpolation.

Remark 6.1 (Second-order Markov process). Even though Y is Markovian, S is not. However, the two-dimensional process U defined by $U_n := (S_n, S_{n+1})$ is Markovian. It lives on a ladder embedded into $\mathbb{Z}^2$. See Figure 6.1.   ◇

In Chapter 5, several scaling limits of the form

$$\lim_{n \to \infty} \mathsf{Law} \left(\frac{S_n}{b_n} \right) = \mathsf{L}$$

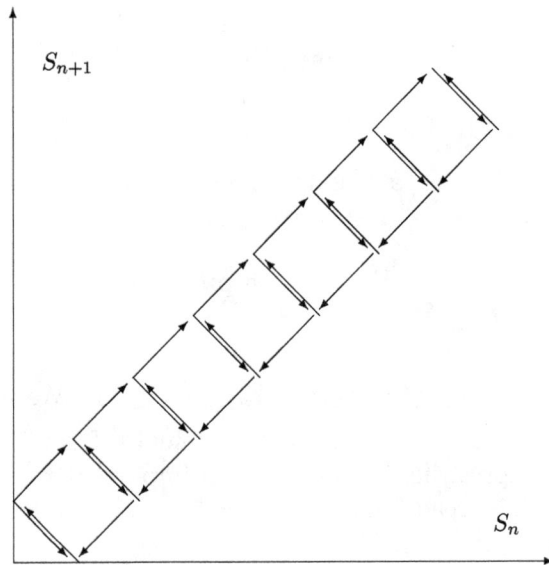

Fig. 6.1.   The process of ordered pairs $U_n := (S_n, S_{n+1})$ is a Markov chain.

---

[1]The reader should not confuse S with S, as the latter is our general notation for state space.

have been established, where $\{b_n\}_{n\geq 1}$ is an appropriate sequence (depending on the sequence of p_n's) tending to infinity and L is a non-degenerate probability law; the focus was on the $\lim_{n\to\infty} p_n = 0$ case.

Let us now review briefly some results from Chapter 5 that concern the one-dimensional distribution of the coin-turning walk.

(i) ("Time-homogeneous case.") Let $p_n = c$ for all $n \geq 1$, where $0 < c < 1$. Then

$$\mathsf{Law}\left(\frac{S_N}{\sqrt{N}}\right) \to \mathsf{Normal}\left(0, \sigma_c^2\right),$$

$$\text{where } \sigma_c^2 := 1 + 2\sum_{i=1}^{\infty} \mathsf{Cov}(Y_i, Y_j) = \frac{1-c}{c}.$$

(ii) ("Lower critical case,") Fix $a > 0$ and let

$$p_n = \frac{a}{n}, \quad n \geq n_0$$

for some $n_0 \in \mathbb{N}$. Then[2]

$$\mathsf{Law}\left(\frac{S_N}{N}\right) \to \mathsf{Beta}(a, a).$$

(iii) ("Lower subcritical case,") Fix $\gamma, a > 0$ and let

$$p_n = \frac{a}{n^\gamma}, \quad n \geq n_0$$

for some $n_0 \in \mathbb{N}$. (Since $\gamma > 1$ corresponds to the supercritical case, we assume that $0 < \gamma < 1$.) Then

$$\mathsf{Law}\left(\frac{S_N}{\sqrt{N^{1+\gamma}}}\right) \to \mathsf{Normal}\left(0, \sigma_{a,\gamma}^2\right), \quad \text{where } \sigma_{a,\gamma}^2 := \frac{1}{a(1+\gamma)}.$$

---

[2]A nice exercise, left to the reader, is to show that when the sequence is precisely $(p_1 = 1/2), p_2 = 1/3, p_3 = 1/4, p_4 = 1/5, \ldots$, $\frac{S_N}{N}$ has precisely discrete uniform law for each N. This fact can be related to Pólya urns — see more on this in Chapter 6.6.

Remark 6.2 (Supercritical cases). Note that if $\sum_n p_n < \infty$, then by the Borel–Cantelli lemma, only finitely many turns will occur a.s.; therefore the X_j's will eventually become all ones or all zeros, and hence

$$\overline{X}_N \to \zeta \text{ a.s.,}$$

where $\zeta \in \{0, 1\}$. By the symmetry of the definition with respect to heads and tails (or, by the bounded convergence theorem), ζ is a Bernoulli(1/2) random variable.

Similarly, if $\sum_n q_n < \infty$, then S will eventually be localized (stuck) at two neighboring integers, again, by the Borel–Cantelli lemma. ◇

These two trivial cases (we call them "lower supercritical" and "upper-supercritical" cases) are not considered, and so we have the following assumption.

Assumption 6.1 (Divergence). In the sequel, we are going to assume that $\sum_n p_n = \infty$ and also $\sum_n q_n = \infty$.

6.1   Mixing

Unlike in Chapter 5 and the previous section of this chapter, we now *do not* randomize the walk with taking Y_1 to be a symmetric random variable. Nevertheless, it is still true for the indicators of turns W_k that $Y_j = Y_i(-1)^{\sum_{i+1}^{j} W_k}$, $j > i$, and that for $e_{i,j} = \prod_{k=i+1}^{j}(1 - 2p_k)$ we have $\mathbb{E}(Y_j \mid Y_i) = Y_i\mathbb{E}(-1)^{\sum_{i+1}^{j} W_k} = e_{i,j}Y_i$, hence $\mathbb{E}(Y_iY_j) = e_{i,j}$.

6.1.1   *Characterization of mixing*

We say that the sequence of random variables $(Y_n)_{n\geq 1}$ satisfies **the mixing condition** if

$$\lim_{j\to\infty} e_{ij} = 0, \quad \forall i \in \mathbb{N}. \tag{6.2}$$

(Recall that by (5.3), e_{ij} may be interpreted as an eigenvalue.) Under mixing, one has that $\lim_{j\to\infty}\mathbb{E}(Y_j \mid Y_i) = 0$, so Y_j "becomes symmetrized" for i fixed and large j. Also, $\lim_{j\to\infty}\mathbb{E}(Y_iY_j) = 0$ and

$\lim_{j \to \infty} \mathbb{E} Y_j = 0$, hence, for fixed $i \geq 2$,

$$\lim_{j \to \infty} \text{Cov}(Y_j, Y_i) = 0, \tag{6.3}$$

in accordance with the usual notion of mixing.

Mixing has a very simple characterization in terms of the sequence $\{p_n\}$, as follows.

Proposition 6.1 (Condition for mixing). *Mixing holds if and only if*

$$\sum_n \min(p_n, q_n) = \infty. \tag{6.4}$$

Proof. Since

$$\min(p_i, q_i) = \begin{cases} p_i, & \text{if } p_i \leq 1/2; \\ q_i = 1 - p_i, & \text{if } p_i > 1/2, \end{cases}$$

we have

$$|e_{i,j}| = \left| \prod_{k=i+1}^{j} (1 - 2p_k) \right|$$

$$= \prod_{i<k\leq j, p_k \leq 1/2} (1 - 2p_k) \times \prod_{i<k\leq j, p_k > 1/2} (1 - 2q_k)$$

$$= \prod_{k=i+1}^{j} (1 - 2\min(p_k, q_k)).$$

When $p_k \neq 1/2$ for all $k \geq 1$, (6.2) and (6.4) are equivalent by a well-known result about infinite products; when $p_k = 1/2$ infinitely often, (6.2) and (6.4) are clearly simultaneously satisfied.

In all other cases, define $k_0 := \max\{k \in \mathbb{N} \mid p_k = 1/2\}$. For $i < k_0$, $e_{i,j} = 0$ for all large j, while for $i \geq k_0$, (6.2) is tantamount to (6.4), just like in the first case. □

6.1.2 Why is mixing a natural assumption?

The mixing condition is stronger than Assumption 6.1 if p_k keeps crossing the line $1/2$ (i.e., $\liminf p_k < 1/2 < \limsup p_k$), while they are equivalent when p_k settles on one side of $1/2$ eventually.

In the first case, Assumption 6.1 is automatically satisfied, as $p_k \geq 1/2$ i.o. and also $q_k \geq 1/2$ i.o. Defining $I := \{i \in \mathbb{N} : p_i \leq 1/2\}$, we see that the mixing condition is nevertheless violated if and only if

$$\sum_i \min(p_i, q_i) = \sum_{i \in I} p_i + \sum_{i \notin I} q_i < \infty,$$

that is, when $\sum_{i \in I} p_i < \infty$ and $\sum_{i \notin I} q_i < \infty$. In this case, recalling that W_i is the indicator of a turn at time i, by Borel–Cantelli,

$$\mathbb{P}\left(\exists n_0 \in \mathbb{N} : W_i = \mathbb{1}_{I^c}(i) \text{ for all } i \geq n_0 \mid \mathcal{F}_1\right) = 1,$$

where $\mathbb{1}_{I^c}$ is the characteristic function of the set $\mathbb{N} \setminus I$. That is, along I, "turning" eventually stops, while along $\mathbb{N} \setminus I$, "staying" eventually stops.

Our conclusion is that when mixing does not hold, the random walk is "eventually deterministic," and thus the setup is less interesting. For example, from the point of view of recurrence, the problem becomes a question about a deterministic process; whether that process takes any integer value infinitely many times depends simply on the set I (as long as $\sum_{i \in I} p_i < \infty$ and $\sum_{i \notin I} q_i < \infty$).

To have a concrete example, let $I = \{2, 4, 6, \ldots\}$ be the set of positive even integers. Then, for large times, the walk will alternate between taking two consecutive steps up and taking two consecutive steps down. This excludes recurrence of course, as the process becomes stuck at some triple of consecutive integers. We summarize the above discussion in Figure 6.2.

Finally, to emphasize the connection with eigenvalues, the reader should recall (5.3) and the discussion in Chapter 2.

6.2   Recurrence/Transience and Scaling Limits

6.2.1   *The law of the nth step for large n*

Recall representation (6.1). When $p_k \leq 1/2$ for all large k, $\rho := \prod_{i=2}^{\infty}(1 - 2p_i)$ is well defined as the terms are in $[0, 1]$ with finitely many exceptions. In particular, when $\sum p_i < \infty$, by Borel–Cantelli,

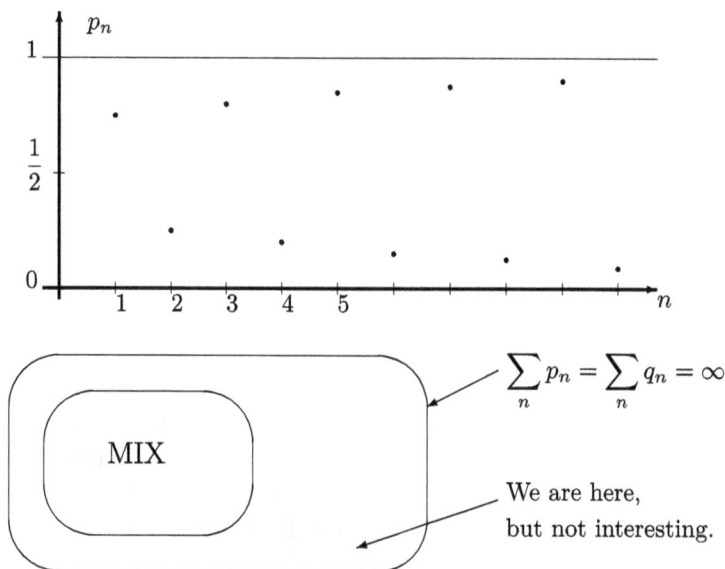

Fig. 6.2. Even if $\sum_n p_n = \sum_n q_n = \infty$ holds, mixing may fail, as it is equivalent to $\sum_n \min(p_n, q_n) < \infty$.

$Y_i = Y$ for all large i, a.s., and just like in Subsection 5.5, we have

$$\mathbb{P}(Y = 1 \mid Y_1 = 1) = \lim_n \mathbb{P}(Y_n = 1 \mid Y_1 = 1) = \frac{1 + \rho}{2},$$

$$\mathbb{P}(Y = 1 \mid Y_1 = -1) = \lim_n \mathbb{P}(Y_n = 1 \mid Y_1 = -1) = \frac{1 - \rho}{2}.$$

This may be generalized as follows.

Theorem 6.1. *Define* $N := \mathrm{card}\{i : p_i > 1/2\} \in \mathbb{N} \cup \{\infty\}$:

(a) *If mixing holds, or if* $\exists i : p_i = 1/2$, *then* $\lim_n \mathbb{P}(Y_n = 1 \mid \mathcal{F}_1) = 1/2$.

(b) *If mixing does not hold, then there are two cases* ($k \in \{-1, 1\}$):

    (i) *if* $N < \infty$, *then* $\lim_{n \to \infty} \mathbb{P}(Y_n = 1 \mid Y_1 = k) = \frac{1}{2}(1 + k\rho)$, *and* $\rho \neq 0$,

    (ii) *if* $N = \infty$, *then* $\mathbb{P}(Y_n = 1 \mid Y_1 = k)$ *has no limit.*

Remark 6.3 (Ergodicity). Part (a) in Theorem 6.1 is interpreted as "mixing[3] implies ergodicity," since $(1/2, 1/2)$ is the invariant distribution for the switching matrix

$$M = \begin{pmatrix} 0 & 1 \\ 1 & 0 \end{pmatrix},$$

and we can consider our model as one where at step n the transition given by M may or may not apply (with probabilities p_n and $1 - p_n$, resp.).   ◇

Remark 6.4 (Speed of convergence). Regarding Theorem 6.1, one may wonder what the speed of convergence is when the limit exists. A closer look at the proof of Theorem 6.1 in Section 6.4 shows that the total variation distance to the limit decays as

$$\exp\left(-2 \sum_{1}^{n} \min\{p_i, 1 - p_i\} \right).$$

We leave the details to the reader.   ◇

6.2.2   *Scaling limits for the walk*

Concerning the results of Chapter 5, a natural question[4] is whether they can be extended to convergence in the process sense, in the spirit of the classical Donsker invariance principle (see, e.g., (Karatzas and Shreve, 1991) for the classical result and its proof). We are now going to answer this question, and moreover, we are also going to consider additional cases, when turns are becoming more and more frequent (i.e., p_n is close to one), such as, for example, $p_n = 1 - c/n$ or $p_n = 1 - n^{-\gamma}$, $0 < \gamma < 1$ for large n.

Note: In the rest of this chapter, for convenience **we assume again that** $p_1 = 1/2$, i.e., we "symmetrize" the setting.

---

[3]In light of (4.4), we can say "merging implies ergodicity" as well.
[4]This question was first proposed by S. O'Rourke

6.2.2.1 *The time-homogeneous case*

As a warm up, we start with the time-homogeneous case.

Theorem 6.2 (Time-homogeneous case). *Assume that $p_n = c$ for $n \geq n_0$. For $n \geq 1$, define the rescaled walk S^n by*

$$S^n(t) := \frac{S_{\lfloor \gamma_c n t \rfloor}}{\sqrt{n}}, \ t \geq 0,$$

where $\gamma_c := \frac{c}{1-c}$, and let $\mathcal{W}$ denote the Wiener measure. Then $\lim_{n \to \infty} \mathsf{Law}(S^n) = \mathcal{W}$ on $C[0, \infty)$.

Remark 6.5 (Alternative argument for Theorem 6.2). We are going to show that Theorem 6.2 follows trivially from our general martingale approximation method of Subsection 6.4.2. However, we note that one can easily get the result intuitively, using that the "turning times" are geometrically distributed. Indeed, assuming that, e.g., $Y_1 = 1$, we can consider the period consisting of the first run of 1's together with the first run of -1's. The second, third, etc. periods are defined similarly, and the piecewise linear "roof-like" processes in these periods are i.i.d. (up to their respective starting values). The classical invariance principle applies to the embedded process considered at each second "turning time." The scaling between the original walk and the embedded one is indicated by the fact that, since the length of each run is geometrically distributed, and those geometric variables are independent, the renewal theorem applies to the lengths of the periods. ◇

Remark 6.6. Theorem 6.2 is also covered by those in Davydov (1970, 1973). The first one treats the "uniformly strong mixing" condition for Markov chains and weak convergence. ◇

6.2.2.2 *Heating regime*

The following theorem gives an invariance principle for the "heating" case, that is, for the case when the p_n are getting close to one. But before that we present an important remark.

Remark 6.7 (Even and odd parts). It turn out that in the heating regime, the right approach is to look at the sums of the two subseries $I = \sum_{\text{odd}} := \sum_{k=1}^{\infty} q_{2k-1}$ and $II = \sum_{\text{even}} := \sum_{k=1}^{\infty} q_{2k}$

separately. If either $I < \infty$ or $II < \infty$, then the invariance principle breaks down.

Indeed, by Borel–Cantelli, after some finite time, every other step turns the coin a.s., and consequently, S is stuck on a set of size three, which rules out the validity of any invariance principle. We conclude that for an invariance principle to hold, it is not enough to assume merely that $\sum_{k=0}^{\infty} q_k = \infty$; one needs to assume that in fact $I = II = \infty$. ⬦

In light of the previous remark, without the loss of generality from now on we work under the following assumption.

Assumption 6.2. $I = II = \infty.$

Before we state the following theorem, we need some more notations, Introduce

$$a_n := \sum_{i=0}^{\infty} \mathsf{Cov}(Y_n, Y_{n+i})$$

$$= 1 + \sum_{i=1}^{\infty} (1 - 2p_{n+1})(1 - 2p_{n+2}) \cdots (1 - 2p_{n+i})$$

$$= 1 + \sum_{i=1}^{\infty} (-1)^i (1 - 2q_{n+1})(1 - 2q_{n+2}) \cdots (1 - 2q_{n+i}), \ n \geq 1,$$

$$(6.5)$$

which is well defined as the sum of a Leibniz series,

$$v_m = \sum_{i=1}^{m} 4a_i^2 p_i q_i, \ m \geq 1, \tag{6.6}$$

and

$$\xi_i := (-1)^{W_i} - \mathbb{E}\left[(-1)^{W_i}\right] = (-1)^{W_i} + 2p_i - 1; \quad \Lambda_n^2 := \sum_{i=1}^{n} a_i^2 \xi_i^2$$

so that $\mathbb{E}\xi_i^2 = \mathbb{V}\mathrm{ar}\left((-1)^{W_i}\right) = 4p_i q_i$ and $\mathbb{E}\Lambda_n^2 = v_n$.

Theorem 6.3 (Invariance principle; heating regime). *Assume that $q_n \to 0$. Besides Assumption 6.2, assume that there exists a $C > 0$ such that at least one of the following two assumptions is* *satisfied:*

$$q_{2m} \geq C \max_{\ell \geq m} q_{2\ell+1}, \quad \forall m \geq m_0 \ (even \ terms \ \text{``dominate''}); \quad (6.7)$$

$$q_{2m+1} \geq C \max_{\ell \geq m+1} q_{2\ell}, \quad \forall m \geq m_0 \ (odd \ terms \ \text{``dominate''}). \quad (6.8)$$

(a) *Define the rescaled walk* S^n *by setting*

$$S^n(t) := \frac{S_{Z(nt)}}{\sqrt{n}}, \quad t \geq 0, \quad (6.9)$$

where $Z(x) := \inf\{n \in \mathbb{N} : v_n \geq x\}$. *Then*

$$\lim_{n \to \infty} \mathrm{Law}(S^n) = \mathcal{W} \text{ on } C[0, \infty), \quad (6.10)$$

where $\mathcal{W}$ *is the Wiener measure.*

(b) *We have* $\lim_{n \to \infty} \Lambda_n = \infty$ *almost surely,*[5] *and further-more,* $\lim_{n \to \infty} \frac{\Lambda_n^2}{\mathbb{E}\Lambda_n^2} = 1$ *in probability. (Hence, in particular,* $\lim_{n \to \infty} v_n = \infty$ *holds too.)*

Remark 6.8 (Equivalent condition). One can rewrite (6.7) by "looking back" too:

$$q_{2m+1} \leq \mathrm{const} \cdot \min\{q_{2\ell}, \ell \leq m\}, \quad \forall m \geq 0,$$

as both are equivalent to saying that $q_n \geq \mathrm{const} \cdot q_r$ for $r > n$ if n is even and r is odd. A similar statement holds for (6.8). ◇

6.2.2.3 *Cooling regime*

When $\lim_{n \to \infty} p_n = 0$, one deals with a so-called "cooling dynamics" as the turns become infrequent. In this case, the scaling limit is not necessarily Brownian motion, as the following theorem shows. Loosely speaking, the order $\mathrm{const} \cdot n^{-1}$ is the critical one in the sense that for sequences of larger order the invariance principle is in force, however at this order or below it, the situation is dramatically different.

Theorem 6.4 (Cooling regime). *Let the process* S^n *be defined by* $S^n(t) := S_{nt}/n$, $t \geq 0$, *where for non-integer values of* nt *we assign* S_{nt} *using the usual linear interpolation. Let* $\mathcal{R}$ *be the process ("random ray") defined by* $\mathcal{R}(t) := tR$, *where* R *is a random variable*

---

[5]Note that the author of Drogin (1972) proves, in fact, *two* invariance principles. The second one uses the function s^2 (our Λ^2) for time change.

equal to ± 1 with equal probabilities. We have the following limits in the process sense:

(1) *Supercritical case:* $\sum_{n=1}^{\infty} p_n < \infty$. *Then* $\lim_{n \to \infty} |S^n(\cdot) - \mathcal{R}(\cdot)|_{\infty} = 0$ *almost surely.*

(2) *Strongly critical case:* $p_n = o(1/n)$ *but* $\sum_{n=1}^{\infty} p_n = \infty$. *Then* $\lim_{n \to \infty} S^n(\cdot) = \mathcal{R}(\cdot)$ *in law.*

(3) *Critical case:* $p_n = c/n$ *for* $n \geq n_0$. *Then* $\lim_{n \to \infty} S^{(n)}$ *is the zigzag process defined further in Section 6.4.1 (the limit is meant in law).*

(4) *Subcritical case: (Cooling but larger order than $1/n$). Let* $p_1 = 1/2$. *Assume that, as* $n \to \infty$,

    (a) $A_n := n p_n \uparrow \infty$,
    (b) $p_n \downarrow 0$.

Then, for the rescaled walk (6.9), the invariance principle (6.10) holds.

We next turn to the case when neither cooling nor heating takes place. This is a bit broader than the "uniformly elliptic" case in Dolgopyat and Sarig (2023), where $0 < \varepsilon < p_n < 1 - \varepsilon$ for *all* indices.

6.2.2.4 *Neither heating nor cooling regime*

The following result generalizes the case when $\lim_{n \to \infty} p_n = a$ with $0 < a < 1$, as well as the time-homogeneous case of Theorem 6.2: the invariance principle holds as long as the p_n are bounded away from both 0 and 1 for large indices.

Theorem 6.5 (Invariance principle; neither heating nor cooling). *Assume that*

$$0 < \liminf_{n \to \infty} p_n \leq \limsup_{n \to \infty} p_n < 1. \tag{6.11}$$

Then for the rescaled walk (6.9), the invariance principle (6.10) holds.

6.2.3   *Validity of the WLLN*

With regard to the weak law of large numbers (by which we mean that $S_n/n \to 0$ in probability), we know that it breaks down at the critical regime. On the other hand, the following result shows that above that order it is always in force.

Theorem 6.6 (WLLN). *Let $p_n \leq 1/2$ for all $n \geq 1$ and assume that $\lim_{n \to \infty} np_n = \infty$. Then $\lim_{n \to \infty} \frac{S_n}{n} = 0$ in probability.*

6.2.4 *Recurrence*

We now turn our attention to the recurrence/transience of the walk and its scaling limit.

Definition 6.2 (Range and effective range). The *range* of the process S is defined as

$$\mathcal{R}_0 := \{x \in \mathbb{Z} : \ S_n = x \text{ for some } n\}.$$

The *effective range* of the process S is

$$\mathcal{R} := \{x \in \mathbb{Z} : \ S_n = x \text{ for infinitely many } n\} \subseteq \mathcal{R}_0.$$

Note that either $\mathcal{R} = \emptyset$ or $\mathcal{R}$ is a connected set in $\mathbb{Z}$.

Definition 6.3 (Localization). We say that the walk *gets stuck* or *localizes* if $\mathcal{R}$ is non-empty and finite. We call S *recurrent*, if $\mathcal{R} = \mathbb{Z}$.

Remark 6.9 (Finite range). Clearly, if the walk gets stuck, then not only $\mathcal{R}$, but also $\mathcal{R}_0$ is finite.

Theorem 6.7 (Dichotomy under mixing). *The mixing condition implies that either*

$$\mathbb{P}(S \text{ is recurrent}) = 1 \quad or \quad \mathbb{P}(S \text{ gets stuck}) = 1.$$

In other words, the effective range is either $\mathbb{Z}$ or it is a finite non-empty set.

Proof. By representation (6.1), one obtains that for $n \geq N$,

$$|S_n - S_N| = \left| \sum_{k=N+1}^{n} (-1)^{W_{N+1} + W_{N+2} + \cdots + W_k} \right|.$$

Consequently, the event $\{S \text{ gets stuck}\}$ is a tail event for the filtration generated by the independent variables $W_1, W_2, \ldots$ and thus has probability either 0 or 1, by Kolmogorov's zero-one law. Assume that the probability of getting stuck is not one, hence it is zero. Recursively define the sequence of stopping times (τ, η) as follows. Let $\tau_0 = 0$ and for $k \geq 1$ let

$$\eta_k = \inf\{j > \tau_k : |e_{\tau_k, j}| \leq 1/2\} \tag{6.12}$$

(recall that $e_{i,j} = \mathbb{E}(Y_i Y_j) = \prod_{k=i+1}^{j}(1-2p_k)$); by mixing, η_k is finite.

Let $a \in \mathbb{Z}_+$. Since S does not localize (gets stuck), we have $\limsup_{n \geq \eta_k} |S_n - S_{\eta_k}| = \infty$, in particular,

$$\tau_{k+1} := \inf\{n \geq \eta_k : |S_n - S_{\eta_k}| = \eta_k + a\}$$

is finite.

Let

$$\Delta := S_{\tau_{k+1}} - S_{\eta_k} = Y_{\eta_k} \sum_{i=\eta_k+1}^{\tau_{k+1}} (-1)^{W_{\eta_k+1}+W_{\eta_k+2}+\cdots+W_i},$$

then $|\Delta| = \eta_k + a$ albeit we do not know the sign of Δ. Next,

$$\mathbb{P}(Y_{\eta_k} = 1 \mid \mathcal{F}_{\tau_k}) = \mathbb{P}(\kappa Y_{\tau_k} = 1 \mid \mathcal{F}_{\tau_k}) = \begin{cases} \mathbb{P}(\kappa = 1), & \text{if } Y_{\tau_k} = 1; \\ \mathbb{P}(\kappa = -1), & \text{if } Y_{\tau_k} = -1, \end{cases}$$

where $\kappa = (-1)^{W_{\tau_k+1}+W_{\tau_k+2}+\cdots+W_{\eta_k}}$. On the other hand, $\kappa = Y_{\eta_k}/Y_{\tau_k} = Y_{\eta_k} Y_{\tau_k}$ so

$$|\mathbb{E}\kappa| = |\mathbb{P}(\kappa = 1) - \mathbb{P}(\kappa = -1)| \leq \frac{1}{2}$$

by (6.12), while $\mathbb{P}(\kappa = 1) + \mathbb{P}(\kappa = -1) = 1$, yielding

$$\mathbb{P}(\kappa = 1) \geq \frac{1}{4}, \quad \mathbb{P}(\kappa = -1) \geq \frac{1}{4}.$$

So $\mathbb{P}(Y_{\eta_k} = 1 \mid \mathcal{F}_{\tau_k}) \geq \frac{1}{4}$; by the same argument, $\mathbb{P}(Y_{\eta_k} = -1 \mid \mathcal{F}_{\tau_k}) \geq \frac{1}{4}$. This, in turn, leads to

$$\mathbb{P}(\Delta = |\Delta| \mid \mathcal{F}_{\tau_k}) \geq \frac{1}{4}, \quad \mathbb{P}(\Delta = -|\Delta| \mid \mathcal{F}_{\tau_k}) \geq \frac{1}{4}.$$

From the definition of τ_{k+1} and the trivial fact that $|S_{\eta_k}| \leq \eta_k$, it follows now that

$$\mathbb{P}(S_{\tau_{k+1}} \leq -a \mid \mathcal{F}_{\tau_k}) \geq \frac{1}{4},$$

so $\mathbb{P}(S_n \leq -a$ infinitely often$) = 1$ by the extended Borel–Cantelli lemma. Since a is arbitrary, $\liminf_n S_n = -\infty$ a.s. By a symmetric argument, $\limsup_n S_n = +\infty$ a.s. as well, yielding that $\mathcal{R} = \mathbb{Z}$ a.s. $\qquad\square$

Next, let us introduce the following mild condition on the walk.

Assumption 6.3 (Spreading). $|S_n| \to \infty$ as $n \to \infty$ in probability.

Remark 6.10. Assumption 6.3 is trivially satisfied when $\sigma_n^2 := \mathsf{Var}(\mathsf{S_n}) \to \infty$ as $n \to \infty$, and the scaling limit

$$\lim_{m \to \infty} \mathbb{P}\left(\frac{S_{n+m}}{\sigma_{n+m}} \in [a,b] \mid \mathcal{F}_n\right) = Q([a,b]), \quad \text{a.s.} \qquad (6.13)$$

holds with $a, b \in \mathbb{R}$, $a \leq b$ and $n \in \mathbb{N}$, and some probability measure Q such that $Q(\{0\}) = 0$. These scaling limits were established in many cases in Chapter 5.

Let us now assume, in addition, mixing. Reformulate (6.13) as

$$\lim_{m \to \infty} \mathbb{P}\left(\frac{S_{n+m} - S_n}{\sigma_{n+m}} \in [a,b] \mid S_n, Y_n\right) = Q([a,b]), \quad \text{a.s.}$$

for all $a < b$. It is easy to see that the conditioning on Y_n could be safely dropped, as the "initial" nth step gets forgotten. $\qquad\diamond$

The next statement follows immediately from Theorem 6.7.

Corollary 6.1 (Spreading and mixing together yield recurrence). *Besides Assumption 6.3, assume also mixing. Then S is recurrent.*

Proof. By Theorem 6.7, no recurrence would imply localization with probability one. In light of Remark 6.9 however, the latter contradicts Assumption 6.3. $\qquad\square$

We have also, in terms of $q_n := 1 - p_n$, the following result.

Corollary 6.2 (Recurrence in terms of q_n). *Assume mixing and furthermore, either*

(a) $\liminf_n q_n > 0$ *or*
(b) $q_n \searrow 0$.

Then S is recurrent.

Proof. Given the dichotomy in Theorem 6.7, one only needs to rule out the possibility that S gets stuck a.s. In case (a), this is trivial: indeed, define $\varepsilon = \frac{1}{2}\liminf_n q_n > 0$ and let n_0 be such that $q_n \geq \varepsilon$ for all $n \geq n_0$. Suppose that $\mathcal{R}$ is finite, and let x be its right-most point. Then eventually the walk will visit x only from the left. Assuming that $n \geq n_0$, this would mean that every time $S_n = x$, with probability at least ε, the walk will continue moving to the right. From the extended Borel–Cantelli lemma, it follows that then $S_{n+1} = x + 1$ for infinitely many ns a.s., yielding a contradiction.

In case (b), again assume that $\mathcal{R}$ is finite, and x is its right-most point. Let $\tau_0 = 0$ and for $k \geq 1$ define[6]

$$\tau_k = \inf\{n > \tau_{k-1} : (S_{n-2}, S_{n-1}, S_n) = (x - 2, x - 1, x)\}$$

which means the walk crosses the edge $[x - 1, x]$ from the left for the first time after it abandoned vertex $x - 1$ to the left in the past. Since q_n is small, it is very likely that the walk will "shuttle" back and forth across the edge before crossing either the edge $[x - 2, x - 1]$ or $[x, x+1]$. Let $\eta \geq 1$ be the number of shuttles, that is, the number of times after τ_k, when the walk crosses $[x - 1, x]$ in either direction before eventually leaving to either $x - 2$ or $x + 1$. Then, if η is odd, then the next crossed edge will be to the right, and if η is even, it will be to the left. By the assumed monotonicity of q_n, we conclude that $\mathbb{P}(\eta = i + 1) \leq \mathbb{P}(\eta = i)$ for all $i \in \mathbb{Z}_+$, and thus

$$\mathbb{P}(\eta \text{ is odd}) - \mathbb{P}(\eta \text{ is even}) = \sum_{k=1}^{\infty} [\mathbb{P}(\eta = 2k - 1) - \mathbb{P}(\eta = 2k)] \geq 0.$$

Therefore,

$$\mathbb{P}(S_{\tau_k + \eta} = x + 1 \mid \mathcal{F}_{\tau_k}) \geq \frac{1}{2},$$

---

[6]To be pedantic, we should also consider the case when $\mathcal{R} = \{x - 1, x\}$, however, it is easy to see that this is impossible since $\sum q_n = \infty$ by mixing.

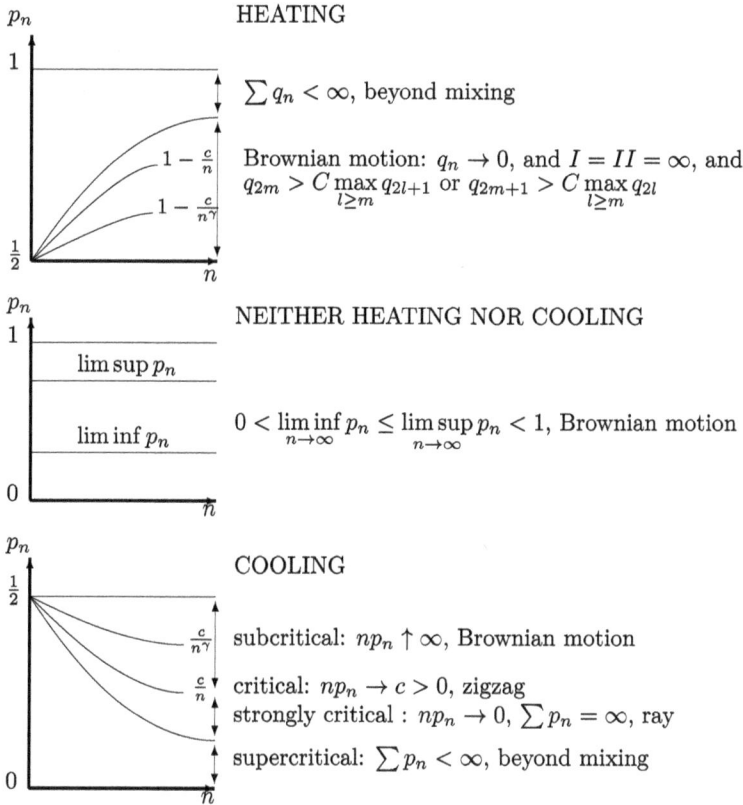

Fig. 6.3. Three regimes of possible convergence.

hence, by the extended Borel–Cantelli lemma, $x + 1$ is also visited infinitely often, contradicting the definition of x. □

Finally, we summarize our scaling results in Figure 6.3.

6.3 Examples and Open Problems

In this section, we compute the scaling $Z(\cdot)$ for a few examples in the cooling regime and the heating regime. We first give two concrete examples for the heating regime. Note that the scaling function $Z(\cdot)$ is the generalized inverse of $v(m) := \sum_{n=1}^{m} 4a_n^2 p_n q_n$. Hence, it suffices to compute $v(m)$ in order to obtain the scaling of $S^{(n)}$.

Example 6.1 (Heating regime). Set $p_n = 1 - \frac{c}{2\,n^\gamma}$, for $n \geq n_0$, where $0 < \gamma < 1$. By Proposition 6.2 in Section 6.4.2, $\mathbb{V}\mathrm{ar}\,(S_m) = (1 + o(1))v(m)$, so we only need to compute $\mathbb{V}\mathrm{ar}\,(S_m)$, and then $Z(\cdot)$ is asymptotically equivalent to the "inverse" of $\mathbb{V}\mathrm{ar}\,(S_m)$. First, note that

$$e_{ij} = \mathsf{Cov}(Y_i, Y_j) = \prod_{k=i+1}^{j}(1 - 2p_k), \quad |e_{ij}| = \prod_{k=i+1}^{j}\left(1 - \frac{c}{k^\gamma}\right),$$

and

$$\mathbb{V}\mathrm{ar}\,(S_n) = n + 2\sum_{i=1}^{n-1}\sum_{j=i+1}^{n} e_{ij} = n + 2\sum_{i=1}^{n-1}\sum_{j=i+1}^{n}(-1)^{i+j}\prod_{k=i+1}^{j}|e_{ij}|.$$

Thus,

$$\mathbb{V}\mathrm{ar}\,(S_n) - \mathbb{V}\mathrm{ar}\,(S_{n-1}) = 1 + 2\sum_{i=1}^{n-1} e_{in} = 1 - 2\sum_{i=1}^{n-1}(-1)^{n-1-i}|e_{in}|.$$

Let us now show that

$$\sum_{i=1}^{n-1}(-1)^{n-1-i}|e_{in}| = \sum_{i=1}^{n-1}(-1)^{n-1-i}\prod_{k=i}^{n}\left(1 - \frac{2}{k^\gamma}\right) = \frac{1}{2} - \frac{c + o(1)}{4n^\gamma}. \tag{6.14}$$

In the case when $i \leq n - n^{\frac{2\gamma+1}{3}}$ (note that $\gamma < \frac{2\gamma+1}{3} < 1$), one has

$$|e_{in}| \leq \prod_{k=n-n^{\frac{2\gamma+1}{3}}}^{n}\left(1 - \frac{c}{k^\gamma}\right) \leq \left(1 - \frac{c}{n^\gamma}\right)^{n^{\frac{2\gamma+1}{3}}} \leq \exp\left(-cn^{\frac{1-\gamma}{3}}\right)$$

yielding

$$\sum_{i=1}^{n-n^{\frac{2\gamma+1}{3}}}|e_{in}| < ne^{-cn^{\frac{1-\gamma}{3}}} = o(n^{-\gamma}). \tag{6.15}$$

For $i \geq n - n^{\frac{2\gamma+1}{3}}$, we have

$$\sum_{i=n-n^{\frac{1+2\gamma}{3}}}^{n-1} (-1)^{n-1-i}|e_{in}|$$

$$= (|e_{n-1,n}| - |e_{n-2,n}|) + (|e_{n-3,n}| - |e_{n-4,n}|) + \cdots$$

$$= \frac{c}{(n-1)^\gamma}|e_{n-1,n}| + \frac{c}{(n-3)^\gamma}|e_{n-3,n}| + \frac{c}{(n-5)^\gamma}|e_{n-5,n}| + \cdots$$

$$= d_1 + d_3 + d_5 + \cdots = \sum_{j=1,\ \text{odd}}^{n^{\frac{(2\gamma+1)}{3}}} d_j, \qquad (6.16)$$

where

$$d_j = \frac{c}{(n-j)^\gamma} \left(1 - \frac{c}{(n-j+1)^\gamma}\right)\left(1 - \frac{c}{(n-j+2)^\gamma}\right) \cdots \left(1 - \frac{c}{n^\gamma}\right),$$

with $1 \leq j \leq n^{\frac{2\gamma+1}{3}}$. Define also the quantities

$$b_j := \kappa(1-\kappa)^j, \quad \text{where } \kappa = \frac{c}{n^\gamma}.$$

Note that

$$d_j \leq \frac{c}{(n-j)^\gamma}\left(1 - \frac{c}{n^\gamma}\right)^j = \left(1 - \frac{j}{n}\right)^{-\gamma} b_j = \left(1 + O\left(n^{-\frac{2-2\gamma}{3}}\right)\right) b_j$$

but

$$d_j \geq \frac{c}{n^\gamma}\left(1 - \frac{c}{(n-j+1)^\gamma}\right)^j = \left(1 - c\frac{\left(1-\frac{j-1}{n}\right)^{-\gamma}-1}{n^\gamma - c}\right)^j b_j$$

$$= \left(1 - O\left(\frac{j}{n^{1+\gamma}}\right)\right)^j b_j = \left(1 - O\left(\frac{j^2}{n^{1+\gamma}}\right)\right) b_j$$

$$= \left(1 - O\left(n^{-\frac{1-\gamma}{3}}\right)\right) b_j.$$

Hence,

$$|b_j - d_j| = b_j \times o(1),$$

implying

$$\sum_{j=1,\ \text{odd}}^{n^{(2\gamma+1)/3}} d_j = (1+o(1)) \sum_{j=1,\ \text{odd}}^{n^{(2\gamma+1)/3}} b_j. \qquad (6.17)$$

At the same time, one has that

$$b_1 + b_3 + \cdots = \kappa(1-\kappa)[1 + (1-\kappa)^2 + (1-\kappa)^4 + \cdots]$$

$$= \frac{\kappa(1-\kappa)}{1-(1-\kappa)^2}$$

$$= \frac{1-\kappa}{2-\kappa} = \frac{1}{2} - \frac{\kappa}{2(2-\kappa)} = \frac{1}{2} - \frac{c+o(1)}{4n^\gamma},$$

so

$$\sum_{j=1,\ \text{odd}}^{n^{(2\gamma+1)/3}} b_j = \sum_{j=1,\ \text{odd}}^{\infty} b_j - O\left((1-\kappa)^{n^{\frac{2\gamma+1}{3}}}\right)$$

$$= \sum_{j=1,\ \text{odd}}^{\infty} b_j - O\left(e^{-cn^{\frac{1-\gamma}{3}}}\right)$$

$$= \frac{1}{2} - \frac{c+o(1)}{4n^\gamma}. \qquad (6.18)$$

Then, combining (6.15), (6.16), (6.17) and (6.18), we obtain (6.14). Hence,

$$\mathbb{V}\text{ar}\,(S_n) - \mathbb{V}\text{ar}\,(S_{n-1}) = 1 - 2\left[\frac{1}{2} - \frac{c+o(1)}{4n^\gamma}\right] = \frac{c+o(1)}{2n^\gamma},$$

and as a result, $\mathbb{V}\text{ar}\,(S_n) = \frac{c}{2(1-\gamma)}n^{1-\gamma} + o(n^{1-\gamma})$.

Our conclusion is that $Z(x) \sim \lfloor (2x(1-\gamma)/c)^{\frac{1}{1-\gamma}} \rfloor$, that is, for the rescaled walk (6.9), the limit in (6.10) holds.

Example 6.2 (Heating regime). Let $p_n = 1 - \frac{c}{n}$, $n \geq n_0$, for some $n_0 \geq 1$. From Lemma 6.1 in Section 6.4, $\lim_{n\to\infty} a_n = 1/2$, hence

$$v_m = \sum_{n=1}^{m} 4a_n^2 p_n q_n = (1+o(1)) \sum_{n=1}^{m}\left(1 - \frac{c}{n}\right)\frac{c}{n} = (c+o(1))\ln n.$$

Thus, for the rescaled walk (6.9), the limit (6.10) holds but now with $Z(x) \sim \lfloor e^{x/c} \rfloor$.

Next is an example for the cooling regime.

Example 6.3 (Subcritical case; cooling regime). If $p_n = \frac{c}{n^\gamma}$ for some $c > 0$, $\gamma \in (0,1)$ and all $n \geq n_0$, then for the rescaled walk (6.9), the invariance principle (6.10) holds. Indeed, similar to the previous examples, one only needs to know the order of $\mathbb{V}\mathrm{ar}\,(S_m)$. By Theorem 5.2, $\mathbb{V}\mathrm{ar}\,(S_n) = (1 + o(1))\frac{n^{1+\gamma}}{c(1+\gamma)}$, so $Z(x) \sim \left\lfloor [c(1 + \gamma)]^{\frac{1}{1+\gamma}} (x)^{\frac{1}{1+\gamma}} \right\rfloor$.

We finally present a few open problems. Recall that PPP denotes Poisson point process.

Problem 6.1 (When p_n is not comparable to $1/n$; different PPP's). What can be said about the case when $\liminf_n np_n = 0$ and $\limsup_n np_n = \infty$? A somewhat related question is whether the following is possible for some situations: the scaling limit is a piecewise deterministic process and the turning points form a PPP but the intensity is different from $\mathrm{const}/x \, \mathrm{d}x$.

Problem 6.2 (Random temporal environment). One can also consider a *random walk in a random temporal environment* (as opposed to the more usual random spatial environment) as follows. Assume now that the p_n are i.i.d. random and follow the same distribution (supported on $[0,1]$). What can one say about the walk in the quenched or in the annealed case? Clearly, when the support of the distribution is compactly embedded in $[0,1]$, the question is not too interesting, however, in general, a.s. $\liminf_n p_n = 0$ and $\limsup_n p_n = 1$.

6.4   Additional Proofs for This Chapter

The rest of this chapter is organized as follows. After presenting two preparations sections on martingale approximation and on a piecewise deterministic process, we give the proofs of the main results.

6.4.1   *Preparation I: The zigzag process*

We now define a stochastic process, which we relate to the critical case in the cooling regime.

Definition 6.4 (Zigzag process). Consider a Poisson point process (PPP) on $[0, \infty)$ with intensity measure $\frac{a}{x}\,dx$ with $a > 0$. (Such a process is known as the *scale-invariant Poisson process*, see Arratia *et al.* (1999).) Once the realization is fixed, the value of the process at $t \geq 0$ is obtained as follows. Starting with the segment containing t and going backwards toward the origin, color the first, third, fifth, etc. segments between the points blue. The second, fourth, etc. will be colored red. Given this Poisson intensity, we have infinitely many segments toward zero (and also toward infinity) almost surely.

Let $\lambda_b(t)$ and $\lambda_r(t)$ denote the Lebesgue measure of the union of blue, resp., red segments between 0 and t. Then we define the *zigzag process* X by

$$X_t := W[\lambda_b(t) - \lambda_r(t)],$$

where W is a random sign ($W = -1$ or $W = 1$ with equal probabilities; also called Rademacher variable) that is independent of the Poisson point process. See Figure 6.4.

It is easy to check directly that the law of the process is invariant under scaling both axes by the same number.

Remark 6.11 (One-dimensional marginals). It is more challenging to check directly that the one-dimensional marginal of the zigzag process at $t \geq 0$ is $t\mathsf{Beta}(a, a)$, although this follows immediately from Theorem 6.4 along with the scaling limit result for the one-dimensional distributions in the previous chapter.

Edward Crane has shown us a nice direct proof for this fact though. The interested reader may enjoy trying to find such a proof him/herself. Hint: Use the scale invariance of the PPP used in the definition, compute the moments and use the uniqueness of solution in the Hausdorff moment problem. (The distribution must of course be supported on $[-t, t]$.)   ◇

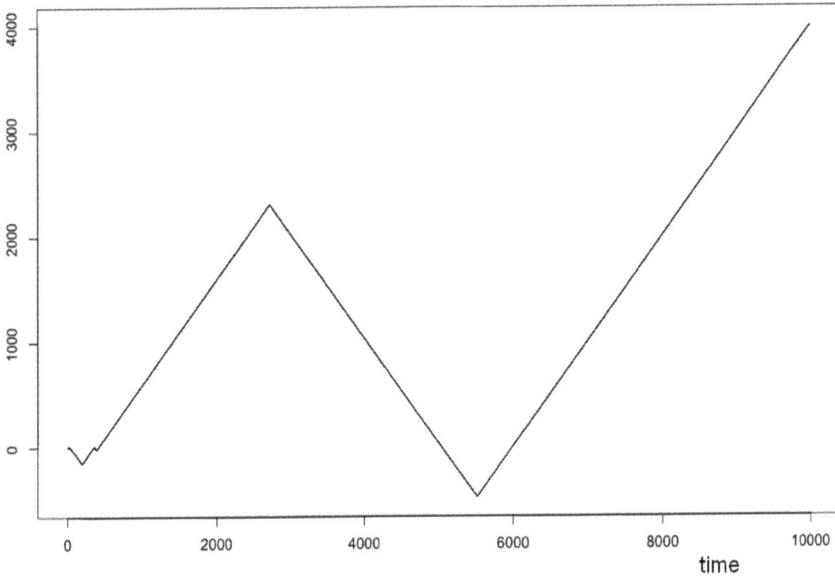

Fig. 6.4. The zigzag process: turning points form a PPP on $[0, \infty)$ with intensity measure $\frac{a}{x} \, dx$. (Obtained by simulating S.)

6.4.2 Preparation II: Approximating the walk with a martingale

We are interested in the scaling limit of the random walk S and, in particular, whether we have a Donsker-style invariance principle, leading eventually to Brownian motion. Following the general principle that "it always helps to find a martingale," in this section we investigate the following important, though still somewhat vague, question.

Question 6.1 (M). *For a given sequence $\{p_n\}_{n \geq 1}$, is the walk S "sufficiently close" to some martingale M?*

After Question (M), the next question is of course the following:

Question 6.2 (INV.M). *Is there an invariance principle for M?*

Focusing now on Question (M) only, we recall from (5.4) the identity $e_{i,j} = \mathbb{E}(Y_j \mid Y_i)/Y_i$ and that for $1 \leq i < j < k$, $e_{i,j}e_{j,k} = e_{i,k}$. With the convention $e_{i,i} := \mathbb{E}(Y_i^2) = 1$, recall the definition of

$a_n = \sum_{i=0}^{\infty} e_{n,n+i}$ from (6.5), assuming that the series is convergent (if $p_n \geq 1/2$ for large n, then it always is; see the following). Then

$$M_n := Y_1 + \cdots Y_{n-1} + a_n Y_n$$

is a martingale. Indeed,

$$\begin{aligned}
\mathbb{E}(M_{n+1} - M_n \mid \mathcal{F}_n) &= \mathbb{E}((1 - a_n)Y_n + a_{n+1}Y_{n+1} \mid \mathcal{F}_n) \\
&= (1 - a_n)Y_n + a_{n+1}\mathbb{E}(Y_{n+1} \mid Y_n) \\
&= [(1 - a_n) + a_{n+1}e_{n,n+1}]Y_n,
\end{aligned}$$

which is identically zero, since $a_{n+1}e_{n,n+1} = a_n - 1$, as

$$a_{n+1}e_{n,n+1} = \sum_{i=1}^{\infty} e_{n,n+1}e_{n+1,n+i} = \sum_{i=1}^{\infty} e_{n,n+i} = a_n - 1.$$

Observe also that

$$\begin{aligned}
\mathbb{V}\mathrm{ar}\,(M_{n+1} - M_n) &= \mathbb{V}\mathrm{ar}\,((1 - a_n)Y_n + a_{n+1}Y_{n+1}) \\
&= (1 - a_n)^2\mathbb{V}\mathrm{ar}\,(Y_n) + a_{n+1}^2\mathbb{V}\mathrm{ar}\,(Y_{n+1}) \\
&\quad + 2(1 - a_n)a_{n+1}\mathrm{Cov}(Y_n, Y_{n+1}) \\
&= a_{n+1}^2 + (1 - a_n)^2 + 2(1 - a_n)a_{n+1}e_{n,n+1} \\
&= a_{n+1}^2 - (1 - a_n)^2 = a_{n+1}^2\left[1 - e_{n,n+1}^2\right] \\
&= 4a_{n+1}^2 p_{n+1}q_{n+1} \qquad\qquad\qquad\qquad (6.19)
\end{aligned}$$

since $\mathbb{V}\mathrm{ar}\,(Y_n) = \mathbb{E}(Y_n^2) = 1$ for each n.

To understand what we mean by being *sufficiently close* to a martingale, recall that the rescaled walk S^n is defined by

$$S^n(t) := \frac{S_{Z(nt)}}{\sqrt{n}} = \frac{M_{Z(nt)} + (1 - a_{Z(nt)})Y_{Z(nt)}}{\sqrt{n}}, \quad t \geq 0,$$

where

$$Z(n) := \inf\{m : \ v_m \geq n\}, \quad n \geq 1.$$

Since $|Y_k| = 1$, if the a_n are not too large, then it suffices to analyze the sequence of the rescaled martingales $M^n(t) := \frac{M_{Z(nt)}}{\sqrt{n}}$ instead of the sequence of the rescaled random walks. Thus, we have the answer in the affirmative to Question (M), provided that

(a) a_n is well defined,
(b) $a_{Z(n)} = o(\sqrt{n})$ (e.g., when a_n remains bounded) as $n \to \infty$. (We dropped t as it is just a constant.)

Proposition 6.2 (Equivalent conditions for (b)). *Set*

$$\sigma_n^2 := \mathbb{V}\text{ar}\,(S_n).$$

Since the martingale differences $M_i - M_{i-1}$ are uncorrelated and centered, one has

$$\mathbb{V}\text{ar}\,(M_n) = \mathbb{E}\left[\left(\sum_{i=1}^{n}[M_i - M_{i-1}]\right)^2\right] = \sum_{i=1}^{n}\mathbb{E}[(M_i - M_{i-1})^2] = v_n,$$

where v_n is defined by (6.6) and $\mathbb{V}\text{ar}\,(M_i - M_{i-1})$ is given by (6.19). Then the conditions

(b.1) $v_n \to \infty, a_n = o\left(\sqrt{v_n}\right)$,
(b.2) $\sigma_n \to \infty, a_n = o(\sigma_n)$

are equivalent, and when they are satisfied, one has $\sqrt{v_n} \sim \sigma_n$.
  Of course, (b.1) $\Leftrightarrow$ (b.2) $\Rightarrow$ (b). Moreover, if $v_n \to \infty$, then the condition $a_n = o\left(\sqrt{v_n}\right)$ is in fact equivalent to (b). The proofs of these statements are provided later. ◇

To answer Question (INV.M), we refer the invariance principle of Drogin.

Proposition D (Part of Theorem 1 in Drogin (1972)). *Let $(X_i)_{i\geq 1}$ be a sequence of square integrable random variables adapted to the filtration $(\mathcal{F}_i)_{i\geq 1}$. Assume that they are martingale differences: $\mathbb{E}(X_i \mid \mathcal{F}_{i-1}) = 0$ and that $v_m := \sum_{i=1}^{m}\mathbb{E}(X_i^2 \mid \mathcal{F}_{i-1}) \to \infty$ a.s. The processes S and S^n, $n \geq 1$ by $S(v_m) = \sum_{i=1}^{m} X_i$, $S(0) := 0$, and by $S^n(t) := S(nt)/\sqrt{n}$, $t \geq 0$, using linear interpolation between integer times. Then the following are equivalent:*

(i) *If $\epsilon > 0$, then*

$$\frac{1}{n}\sum_{i=1}^{Z(n)} X_i^2 1_{\{X_i^2 > n\epsilon\}} \to_{L_1} 0 \quad \text{as } n \to \infty. \tag{6.20}$$

(ii) *As $n \to \infty$, the law of S^n converges to the Wiener measure and*

$$\frac{v_{Z(n)}}{n} \to_{L_1} 1.$$

Note that, in our setting, both v_m and $Z(n)$ are deterministic. To summarize the discussion on Questions (M) and (INV.M) above, in our setting, to prove that the limiting process is Brownian motion, we need to check the following conditions:

(a) *a_n is well defined.*
(b) *$a_{Z(n)} = o(\sqrt{n})$, or equivalently, $a_n = o(\sigma_n)$ (given $\sigma_n \to \infty$), as $n \to \infty$.*
(c) *$v_n \to \infty$ and (6.20) holds.*

Here (a), (b) guarantee an answer in the affirmative to (M), and (c) guarantees an answer in the affirmative to (INV.M).

6.4.3   Some specific cases

The first two cases we are looking at are in the cooling regime, the last one is in the heating regime. We use the conditions discussed in the last paragraph in Proposition 6.2.

6.4.3.1   Cooling, critical

Let $p_n = c/n$ for large n. If $c \geq 1/2$, then (a) fails to hold because then $a_n = \infty$. Otherwise, a_n is of order n^{1-2c}, and $\sqrt{v_n}$ is of the same order, and thus (b) fails to hold. In both cases, the answer to Question (M) is negative.

6.4.3.2   Cooling, subcritical

Let $p_n \leq 1/2$ for all[7] $n \geq 1$ and $p_n = c/n^\gamma$ for n large, where $0 < \gamma < 1$. In this case, the answers to (M) and to (INV.M) are both in the affirmative, and one can compute that $a_n = \frac{n^\gamma}{2c}(1 + o(1))$.

---

[7]We may assume this without the loss of generality, as the validity of the invariance principle does not depend on a finite number of terms.

6.4.3.3 *Cooling, subcritical; the necessity of* $\liminf_{n\to\infty} \frac{p_n}{p_{n+1}} > 0$

One can see that assumption (a) in Theorem 6.4(4) guarantees that

$$\liminf_{n\to\infty} \frac{p_n}{p_{n+1}} > 0.$$

The following example shows indeed the necessity of this bound, that is, by showing that the property that $a_n^2 = o(v_n)$ can break down if this $\liminf$ is zero. Indeed, let

$$p_i := \frac{\ln k}{2 \cdot k!} \quad \text{for } k! < i \le (k+1)!, \quad k = 1, 2, \ldots.$$

Then

$$\prod_{i=k!+1}^{(k+1)!} (1 - 2p_i) = \left(1 - \frac{\ln k}{k!}\right)^{k \cdot k!} = (1 + o(1))e^{-k \ln k} = \frac{1 + o(1)}{k^k}$$

and $\sum_k \frac{(k+1)! - k!}{k^k} < \infty$, so a_n is well defined. Moreover,

$$a_{m!} = \sum_{i=0}^{\infty} e_{m!, m!+i} \ge 1 + \sum_{i=1}^{(m+1)! - m!} (1 - 2p_{m!+1}) \cdots (1 - 2p_{m!+i})$$

$$= 1 + \left[1 - \frac{\ln m}{m!}\right] + \left[1 - \frac{\ln m}{m!}\right]^2 + \cdots + \left[1 - \frac{\ln m}{m!}\right]^{(m+1)! - m!}$$

$$= \frac{1 - O\left(e^{-m \ln m}\right)}{1 - \left(1 - \frac{\ln m}{m!}\right)} = (1 + o(1)) \frac{m!}{\ln m}.$$

At the same time, one has that

$$\frac{v_{m!}}{4} = \sum_{i=1}^{m!} a_i^2 p_i q_i = \sum_{k=0}^{m-1} \sum_{i=k!+1}^{(k+1)!} a_i^2 p_i q_i \le \sum_{k=0}^{m-1} \left[\sum_{i=k!+1}^{(k+1)!} a_i^2 \frac{\ln k}{k!}\right]$$

$$\le \sum_{k=0}^{m-1} (1 + o(1)) \frac{k!}{\ln k} \le \frac{(1 + o(1))(m-1)!}{\ln(m-1)} \le \frac{1 + o(1)}{m} \cdot \frac{m!}{\ln m}$$

$$= o\left(a_{m!}^2\right),$$

since for $k! < i \le (k+1)!$,

$$a_i \le \sum_{j=0}^{(k+1)!-k!} \left[1 - \frac{\ln k}{k!}\right]^j$$

$$+ \left[1 - \frac{\ln k}{k!}\right]^{(k+1)!-k!} \cdot \sum_{j=0}^{(k+2)!-(k+1)!} \left[1 - \frac{\ln(k+1)}{(k+1)!}\right]^j$$

$$+ \left[1 - \frac{\ln k}{k!}\right]^{(k+1)!-k!} \cdot \left[1 - \frac{\ln(k+1)}{(k+1)!}\right]^{(k+1)!-k!}$$

$$\cdot \sum_{j=0}^{(k+3)!-(k+2)!} \left[1 - \frac{\ln(k+2)}{(k+2)!}\right]^j + \cdots.$$

Hence,

$$a_i \le (1+o(1))\frac{k!}{\ln k} + e^{-k \ln k}(k+2)!$$

$$+ e^{-k \ln k} e^{-(k+1)\ln(k+1)}(k+3)! + \cdots$$

$$\le (1+o(1))\frac{k!}{\ln k} + \frac{(k+2)!}{k^k} + \frac{(k+3)!}{(k+1)^{m+1}} + \frac{(k+4)!}{(k+2)^{k+2}} + \cdots$$

$$= (1+o(1))\frac{k!}{\ln k}.$$

In addition, it is worth noting that with these p_i's, the assumption (4)(a) in Theorem 6.4 is violated too, since for $i = (m+1)!$, one has

$$ip_i = (m+1)! \, p_{(m+1)!} = (m+1)!\frac{\ln m}{2 \cdot m!} = \left[\frac{1}{2} + \frac{m}{2}\right]\ln m,$$

while

$$(i+1)p_{i+1} = [(m+1)!+1]\frac{\ln(m+1)}{2 \cdot (m+1)!} = \left[\frac{1}{2} + o(1)\right]\ln m \ll ip_i.$$

6.4.3.4   *Heating*

Let $p_n = 1 - q_n$ and $q_n \to 0$ but $\sum q_n = \infty$. We have

$$a_n = 1 + \sum_{i=1}^{\infty}(-1)^i(1-2q_{n+1})(1-2q_{n+2})\cdots(1-2q_{n+i}),$$

and, since $1 - 2p_n = 2q_n - 1 < 0$ for large n, using the Leibniz criterion, along with the assumption that $\sum q_n = \infty$, it follows that a_n is well defined. The validity of the martingale approximation follows from the fact that $a_n \leq 1$ but $v_n \to \infty$ as $n \to \infty$; see the proof of Theorem 6.3.

6.4.4 *Proof of Theorem 6.1*

Clearly, if $p_i = 1/2$ for some $i \in \mathbb{N}$, then the process "gets symmetrized" from time i on (and $\rho = 0$), and the statement is trivial. We thus assume in the rest of the proof that $p_i \neq 1/2$, $\forall i \in \mathbb{N}$.

Furthermore, we handle the conditional probability $\mathbb{P}(\cdot \mid Y_1 = 1)$ only, the argument for $\mathbb{P}(\cdot \mid Y_1 = -1)$ is similar. In terms of the W_i, one has $Y_n := (-1)^{\sum_{i=1}^{n} W_i}$, where $W_1, W_2, W_3, \ldots$ are independent Bernoulli variables with parameters $p_1, p_2, p_3, \ldots$, respectively, and we handle the $p_1 = 0$ (i.e., $W_1 \equiv 0$) case. In particular, $\prod_{i=1}^{n}(1 - 2p_i) = \prod_{i=2}^{n}(1 - 2p_i)$.

Let $x_n := \mathbb{P}(Y_n = 1)$. We have the recursion

$$x_{n+1} = p_n(1 - x_n) + (1 - p_n)x_n, \quad n \geq 1,$$

$$x_1 = 1,$$

and the substitution $y_n := x_n - 1/2$ yields $y_{n+1} = (1 - 2p_n)y_n$ with $y_1 = 1/2$. Hence,

$$y_{n+1} = \frac{1}{2}\prod_{i=1}^{n}(1 - 2p_i). \tag{6.21}$$

Case 1: $N = \infty$. We have to prove that x_n converges to $1/2$ or has no limit, according to whether $\min(p_i, q_i)$ is summable or not.

Let $N_n := \text{card}\{i \leq n : p_i > 1/2\}$; then $\lim_{n \to \infty} N_n = \infty$. Since

$$\min(p_i, q_i) = \begin{cases} p_i. & \text{if } p_i < 1/2; \\ q_i = (1 - p_i), & \text{if } p_i > 1/2, \end{cases}$$

we have

$$\prod_{i=1}^{n}(1 - 2p_i) = \prod_{i \leq n, p_i \leq 1/2}(1 - 2p_i) \times \prod_{i \leq n, p_i > 1/2}(1 - 2p_i)$$

$$= (-1)^{N_n} \prod_{i\leq n, p_i \leq 1/2} (1-2p_i) \times \prod_{i\leq n, p_i > 1/2} (1-2q_i)$$

$$= (-1)^{N_n} \prod_{i=1}^{n} (1 - 2\min(p_i, q_i)).$$

Given that $\lim_n (-1)^{N_n}$ does not exist, there are two cases:

(i) The right-hand side converges because the product (without the $(-1)^{N_n}$ factor) converges to zero and mixing holds ($\sum_{i=1}^{\infty} \min(p_i, q_i) = \infty$), in which case $\lim_n y_n = 0$ and $\lim_n x_n = 1/2$.
(ii) The right-hand side has no limit and mixing does not hold in which case y_n (hence x_n) has no limit.

Case 2: $N < \infty$. Let us assume first that $N = 0$, that is, $p_i < 1/2, i \geq 1$. If $\sum_{i=1}^{\infty} p_i = \infty$, then in (6.21) we have $\prod_{i=1}^{n}(1-2p_i) \downarrow 0$, implying $\lim_n y_n = 0$ and $\lim_n x_n = 1/2$. If $\sum_{i=1}^{\infty} p_i < \infty$, then $\rho = \prod_{i=1}^{\infty}(1-2p_i) > 0$ and $\lim_n y_n = \frac{1}{2}\rho$, that is, $\lim_n x_n = \frac{1}{2}(1+\rho)$.

In the general case, for large i, $\min(p_i, q_i) = p_i < 1/2$, and mixing is tantamount to $\sum_{i=1}^{\infty} p_i = \infty$. The proof is very similar as before, using the fact that the product has positive terms for large enough indices.

6.4.5 *Proof of Theorem* **6.2**

The martingale method is applicable in this case too. Indeed, direct computation gives $a_n = \frac{1}{2p}$, $\forall n$, and $v_n = \sum_n 4a_i^2 p_i q_i = \frac{1-p}{p}n$. Hence, $a_n = o(\sqrt{v_n})$, $a_i^2 \xi_i^2$ are bounded, so (6.20) holds, and thus we can apply Proposition D1972, yielding the answer to (INV.M) in the affirmative. □

6.4.6 *Proof of Theorem* **6.3**

We first prove the statement under the more restrictive assumption that

$$\limsup_{n\to\infty} \frac{q_{n+1}}{q_n} < \infty, \tag{6.22}$$

and then we upgrade it for showing the statement under the condition appearing in the theorem.

6.4.6.1 *STEP 1*

We start with a simple lemma.

Lemma 6.1. *Assume that for the non-negative sequence (q_n) such that*

- $q_n \to 0$,
- $\sum_n q_n = \infty$,
- *the condition (6.22) holds.*

Then $\liminf_{n \to \infty} a_n > 0$, where the a_n are defined by (6.5). Moreover, if $\lim_{n \to \infty} \frac{q_{n+1}}{q_n} = 1$, then $\lim_{n \to \infty} a_n = 1/2$.

Remark 6.12. The condition that $q_n \to 0$ is really necessary in Lemma 6.1. Indeed, fix $c_1, c_2 > 0$, $c_1 \neq c_2$ and let $q_n = c_1/n$ if n is odd and $q_n = c_2/n$ if n is even. Then $q_{n+1}/q_n \not\to 1$, though (6.22) still holds. In this case, $a_n \not\to 1/2$ rather (as it is not hard to show) $\lim_{k \to \infty} a_{2k} = \frac{c_1}{c_1+c_2} \neq \frac{c_2}{c_1+c_2} = \lim_{k \to \infty} a_{2k+1}$. ◇

Proof. Fix some n, and for $m \geq n$, let

$$w_m = \prod_{j=n+1}^{m} (1 - 2q_j), \quad m > n, \quad w_n = 1,$$

and note that $w_m \downarrow 0$ as $m \to \infty$ due to $\sum q_i = \infty$. Then

$$a_n = \sum_{i=0}^{\infty} (-1)^i w_{n+i} = \sum_{k=0}^{\infty} (w_{n+2k} - w_{n+2k+1}) = \sum_{k=0}^{\infty} 2q_{n+2k+1} w_{n+2k}.$$

Now, take any finite $c > \limsup_n q_{n+1}/q_n$, and assume that n is so large that the quantity $q_{\ell+1}/q_\ell < c$ for all $\ell \geq n$. Then

$$
\begin{aligned}
w_{n+2k} - w_{n+2k+2} &= 2(q_{n+2k+1} + q_{n+2k+2} \\
&\quad - 2q_{n+2k+1}q_{n+2k+2}) \cdot w_{n+2k} \\
&\leq 2(q_{n+2k+1} + q_{n+2k+2}) \cdot w_{n+2k} \\
&\leq (1 + c) \cdot 2q_{n+2k+1}w_{n+2k}. \quad (6.23)
\end{aligned}
$$

As a result,

$$(1+c)a_n = \sum_{k=0}^{\infty}(1+c)\cdot 2q_{n+2k+1}\,w_{n+2k}$$

$$\geq \sum_{k=0}^{\infty}(w_{n+2k} - w_{n+2k+2}) = w_n = 1 > 0,$$

where the telescopic sum converges due to the fact that $w_m \to 0$. Since $c > \limsup_n q_{n+1}/q_n$ is arbitrary, we can even conclude that

$$\liminf_{n\to\infty} a_n \geq \frac{1}{1 + \limsup_{n\to\infty} q_{n+1}/q_n}. \tag{6.24}$$

To prove the second part of the claim, observe that from (6.24) we already have $\liminf_n a_n \geq 1/2$. To show the counterpart of this, fix an arbitrary $\varepsilon > 0$ and let n be so large that $\frac{q_{n+2k+2}}{q_{n+2k+1}} \geq 1 - \varepsilon/2$ and $q_{n+2k+2} \leq \varepsilon/4$ for all $k \geq 0$. Then

$$q_{n+2k+1} + q_{n+2k+2} - 2q_{n+2k+1}q_{n+2k+2}$$

$$= \left(1 + \frac{q_{n+2k+2}}{q_{n+2k+1}} - 2q_{n+2k+2}\right)q_{n+2k+1} \geq (2 - \varepsilon)\cdot q_{n+2k+1},$$

given that $\frac{q_{n+2k+2}}{q_{n+2k+1}} \to 1$ and $q_{n+2k+2} \to 0$. Hence, (see (6.23))

$$w_{n+2k} - w_{n+2k+2} \geq (2 - \varepsilon)\cdot 2q_{n+2k+1}w_{n+2k}$$

and

$$(2 - \varepsilon)a_n = \sum_{k=0}^{\infty}(2 - \varepsilon)\cdot 2q_{n+2k+1}\,w_{n+2k}$$

$$\leq \sum_{k=0}^{\infty}(w_{n+2k} - w_{n+2k+2}) = w_n = 1.$$

Since $\varepsilon > 0$ is arbitrary, we conclude that $\limsup_n a_n \leq 1/2$, which completes the proof. $\qquad\square$

We now continue the proof of the theorem under the assumption that (6.22) holds.

Proof of Theorem 6.3(a). Note that all answers to Questions (M) and (INV.M) of Section 6.4.2 are checked and in the affirmative (as a_n is well defined and stays bounded), except (6.20). Since in our case $X_i = a_i \xi_i Y_{i-1}$ and $|Y_i| = 1$, what we need is to show that

$$\lim_{n \to \infty} \frac{1}{n} \sum_{i=1}^{Z(n)} a_i^2 \xi_i^2 \mathbb{1}_{\{a_i^2 \xi_i^2 > n\epsilon\}} = 0. \tag{6.25}$$

(Note that $Z(n)$ in our case is deterministic and so is v_m.) Since

$$\xi_i^2 = [(-1)^{W_i} + (2p_i - 1)]^2 \le 4, \quad \text{and} \quad |a_i| \le 1,$$

as a_i is a Leibniz series, all but finitely many terms in the sum in (6.25) are zero, proving (6.25). We conclude that (6.20) holds.

Next, a direct computation shows that $v_m = 4 \sum_{i=1}^{m} a_i^2 p_i q_i$. Then

$$\lim_{m \to \infty} v_m = \sum_{i=1}^{\infty} 4 a_i^2 p_i q_i = \infty$$

follows from Lemma 6.1 and from the assumptions $p_n \to 1$ and $\sum q_n = \infty$. The proof of (a) is thus complete.

Proof of Theorem 6.3(b). We first prove that $\Lambda_n^2 = \sum_{i=1}^{n} a_i^2 \xi_i^2 \to \infty$. Recall that $\xi_i = (-1)^{W_i} - \mathbb{E}(-1)^{W_i} = 2p_i - 1 + (-1)^{W_i}$ satisfies $\mathbb{E}\xi_i^2 = \mathbb{V}\mathrm{ar}\,((-1)^{W_i}) = 4p_i q_i$. Let also $U_i := a_i^2 \xi_i^2 \in [0, 4]$.

Since the W_i are independent, so are the ξ_i, and hence, for Λ_n^2, the three series theorem applies: the non-negative series $\sum_i U_i$ diverges if for some $A > 0$, $\sum_i \mathbb{E}[U_i; |U_i| \le A]$ diverges. But for $A > 4$,

$$\sum_i \mathbb{E}[U_i; |U_i| \le A] = \sum_i \mathbb{E}(U_i) = \sum_i a_i^2 p_i q_i = \infty,$$

as a_i is bounded away from zero, $p_i \to 1$ and $\sum q_i = \infty$.

Alternatively, let $\epsilon > 0$. Then $p_i \to 1$ and $\sum q_i = \infty$ along with the second Borel–Cantelli lemma guarantee that $\xi_i = 2p_i - 1 + (-1)^{W_i} \ge 2 - \epsilon$ for infinitely many i's almost surely. We are done because the a_i are bounded away from zero.

For the second statement, by using Chebyshev's inequality, it is enough to show that

$$\lim_{n \to \infty} \frac{\mathbb{V}\mathrm{ar}\,(\Lambda_n^2)}{(\mathbb{E}\Lambda_n^2)^2} = 0. \tag{6.26}$$

Since $a_n, p_n, q_n \in [0,1]$,

$$\mathbb{V}\mathrm{ar}\,(\Lambda_n^2) = 4\sum_{i=1}^{n} a_i^4 p_i q_i (p_i - q_i)^2 \leq 4\sum_{i=1}^{n} q_i. \tag{6.27}$$

Moreover, for large n's,

$$\mathbb{E}\Lambda_n^2 = v_n = 4\sum_{i=1}^{n} a_i^2\, p_i\, q_i \geq c\sum_{i=1}^{n} q_i \tag{6.28}$$

for some $c > 0$, since $\liminf_{i\to\infty} a_i > 0$ by Lemma 6.1 and $p_i \to 1$. Given that $\sum_{i=1}^{n} q_i \to \infty$, (6.27) and (6.28) together yield (6.26), thus completing the proof of the statement. $\qquad\square$

6.4.6.2   *STEP 2*

We now upgrade the result obtained in STEP 1, by dropping the restriction that (6.22) holds. We need the following.

Lemma 6.2 (Comparison with "regular" sequences). *Let* $0 \leq q_n \leq 1/2$, $n \geq 1$:

(i) *Assume that there exists a sequence* $q_n^* \to 0$ *such that* q_n^* *is not summable, regular in the sense that (6.22) holds, and* $q_n \leq q_n^*$ *for even* n*, while* $q_n \geq q_n^*$ *for odd* n*. Then* $\liminf_{k\to\infty} a_{2k} > 0$.
(ii) *Assume that there exists a sequence* $\tilde{q}_k \to 0$ *such that* $\tilde{q}_n$ *is not summable, regular in the sense that (6.22) holds, and* $q_n \leq \tilde{q}_n$ *for odd* n*, while* $q_n \geq \tilde{q}_n$ *for even* n*. Then* $\liminf_{k\to\infty} a_{2k+1} > 0$.

Proof of Lemma 6.2. Since $0 \leq q_n \leq 1/2$ for $n \geq 1$, it is easy to check the following (for example, by observing that for $k > n$, the coefficients of q_k in a_n form a Leibniz series as well):

- Let $n = 2k$. Then a_n is decreasing[8] in all q_i for which i is even and increasing in all q_i for which i is odd.
- Let $n = 2k+1$. Then a_n is increasing in all q_i for which i is even and decreasing in all q_i for which i is odd.

---

[8]The terms *increasing* and *decreasing* are not used in the strict sense.

Turning to the proof of (i) (a similar proof works for (ii), which we omit), note that, because of its monotonicity and non-summability (use $I = \infty$ and $q_{2k}^* \geq q_{2k}$), Step 1 yields that (q_n^*) is such that $\liminf a_n > 0$, and in particular, $\liminf_k a_{2k} > 0$. Hence, by the first bullet point above, $\liminf_k a_{2k} > 0$ also for (q_n), proving (i). $\qquad\square$

Proof of Theorem 6.3. First, without the loss of generality, we assume that $m_0 = 1$ (changing a finite number of terms does not change the validity of the invariance principle). Similarly, we may and will assume that $q_n \leq 1/2$ for all $n \geq 1$, as we assume anyway that $q_n \to 0$.

We only need that $v_n = 4\sum_{i=1}^n a_i^2 p_i q_i \to \infty$, what is left is very similar to Step 1. This follows from $p_n \to 1$ and Assumption 6.2, provided that either $\liminf_k a_{2k} > 0$ or $\liminf_k a_{2k+1} > 0$. By Lemma 6.2, it is sufficient to construct either a sequence (q_n^*) or a sequence $(\tilde{q}_n)$ satisfying the properties in the lemma. These sequences will be automatically divergent, given Assumption 6.2 and that (q_n^*) resp. $(\tilde{q}_n)$ dominate (q_n) for even resp. odd n's. Now, assume, for example, (6.7) (assuming (6.8) leads to a similar argument). Define

$$\tilde{q}_{2m} := C \max\{q_{2\ell+1}, \ell \geq m\}, \quad m \geq 1,$$

$$\tilde{q}_{2m+1} := \max\{q_{2\ell+1}, \ell \geq m\}, \quad m \geq 0.$$

Then $(\tilde{q}_n)$ is regular because $\frac{\tilde{q}_{n+1}}{\tilde{q}_n} \leq \max\{C^{-1}, C\}$ for all $n \geq 1$, and trivially $q_{2m} \geq \tilde{q}_{2m}$ and $\tilde{q}_{2m+1} \geq q_{2m+1}$. Hence, $\liminf_k a_{2k+1} > 0$ by Lemma 6.2(ii). $\qquad\square$

Remark 6.13 (One of the two subsequences can be arbitrary). Chose an arbitrary "odd" subsequence, satisfying the conditions that it tends to zero and yet not summable. Then take a sufficiently large "even" subsequence that dominates it in the sense of (6.7) but still tends to zero (for example, let $q_{2n} := 1/\sqrt{2n}$ and $q_{2n+1} := 1/(2n+1)$). Then (6.7) holds, while the condition $\limsup_n q_{n+1}/q_n < \infty$ (cf. (6.22) in the proof) fails to hold, as $\lim_n q_{2n}/q_{2n+1} = \infty$.

By the same token, one can first chose an arbitrary non-summable "even" sequence, with the terms tending to zero and then a dominating "odd" one. $\qquad\diamond$

6.4.7   *Proof of Proposition 6.2*

Recall that $S_n = M_n + (1 - a_n)Y_n$, hence

$$\mathbb{V}\text{ar}\,(S_n) = \mathbb{V}\text{ar}\,(M_n) + (1 - a_n)^2 + 2(1 - a_n)\text{Cov}(M_n, Y_n),$$

where, by Cauchy–Schwarz, $|\text{Cov}(M_n, Y_n)| \leq \sqrt{\mathbb{E}(M_n^2)} = \sqrt{v_n}$, so

$$\begin{aligned}
\sigma_n^2 - v_n &= (1 - a_n)(1 - a_n + 2\text{Cov}(M_n, Y_n)) \\
&= (1 - a_n)(1 - a_n + A_n\sqrt{v_n}),
\end{aligned}$$

where $|A_n| \leq 1$. Then

$$\frac{\sigma_n^2}{v_n} - 1 = \frac{1 - a_n}{\sqrt{v_n}} \cdot \left(\frac{1 - a_n}{\sqrt{v_n}} + A_n\right),$$

if $v_n \to \infty$ and $a_n = o(\sqrt{v_n})$ as $n \to \infty$, hence $\sqrt{v_n} \sim \sigma_n$ follows.
    Similarly, we have

$$1 - \frac{v_n}{\sigma_n^2} = \frac{1 - a_n}{\sigma_n}\left(\frac{1 - a_n}{\sigma_n} + A_n\sqrt{v_n}/\sigma_n\right).$$

Using the short hands $w_n := \frac{\sqrt{v_n}}{\sigma}_n$ and $b_n := \frac{1-a_n}{\sigma_n}$, one obtains the quadratic equation $w_n^2 + b_n A_n w_n + b_n^2 - 1 = 0$, where $b_n \to 0$. Hence,

$$w_n = \frac{-b_n A_n \pm \sqrt{b_n^2 A_n^2 + 4(1 - b_n^2)}}{2},$$

but of course $w_n \geq 0$. Therefore, $b_n \to 0$ implies that $w_n \to 1$, that is, $\sqrt{v_n} \sim \sigma_n$. This is clearly the case when $\sigma_n \to \infty$ and $a_n = o(\sigma_n)$ as $n \to \infty$.   □

6.4.8   *Proof of Theorem 6.4: strongly critical case*

First, it is easy to see that if X is a symmetric random variable, concentrated on $[-t, t]$, then $\mathbb{V}\text{ar}\,(X) \leq t^2$, with equality if and only if the law of X is $\frac{1}{2}(\delta_{-t} + \delta_t)$.
    Now assume that $\lim_{n\to\infty} np_n = 0$. By a well-known criterion for tightness (see Theorem 4.10 in Karatzas and Shreve (1991)),

the laws of the $S^{(n)}$ are tight on $C([0, \infty))$ if besides the condition $\lim_{\eta \to +\infty} \sup_{n \geq 1} \mathbb{P}(S^{(n)}(0) > \eta) = 0$, one also has

$$
\limsup_{\delta \downarrow 0} \mathbb{P}_{n \geq 1} \left(\max_{\substack{|t-s| \leq \delta \\ 0 \leq t,s \leq T}} |S^{(n)}(t) - S^{(n)}(s)| > \epsilon \right) = 0, \quad \forall \epsilon > 0, \ T > 0.
$$

$$(6.29)$$

Since $S^{(n)}(0) = 0$, $n \geq 1$, the first condition clearly holds. The second one is satisfied by the uniform Lipschitzness: $|S^{(n)}(t) - S^{(n)}(s)| \leq |t - s|$, $n \geq 1$.

Given tightness, it is sufficient to show that the limit at time $t > 0$ is $\frac{1}{2}(\delta_{-t} + \delta_t)$, that is, it satisfies $\mathbb{V}\text{ar}\,(X) \geq t^2$. Indeed, the only continuous functions f on $(0, T)$ satisfying $|f(t)| = t$ are $f(t) = t$ and $f(t) = -t$. For simplicity, we work with $t = 1$ (otherwise, use a simple scaling), that is, we show that every partial limit at time $t = 1$ is such that its variance is at least one.

To achieve this, fix $N \geq 1$ and recall from Subsection 5.4 (see in particular (5.18)) that

$$
\mathbb{V}\text{ar}\left(\frac{S_N}{N}\right) = \frac{1}{N} + \frac{2}{N^2} \sum_{1 \leq i_1 < i_2 \leq N} e_{i_1, i_2}.
$$

This quantity is monotone decreasing in all p_n's as long as they are all less or equal than $1/2$ because the same holds for each fixed $e_{i,j}$. Fix $\epsilon > 0$ and let $N = N(\epsilon)$ be such that $\epsilon/N \leq 1/2$ and that also $\epsilon/n > p_n$ holds for all $n > N$. Define $\hat{p}_n$ so that it coincides with p_n for $n \leq N$ and $\hat{p}_n = \epsilon/n$ for $n > N$. By monotonicity,

$$
\mathbb{V}\text{ar}\left(\frac{S_n}{n}\right) \geq \mathbb{V}\text{ar}\left(\frac{\hat{S}_n}{n}\right), \quad n \geq 1,
$$

where $\hat{S}$ is the walk for the sequence $(\hat{p}_n)$. Using this comparison along with (5.8), it follows that

$$
\lim_{n \to \infty} \mathbb{V}\text{ar}\left(\frac{\hat{S}_n}{n}\right) = \frac{1}{2\epsilon + 1} \implies \liminf_{n \to \infty} \mathbb{V}\text{ar}\left(\frac{S_n}{n}\right) \geq \frac{1}{2\epsilon + 1}.
$$

Since $\epsilon > 0$ was arbitrary,

$$
\liminf_{n \to \infty} \mathbb{V}\text{ar}\left(\frac{S_n}{n}\right) \geq 1.
$$

Now, if $S_{n_j}/n_j \to X$ in law, then

$$\lim_{j \to \infty} \mathbb{V}\mathrm{ar}\left(\frac{S_{n_j}}{n_j}\right) = \mathbb{V}\mathrm{ar}\,(X)$$

because $\mathbb{E}(S_n) = 0$ and the variables are all supported in $[-1, 1]$ (and so the test function $f(x) = x^2$ is admissible). From the last two displayed formula, we have that $\mathbb{V}\mathrm{ar}\,(X) \geq 1$ and we are done.   □

6.4.9   *Proof of Theorem 6.4: supercritical case*

By the Borel–Cantelli lemma, for almost every ω, either (a) $S_n(\omega) = 1$ for all large n or (b) $S_n(\omega) = -1$ for all large n. As $n \to \infty$, in case (a), the path converges uniformly to a straight line with slope 1; in the second case, it converges uniformly to a straight line with slope -1, as claimed.   □

6.4.10   *Proof of Theorem 6.4: critical case*

Fix $T > 0$, denote by $\mathcal{M}_T$ the set of all locally finite point measures on the interval $(0, T]$ and denote by $N^{(n)} = N^{(n,T)}$ the laws of the point processes induced by the turns of the walk $S^{(n)}$, $n = 1, 2, \ldots$, on the time interval $(0, T]$.

Let $t \in (0, T)$; we **assign a continuous (zigzagged) path** to each point measure. We do it so that we use the constraint that the path increases at[9] t. (Since we build the path "backwards" from t to the origin, we need this arbitrary but fixed constraint.) The construction is spelled out in the following definition.

Definition 6.5 (Assigning paths). Define the map $\Phi_t : \mathcal{M}_T \to C[0, T]$ as follows:

- First, label the (countably many) atoms on $(0, t]$ from right to left as $a_1, a_2, \ldots$, i.e., the closest one on the left to t as a_1, the second closest as a_2, etc., and note that $t = a_1$ is possible; also label the atoms on $(t, T]$, from the closest to the furthest as $b_1, b_2, \ldots$.

---

[9]That is, it increases on $[t, t + \epsilon]$ for some small $\epsilon > 0$.

- Assign "+" sign to the intervals (the union of which is denoted by S_t^+)

$$\ldots [a_7, a_6), [a_5, a_4), [a_3, a_2), [a_1, b_1), [b_2, b_3), [b_4, b_5), [b_6, b_7), \ldots.$$

- Assign "−" sign to the intervals (the union of which is denoted by S_t^-)

$$\ldots [a_8, a_7), [a_6, a_5), [a_4, a_3), [a_2, a_1), [b_1, b_2), [b_3, b_4), [b_5, b_6), \ldots.$$

Let $\mu \in \mathcal{M}_T$. For $0 < r \le T$, define

$$\Phi_t(\mu)(r) := L((0, r] \cap S_t^+) - L((0, r] \cap S_t^-), \quad \text{with } \Phi_t(\mu)(0) := 0, \tag{6.30}$$

where L is the Lebesgue measure on the real line. Then $\Phi_t(\mu)(\cdot)$ is well defined and continuous on $[0, T]$. Intuitively, it describes the difference between the total length of increasing parts and the total length of decreasing parts, assuming an increase at t. Clearly,

$$|\Phi_t(\mu)(r)| \le r, \quad 0 < r \le T. \tag{6.31}$$

Remark 6.14. The case $t = 0$ is excluded, i.e., one cannot set the path $\Phi_t(\mu)(\cdot)$ to first increase at $t = 0$, as our point measures may not be locally finite around 0. For instance, we show that N_n converges to a limiting Poisson Point Process (PPP) N, and this N explodes at 0. However, for $t > 0$, $\Phi_t(r) \to 0$, as $r \to 0$.

We now turn to the case of a PPP with intensity $\frac{c}{x}$ (we replaced the constant a of Theorem 6.4 by c in the proof to avoid confusion).

Proposition 6.3. *(Turning points $\to$ PPP with intensity $\frac{c}{x}$) Given $0 < a < b < \infty$, $c > 0$, set $p_n = \frac{c}{n} \wedge 1$, and denote the number of turns from step $\lceil an \rceil + 1$ to step $\lceil bn \rceil$ by $N^{(n)}((a, b])$. Denoting $\mu_{c;a,b} := c \ln(b/a) = \int_a^b \frac{c}{x}\, dx$, one has*

(i) *for $k \ge 0$, $0 < a < b$, as $n \to \infty$,*

$$\mathbb{P}\left(N^{(n)}((a, b]) = k\right) = \exp(-\mu_{c;a,b}) \frac{\mu_{c;a,b}^k}{k!} + O\left(\frac{1}{n}\right) 49, \tag{6.32}$$

$$\mathsf{Law}(N^{(n)}((a, b])) \xrightarrow{n \to \infty} \mathsf{Poiss}(\mu_{c;a,b}); \tag{6.33}$$

(ii) *given $0 < t_1 < t_2 < \cdots < t_l < \infty$, the random variables*

$$N^{(n)}((t_1, t_2]), N^{(n)}((t_2, t_3]), \ldots, N^{(n)}((t_{l-1}, t_l])$$

are independent (independent increments), and

$$\mathsf{Law}\left(N^{(n)}((t_1, t_2]), N^{(n)}((t_2, t_3]), \ldots, N^{(n)}((t_{l-1}, t_l])\right)$$

$$\xrightarrow{n \to \infty} \mathsf{Poiss}(c)\left((\mu_{c;t_1,t_2}), (\mu_{c;t_2,t_3}) \cdots, (\mu_{c;t_{l-1},t_l})\right),$$

where $\mathsf{Poiss}(c) = \mathsf{Poiss}((0,\infty), \frac{c}{x}\,dx)$ *is the law of the PPP with intensity $\frac{c}{x}\,dx$ on $(0,\infty)$.*

Proof (of Proposition 6.3). The strategy of the proof. We first prove part (i). Once that is done, since the turns from step $\lceil t_i n \rceil + 1$ to step $\lceil t_j n \rceil$ and from $\lceil t_l n \rceil + 1$ to $\lceil t_j n \rceil$ are independent for any $0 < t_i < t_j \le t_l < t_r < \infty$, part (ii) immediately follows.

Regarding part (i), we only need to prove equation (6.32), and then (6.33) easily follows. In fact, we only give here the proof (in three steps) of (6.32) for a, b integers, i.e., $\lceil an \rceil = an$, $\lceil bn \rceil = bn$, for n large enough; the proof for general $0 < a < b$ can then be easily adjusted.

Step 1: Given $c > 0$, and n large enough, define

$$\Pi_{c,n} := \mathbb{P}(\text{no turn between } an + 1 \text{ and } bn).$$

We now provide an estimate for $\Pi_{c,n}$, namely,

$$\Pi_{c,n} = \exp(-\mu_{c;a,b}) + O\left(\frac{1}{n}\right). \tag{6.34}$$

Indeed,

$$\Pi_{c,n} = \frac{an + (1-c)}{an + 1} \cdot \frac{an + 1 + (1-c)}{an + 2}$$

$$\cdot \frac{an + 2 + (1-c)}{an + 3} \cdots \frac{bn - 1 + (1-c)}{bn}$$

$$= \left(\frac{an + (1-c)}{an} \cdot \frac{an + 1 + (1-c)}{an + 1} \cdots \frac{bn - 1 + (1-c)}{bn - 1}\right)$$

$$\times \left(\frac{an}{an+1} \cdot \frac{an+1}{an+2} \cdot \dots \cdot \frac{bn-1}{bn} \right)$$

$$= \frac{a}{b} \cdot \left(\frac{an+(1-c)}{an} \cdot \frac{an+1+(1-c)}{an+1} \dots \frac{bn-1+(1-c)}{bn-1} \right)$$

$$= \frac{a}{b} \exp \left(\sum_{i=1}^{bn-an} \ln(an+i-c) - \ln(an+i-1) \right)$$

$$= \frac{a}{b} \exp \left(\sum_{i=1}^{bn-an} \int_{an+i-1}^{an+i-c} \frac{dx}{x} \right).$$

The exponent tends to $(1-c)\ln\frac{b}{a}$, and so $\lim_{n\to\infty} \Pi_{c,n} = \exp(-c\ln(b/a)) = \exp(-\mu_{c;a,b})$. Indeed,

$$\frac{1-c}{an+i-c} \le \int_{an+i-1}^{an+i-c} \frac{1}{x}\,dx \le \frac{1-c}{an+i-1},$$

hence

$$\sum_{i=1}^{bn-an} \frac{1-c}{an+i-c} \le \sum_{i=1}^{bn-an} \int_{an+i-1}^{an+i-c} \frac{1}{x}\,dx \le \sum_{i=1}^{bn-an} \frac{1-c}{an+i-1},$$

where $\lim_{n\to\infty} \sum_{i=1}^{bn-an} \frac{1}{an+i-c} = \lim_{n\to\infty} \sum_{i=1}^{bn-an} \frac{1}{an+i-1} = \ln(b/a)$ leading to (6.34).

Step 2: We now estimate

$$\mathbb{P}(N^{(n)}((a,b])=1)$$
$$= \mathbb{P}(\text{there is one turn from step } an+1 \text{ to step } bn).$$

Note that the turning step can happen at step $an+i$, for $i = 1,2,\dots,bn-an$, with corresponding probabilities $(\frac{an+1-c}{an+1} \cdot \frac{an+2-c}{an+2} \cdot \dots \cdot \frac{bn-c}{bn}) \cdot \frac{c}{an+i} = \Pi_{c,n} \cdot \frac{c}{an+i}$, $i=0,1,\dots,bn-an-1$. Thus,

$$\mathbb{P}(N^{(n)}((a,b])=1) = \Pi_{c,n} \sum_{i=0}^{bn-an-1} \frac{c}{an+i} = \Pi_{c,n} \cdot c \cdot \Delta_n,$$

where

$$\Delta_n = \sum_{i=0}^{bn-an-1} \frac{1}{an+i}.$$

Since

$$\ln\frac{b}{a} = \int_{an}^{bn} \frac{\mathrm{d}x}{x} \le \Delta_n \le \int_{an}^{bn} \frac{\mathrm{d}x}{x} + \left(\frac{1}{an} - \frac{1}{an+1}\right)(bn-an)$$

$$= \ln\frac{b}{a} + \frac{(b-a)}{a(an+1)},$$

one has

$$\Delta_n = \ln\frac{b}{a} + \mathcal{O}(1/n), \tag{6.35}$$

and then (6.34), (6.35) give

$$\mathbb{P}\left(N^{(n)}((a,b]) = 1\right) = \frac{\mu_{c;a,b}}{1!} e^{-\mu_{c;a,b}} + \mathcal{O}\left(\frac{1}{n}\right).$$

Step 3: We verify (6.32) using induction, and so we assume that

$$\mathbb{P}\left(N^{(n)}((a,b]) = k\right) = \exp(-\mu_{c;a,b})\frac{\mu_{c;a,b}^k}{k!} + \mathcal{O}\left(\frac{1}{n}\right) \tag{6.36}$$

and show that k can be replaced by $k+1$ as well. On the the event $\{N^{(n)}((a,b]) = k\}$, there should be k turns from step $an+1$ to step $bn+1$, say the turns happen at $an+i_1, an+i_2, \dots, an+i_k$, where $i_1, \dots, i_k$ is an increasing sequence taking values in $\{0, 1, \dots, bn - an - 1\}$. Similar to the $k=1$ case, the probability for this to happen is

$$p = \Pi_{c,n} \cdot \left(\frac{c}{an+i_1}\frac{c}{an+i_2}\cdots\frac{c}{an+i_k}\right).$$

Then $\mathbb{P}(N^{(n)}((a,b]) = k)$ is the sum of all such terms, i.e.,

$$\mathbb{P}(N^{(n)}((a,b]) = k)$$

$$= \Pi_{c,n} \cdot c^k \cdot \sum_{0 \le i_1 < \cdots < i_k \le bn-an-1} \frac{1}{an+i_1}\frac{1}{an+i_2}\cdots\frac{1}{an+i_k}.$$

By assumption (6.36) and the estimate (6.34), we have

$$\sum_{0 \le i_1 < \cdots < i_k \le bn-an-1} \frac{1}{an+i_1} \frac{1}{an+i_2} \cdots \frac{1}{an+i_k}$$

$$= \frac{(c\ln(\frac{b}{a}))^k}{k!} + \mathcal{O}\left(\frac{1}{n}\right)$$

$$= \frac{\mu_{c;a,b}^k}{k!} + \mathcal{O}\left(\frac{1}{n}\right). \tag{6.37}$$

Similarly,

$$\mathbb{P}(N^{(n)}((a,b]) = k+1)$$

$$= \Pi_{c,n} \sum_{0 \le i_1 < \cdots < i_{k+1} \le bn-an-1} \frac{c}{an+i_1} \frac{c}{an+i_2} \cdots \frac{c}{an+i_{k+1}},$$

where the sequence $i_1 < i_2 < \cdots < i_k < i_{k+1}$ takes values in $\{0, 1, \ldots, bn - an - 1\}$. Now

$$\frac{c}{an+j} \mathbb{P}(N^{(n)}((a,b]) = k)$$

$$= \Pi_{c,n} \sum_{0 \le i_1 < i_2 < \cdots < i_k \le bn-an-1} \frac{c}{an+i_1} \frac{c}{an+i_2} \cdots \frac{c}{an+i_k} \frac{c}{an+j},$$

for $j = 0, 1, \ldots, bn - an - 1$. Now, consider the sum

$$\left(\sum_{j=0}^{bn-an-1} \frac{c}{an+j} \right) \mathbb{P}\left(N^{(n)}((a,b]) = k \right)$$

$$= \sum_{j=0}^{bn-an-1} \left(\Pi_{c,n} \cdot \sum_{0 \le i_1 < \cdots < i_k \le bn-an-1} \frac{c}{an+i_1} \frac{c}{an+i_2} \right.$$

$$\left. \times \cdots \frac{c}{an+i_k} \frac{c}{an+j} \right). \tag{6.38}$$

In each sum on the right-hand side, there are two different kinds of terms: terms of the type

$$\frac{c}{an+i_1}\frac{c}{an+i_2}\cdots\frac{c}{an+i_k}\frac{c}{an+i_{k+1}},$$

where $i_m, m = 1, 2, \ldots, k+1$ are all different (no repetitions), and terms of the type

$$\frac{c}{an+i_1}\frac{c}{an+i_1}\frac{c}{an+i_2}\cdots\frac{c}{an+i_k},$$

where $i_m, m = 1, 2, \ldots, k$ are all different (one repetition). We then rearrange the right-hand side: sum the "non-repeating" terms as one group, denoted by I; sum the "once repeating" ones where the term $\frac{c}{an+j}$ is the one repeated by I_j, $j = 0, 1, \ldots, bn - an - 1$, and we estimate I, I_j separately:

$$I = (k+1) \cdot \left(\Pi_{c,n} \sum_{i_1<i_2<\cdots<i_k<i_{k+1}} \frac{c}{an+i_1}\frac{c}{an+i_2} \right.$$

$$\left. \times \cdots \frac{c}{an+i_k}\frac{c}{an+i_{k+1}} \right)$$

$$= (k+1) \cdot \mathbb{P}\left(N^{(n)}((a,b]) = k+1 \right),$$

since each product $\frac{c}{an+i_1}\frac{c}{an+i_2}\cdots\frac{c}{an+i_k}\frac{c}{an+i_{k+1}}$ appears $k+1$ times in sum I. Further,

$$I_j = \frac{c^2}{(an+j)^2}\left(\Pi_{c,n} \sum_{\substack{0\leq i_1<i_2<\cdots<i_k\leq bn-an-1 \\ i_m \neq j}} \frac{c}{an+i_1}\frac{c}{an+i_2}\cdots\frac{c}{an+i_k} \right)$$

$$\leq I_0 = \frac{c^2}{(an)^2}\left(\Pi_{c,n} \sum_{1\leq i_1<i_2<\cdots<i_k\leq bn-an-1} \frac{c}{an+i_1}\frac{c}{an+i_2}\cdots\frac{c}{an+i_k} \right)$$

$$\leq \frac{c^2}{(an)^2}\mathbb{P}\left(N^{(n)}((a,b]) = k \right) = \frac{c^2}{(an)^2}\left(\exp(-\mu_{c;a,b})\frac{\mu_{c;a,b}^k}{k!} + \mathcal{O}\left(\frac{1}{n}\right) \right),$$

hence

$$\sum_{j=0}^{bn-an-1} I_j \le (bn - an) \cdot (I_0) \le \frac{bn - an}{(an)^2} \cdot \left(\frac{\mu_{c;a,b}^k}{k!} e^{-\mu_{c;a,b}} + \mathcal{O}\left(\frac{1}{n}\right) \right)$$

$$= \mathcal{O}\left(\frac{1}{n}\right).$$

Now, by estimations of I, I_js, (6.38) is written as

$$(k+1) \cdot \mathbb{P}\left(N^{(n)}((a,b]) = k + 1 \right) + \mathcal{O}\left(\frac{1}{n}\right)$$

$$= I + \sum_{j=1}^{bn-an-1} I_j$$

$$= \left(\sum_{j=0}^{bn-an-1} \frac{c}{an+j} \right) \mathbb{P}\left(N^{(n)}((a,b]) = k \right)$$

$$= \left(\mu_{c;a,b} + \mathcal{O}\left(\frac{1}{n}\right) \right) \cdot \left(\exp(-\mu_{c;a,b}) \frac{\mu_{c;a,b}^k}{k!} + \mathcal{O}\left(\frac{1}{n}\right) \right)$$

$$= \frac{\mu_{c;a,b}^{k+1}}{k!} e^{-\mu_{c;a,b}} + \mathcal{O}\left(\frac{1}{n}\right),$$

and we conclude that (6.36) holds with k replaced by $k + 1$. This completes the proof of Step 3 and of the proposition altogether. □

Note: We use the endpoints $\lceil an \rceil + 1, \lceil bn \rceil$ because $\frac{\lceil an \rceil + 1}{n} \to a^+$, $\frac{\lceil bn \rceil}{n} \to b$, so the above limit represents the number of turns in (a,b] in the scaling limit.

In the sequel, we consider measures equipped with both the weak and the vague topologies. When we consider laws on $C([0,T], |\cdot|_{[0,T]})$ where $|\cdot|_{[0,T]}$ denotes supremum norm, weak convergence is denoted by $\xrightarrow{w}$. When one uses vague topology for measures and random measures are considered, $\mathcal{X}_n \overset{vd}{\to} \mathcal{X}$ will be used for convergence in distribution.

Proposition 6.4 (Convergence for point measures and paths). *Let $0 < t < T$. Then*

(i) *as $n \to \infty$, $N^{(n)} \overset{vd}{\to} \mathsf{Poiss}(c)$ on $\mathcal{M}_T$ equipped with the vague topology, where $\mathsf{Poiss}(c)$ is the PPP on $(0, T]$ with intensity $\frac{c}{x}\,\mathrm{d}x$,*

(ii) *$\Phi_t : \mathcal{M}_T \to C[0, T]$ is a continuous and uniformly bounded functional, when the former space is equipped with the vague topology and the latter with the supremum norm $|.|_{[0,T]}$,*

(iii) *as $n \to \infty$, $\Phi_t(N^{(n)}) \overset{w}{\to} \Phi_t(\mathsf{Poiss}(c))$ on $C([0, T], |\cdot|)$.*

Proof (of Proposition 6.4). (i) In order to use Lemma 1.4, one needs to define a new metric on $(0, T]$ by $\rho(x, y) := |1/x - 1/y|$. Then $\Delta := ((0, T], \rho)$ is a complete separable metric space; note that $(0, \epsilon]$ is not bounded under ρ. Setting $\mathcal{I} := \{(a, b], 0 < a < b \le T\}$, it is obvious that $\mathcal{I}$ is a semi-ring of bounded Borel sets in Δ, and $\mu(\partial(a, b]) = \mu(\{a\} \cup \{b\}) = 0$, hence $\mathcal{I} \subset \widehat{\Delta}_{E\mathsf{Poiss}(c)}$, where $\widehat{\Delta}_{E\mathsf{Poiss}(c)}$ is the class of all bounded sets $A \subset \Delta$ with $E\mathsf{Poiss}(c)(\partial A) = 0$. Then by Lemma 1.4, we only need to prove $N^{(n)}(f) \overset{d}{\to} \mathsf{Poiss}(c)(f)$, for any $f \in \widehat{\mathcal{I}}_+$, i.e., any f with $f = \sum_{i=1}^{k} c_i \mathbb{1}_{(a_i, b_i]}$, where $(a_i, b_i] \in \mathcal{I}$ and $a_i > 0$. Note that f is undefined on $(0, \min a_i]$. Then $N^{(n)} \overset{vd}{\to} \mathsf{Poiss}(c)$ on $(0, T]$ follows from $N^{(n)}(\mathbb{1}_{(a,b]}) \overset{d}{\to} \mathsf{Poiss}(c)(\mathbb{1}_{(a,b]})$ for $0 < a < b \le T$, which in turn follows from Proposition 6.3.

(ii) Assume that $\mu_n, \mu \in \mathcal{M}_T$ and $\mu_n \overset{v}{\to} \mu$. Then for any $\varepsilon > 0$ small enough, $\mu_n \overset{v}{\to} \mu$ on $[\varepsilon, T]$. Since μ is locally finite, it has finitely many atoms on $[\varepsilon, T]$, say $\varepsilon \le x_1 \le \cdots \le x_l \le T$. It easy to see that $\exists\, n_0$ such that for any $n \ge n_0$, μ_n also has l atoms there. Moreover, $\exists\, K = K(\varepsilon, l) \ge n_0$, such that, for any $n \ge K$,

$$\left| x_i^{(n)} - x_i \right| \le \frac{\varepsilon}{2(l+2)^2}, \quad \text{for all } i = 1, 2, \ldots, l.$$

By (6.31), $|\Phi_t(\mu_n)(\varepsilon) - \Phi_t(\mu)(\varepsilon)| \le 2\varepsilon$, and by definition (6.30), we have

$$|\Phi_t(\mu_n)(t) - \Phi_t(\mu)(t)| \le \frac{l+2}{(l+2)^2}\varepsilon, \quad \text{so}$$

$$|\Phi_t(\mu_n) - \Phi_t(\mu)|_{[0,T]} \le \frac{(l+2)^2}{(l+2)^2}\varepsilon = \varepsilon, \quad n \ge K.$$

Hence, Φ_t is continuous. Moreover, $|\Phi_t(\mu)|_{[0,T]} \le T$, so Φ_t is also uniformly bounded.

Finally, (iii) immediately follows from (i), (ii) and Lemma 1.4, completing the proof of Proposition 6.4. □

Having Proposition 6.4 at our disposal, it is now easy to prove that the processes $S^{(n)}$ in the statement of the theorem converge in law to the zigzag process, by checking the convergence of the finite dimensional distributions, and then tightness.

Convergence of fidi's: Given $0 < t_1 < t_2 < \cdots < t_k$, to check that the law of $(S_{t_1}^{(n)}, \ldots, S_{t_k}^{(n)})$ converges as $n \to \infty$, let $A_1, A_2, \ldots, A_k \subset \mathbb{R}$ be Borel sets, and denote

$$\vec{A} := (A_1, \ldots, A_k), \quad -\vec{A} := (-A_1, \ldots, -A_k),$$
$$(S_{\vec{t}}^{(n)} \in \vec{A}) := \left(S_{t_1}^{(n)} \in A_1, \ldots, S_{t_k}^{(n)} \in A_k\right),$$
$$(\Phi_t(u)_{\vec{t}} \in \vec{A}) := (\Phi_t(u)_{t_1} \in A_1, \ldots, \Phi_t(u)_{t_k} \in A_k).$$

Moreover, $\{S_s^{(n)} = +\}$ ($\{S_s^{(n)} = -\}$) denotes the event that the zigzag path is increasing (decreasing) at s^+, by which we mean that there exists a small interval $[s, s+\epsilon]$ such that $S^{(n)}$ has slope 1 (-1) on $(s, s+\epsilon)$. Then

$$\mathbb{P}\left(S_{\vec{t}}^{(n)} \in \vec{A}\right) = \mathbb{P}\left(S_{\vec{t}}^{(n)} \in \vec{A} \mid S^{(n)}(t_1) = +\right)\mathbb{P}\left(S^{(n)}(t_1) = +\right)$$
$$+ \mathbb{P}\left(S_{\vec{t}}^{(n)} \in \vec{A} \mid S^{(n)}(t_1) = -\right)\mathbb{P}\left(S^{(n)}(t_1) = -\right),$$

where, by symmetry, $\mathbb{P}\left(S^{(n)}(t_1) = +\right) = \mathbb{P}\left(S^{(n)}(t_1) = -\right) = \frac{1}{2}$ and

$$\mathbb{P}\left(S^{(n)}_{\vec{t}} \in \vec{A} \mid S^{(n)}(t_1) = +\right) = \mathbb{P}\left(\Phi_{t_1}(N^{(n)})_{\vec{t}} \in \vec{A}\right).$$

By Proposition 6.4, $\Phi_{t_1}(N^{(n)}) \overset{w}{\to} \Phi_{t_1}(\mathsf{Poiss}(c))$ on $C[0, t_k]$; composing with projections yields

$$\mathbb{P}\left(S^{(n)}_{\vec{t}} \in \vec{A} \mid S^{(n)}(t_1) = +\right) \overset{n\to\infty}{\longrightarrow} \mathbb{P}\left(\Phi_{t_1}(\mathsf{Poiss}(c))_{\vec{t}} \in \vec{A}\right).$$

Similarly,

$$\mathbb{P}\left(S^{(n)}_{\vec{t}} \in \vec{A} \mid S^{(n)}(t_1) = -\right) = \mathbb{P}\left(-S^{(n)}_{\vec{t}} \in -\vec{A} \mid -S^{(n)}(t_1) = +\right)$$

tends to $\mathbb{P}\left(\Phi_{t_1}(\mathsf{Poiss}(c))_{\vec{t}} \in -\vec{A}\right)$ as $n \to \infty$, hence

$$\mathbb{P}\left(S^{(n)}_{\vec{t}} \in \vec{A}\right)$$

$$\overset{n\to\infty}{\longrightarrow} \frac{1}{2}\left(\Phi_{t_1}(\mathsf{Poiss}(c))_{\vec{t}} \in \vec{A}\right) + \frac{1}{2}\left(\Phi_{t_1}(\mathsf{Poiss}(c))_{\vec{t}} \in -\vec{A}\right).$$

Tightness: We repeat the argument in the proof of the strongly critical case in Theorem 6.4 here. Use that $\lim_{\eta\to+\infty} \sup_{n\geq 1} \mathbb{P}(S^{(n)}(0) > \eta) = 0$ together with

$$\limsup_{\delta\downarrow 0}_{n\geq 1} \mathbb{P}\left(\max_{\substack{|t-s|\leq\delta \\ 0\leq t,s<T}} |S^{(n)}(t) - S^{(n)}(s)| > \epsilon\right) = 0, \quad \forall\epsilon,\ T > 0,$$

are sufficient for tightness on $C([0, \infty))$. These are indeed satisfied because $S^{(n)}(0) = 0$, $n \geq 1$, and because of the uniform Lipschitzness. This completes the proof of the theorem in the critical case. $\square$

Note: One can use any Φ_s, $s > 0$, instead of Φ_{t_1} (again, $s = 0$ is excluded), without causing too much change; then

$$\mathbb{P}\left(X^{(n)}_{\vec{t}} \in \vec{A} \mid X^{(n)}(s) = +\right) \overset{n\to\infty}{\longrightarrow} \mathbb{P}\left(\Phi_s(\mathsf{Poiss}(c))_{\vec{t}} \in \vec{A}\right).$$

Remark 6.15. We can also generalize the condition $A_n := np_n = c$ a bit, namely, one can mimic the proof in Proposition 6.3 to show the following. If the A_n are stable in the sense that

$$\sum_{k=an}^{bn} \frac{A_k - c}{k} \overset{n \to \infty}{\Longrightarrow} 0, \quad \text{that is,} \quad \sum_{k=an}^{bn} \frac{A_k}{k} \overset{n \to \infty}{\Longrightarrow} c \ln(b/a),$$

$$\forall 0 < a < b < \infty,$$

then the turns $N^{(n)}$ tend to a PPP with intensity $\lambda(x) = \frac{c}{x} \, dx$. Hence, the law of $S^{(n)}$ tends to that of the same zigzag process, i.e., we have the same scaling limit. This includes, for example, the following cases:

- $A_n \equiv c$ for all large n,
- $\lim_{n\to\infty} A_n = c$,
- A_n is periodic with average c,

where c is a positive constant. ◇

6.4.11   *Proof of Theorem 6.4: subcritical case*

Following the martingale approximation approach and again to prove all conditions at the end of Section 6.4.2, we prove the result in the following steps:

(i) The $a_n \geq 1$ are well defined; furthermore, $a_n = o(n)$.
(ii) $\lim_{m\to\infty} v_m = \infty$.
(iii) $a_n^2 = o(v_n)$.
(iv) As $n \to \infty$,

$$\frac{1}{n} \sum_{i=1}^{Z(n)} a_i^2 \xi_i^2 \mathbb{1}_{\{a_i^2 \xi_i^2 > n\varepsilon\}} \overset{L^1}{\longrightarrow} 0. \tag{6.39}$$

Step (i). Since $1 - x \leq e^{-x}$, $x > 0$, and A_n is a monotone increasing sequence, we have

$$e_{n,n+i} = \prod_{k=n+1}^{n+i} (1 - 2p_k) \leq e^{-(2p_{n+1}+\cdots+2p_{n+i})}$$

$$= e^{-2\left(\frac{A_{n+1}}{n+1} + \cdots + \frac{A_{n+i}}{n+i}\right)} \leq e^{-2A_{n+1}\left(\frac{1}{n+1} + \cdots + \frac{1}{n+i}\right)}$$

$$\leq e^{-2A_{n+1}\int_1^i \frac{dx}{n+x}} = e^{-2A_{n+1}\ln\frac{n+i}{n+1}} = \left(\frac{n+1}{n+i}\right)^{2A_{n+1}}, \quad (6.40)$$

since $\sum_{j=a}^b \frac{1}{j} \geq \int_a^b \frac{dx}{x}$ for all integers a, b with $b > a \geq 2$. So,

$$a_n = 1 + \sum_{i=1}^\infty e_{n,n+i} \leq 1 + \sum_{i=1}^\infty \left(\frac{n+1}{n+i}\right)^{2A_{n+1}}$$

$$\leq 1 + \int_0^\infty \left(\frac{n+1}{n+x}\right)^{2A_{n+1}} dx$$

$$= 1 + \frac{(n+1)^{2A_{n+1}}}{2A_{n+1} - 1} \frac{1}{n^{2A_{n+1}-1}} = 1 + \frac{n}{2A_{n+1} - 1}\left(1 + \frac{1}{n}\right)^{2A_{n+1}}$$

$$= 1 + \frac{n}{2A_{n+1} - 1}\left(1 + \frac{1}{n}\right)^{n \cdot 2p_{n+1}}\left(1 + \frac{1}{n}\right)^{2p_{n+1}}$$

$$= 1 + \frac{n}{2A_{n+1} - 1}e^{2p_{n+1}}(1 + o(1)) = 1 + \frac{n(1 + O(p_{n+1}))(1 + o(1))}{2A_{n+1} - 1}$$

for large n. Since $A_{n+1} \to \infty$, we have $a_n = o(n)$.

Step (ii). There exists an $N \geq 1$ such that for all $n \geq N$ we have $p_n \geq \frac{1}{n}$ and $q_n \geq \frac{1}{4}$. Also, $a_n \geq 1$. Hence, for m large enough, $v_m = \sum_{n=1}^m 4a_n^2 p_n q_n \geq \sum_{n=N}^m 4p_n q_n \geq \sum_{n=N}^m \frac{1}{n} \to \infty$ as $m \to \infty$.

Step (iii). Since $p_n \downarrow 0$, one has

$$p_n a_n = p_n \left[1 + (1 - 2p_{n+1}) + (1 - 2p_{n+1})(1 - 2p_{n+2}) + \cdots\right]$$

$$\geq p_n \sum_{k=0}^\infty (1 - 2p_n)^k = \frac{1}{2}.$$

From its definition it follows that v_n is monotone, and we also know that $v_n \to \infty$. Hence, by the Stolz–Cesàro theorem,[10] we have

$$\limsup_{n\to\infty} \frac{a_n^2}{v_n} \leq \limsup_{n\to\infty} \frac{a_n^2 - a_{n-1}^2}{v_n - v_{n-1}} = \limsup_{n\to\infty} \frac{(a_n + a_{n-1})(a_n - a_{n-1})}{4p_n q_n a_n^2}$$

$$\leq \limsup_{n\to\infty} \frac{(a_n + a_{n-1})(a_n - a_{n-1})}{2a_n}$$

$$\leq \frac{1}{2} \limsup_{n\to\infty} (a_n - a_{n-1}), \tag{6.41}$$

since $4p_n q_n a_n^2 = (2p_n a_n) \cdot q_n \cdot 2a_n$, and $q_n \to 1$, $p_n a_n \geq 1/2$, $a_{n-1} \leq a_n$. Next,

$$a_n - a_{n-1} = \sum_{i=1}^{\infty} [e_{n,n+i} - e_{n-1,n-1+i}]$$

$$= \sum_{i=1}^{\infty} [e_{n,n+i-1}(1 - 2p_{n+i}) - (1 - 2p_n)e_{n,n-1+i}]$$

$$= 2 \sum_{i=1}^{\infty} (p_n - p_{n+i})e_{n,n+i-1}. \tag{6.42}$$

We have (e.g., by integrating by parts)

$$\sum_{i=1}^{\infty} \frac{i}{(n-1+i)^{2A_{n+1}+1}} \leq \int_0^{\infty} \frac{x\,dx}{(n-1+x)^{2A_{n+1}+1}}$$

$$= \frac{1}{2A_{n+1}(2A_{n+1} - 1)(n-1)^{2A_{n+1}-1}}.$$

---

[10]This is the discrete version of L'Hospital's rule — see, e.g., Problem 70 in Pólya and Szegő (1998).

From the monotonicity of p_n and np_n, we get $p_n \geq p_{n+i}$ and $\frac{p_{n+i}}{p_n} \geq \frac{n}{n+i}$. Then, from (6.40) and (6.42), it follows that[11]

$$0 \leq \frac{a_n - a_{n-1}}{2p_n} = \sum_{i=1}^{\infty} \left(1 - \frac{p_{n+i}}{p_n} \right) e_{n,n+i-1}$$

$$\leq \sum_{i=1}^{\infty} \frac{i}{n+i} \cdot \left(\frac{n+1}{n+i-1} \right)^{2A_{n+1}}$$

$$\leq \sum_{i=1}^{\infty} \frac{(n+1)^{2A_{n+1}} \cdot i}{(n-1+i)^{2A_{n+1}+1}} \leq \frac{(n+1)^{2A_{n+1}}}{2A_{n+1}(2A_{n+1}-1)(n-1)^{2A_{n+1}-1}}$$

$$= \frac{(n-1)\left(1 + \frac{1}{n-1}\right)^{2A_{n+1}}}{2A_{n+1}(2A_{n+1}-1)} = \frac{(n-1)(1+O(p_n))}{4A_{n+1}^2(1+o(1))}.$$

Hence,

$$0 \leq a_n - a_{n-1} \leq 2p_n \frac{n+o(n)}{4A_{n+1}^2} = \frac{A_n(1+o(1))}{2A_{n+1}^2} \leq \frac{1+o(1)}{2A_{n+1}} \to 0,$$

so the right-hand side of (6.41) tends to zero.

Step (iv). We show how, in our case, (iii) implies (iv). Since a_i increases in i, and $|\xi_i| \leq 2$ gives $a_i^2 \xi_i^2 \leq 4a_i^2$, we have

$$\{i : a_i^2 \xi_i^2 \geq \varepsilon n\} \subset \{i : 4a_i^2 \geq \varepsilon n\} = \{i : i \geq (f^2(n))^{-1}(\varepsilon n/4)\},$$

where $f(\cdot)$ is the linear interpolation such that $f(i) = a_i$, and here v can be treated also as a positive strictly increasing function on $[0, \infty)$ with $v(m) = v_m$, so both $(f^2)^{-1}, v^{-1}$ are well defined, positive and strictly increasing. Using that $Z(n) = v^{-1}(n)$, Drogin's condition (6.39) will be verified if we show that

$$v^{-1}(n) < (f^2(n))^{-1}(\varepsilon n/4), \tag{6.43}$$

---

[11]The last equality is elementary: $\left(1 + \frac{1}{n-1}\right)^{2A_{n+1}} = (1 + \frac{1}{n-1})^{(n-1)2p_{n+1}} \cdot (1 + \frac{1}{n-1})^{4p_{n+1}} = O(e^{2p_{n+1}}) = O(1 + 2p_{n+1}) = O(1 + p_n).$

for n large enough, because then, for n large enough, $a_i^2 \xi_i^2 < \varepsilon n$ for $i \leq Z(n)$, that is,

$$\mathbb{1}_{\{a_i^2 \xi_i^2 > n\varepsilon\}} = 0, \ 1 \leq i \leq Z(n).$$

Since $a_m^2 = o(v_m)$, i.e., $f^2(x) = o(v(x))$, for this ε, there is an M such that for $l \geq M$, $f^2(l)/v(l) < \varepsilon/4$, and for such an M, there is an N such that for $x \geq N$ we have $v^{-1}(x) \geq M$. Hence,

$$\frac{f^2(v^{-1}(x))}{v(v^{-1}(x))} = \frac{f^2(v^{-1}(x))}{x} < \frac{\varepsilon}{4}, \quad \forall x \geq N,$$

that is, (6.43) holds for $n \geq N$. This completes the proof of (iv) and that of the theorem altogether.

6.4.12 Proof of Theorem 6.5

We again use the martingale approximation approach of Section 6.4.2. Note that

$$a_n = 1 + \sum_{i=0}^{\infty} \prod_{k=n+1}^{n+i} (1 - 2p_k). \tag{6.44}$$

Without the loss of generality, we may assume that $0 < a < p_n < b < 1$. Then $r := \max\{|2a - 1|, |2b - 1|\} < 1$, and

$$\left| \prod_{k=n+1}^{n+i} (1 - 2p_k) \right| \leq r^i,$$

which is why the sum in (6.44) is well defined, that is, the a_n are well defined, for all $n \geq 1$. Furthermore,

$$1 + \sum_{i=0}^{\infty} \prod_{k=n+1}^{n+i} (1 - 2p_k) \leq 1 + \sum_{i=1}^{\infty} \prod_{k=n+1}^{n+i} |1 - 2p_k| \leq 1 + \sum_{i=1}^{\infty} r^i = \frac{1}{1 - r},$$

which gives $|a_n| \leq \frac{1}{1-r}$ for all n.

Next, we prove that $v(m) \overset{m\to\infty}{\longrightarrow} \infty$ or, equivalently, that $\sigma_n \overset{n\to\infty}{\longrightarrow} \infty$:

(i) If $p_n \leq 1/2$, $\forall n$, then $a_n > 1$, $\forall n$, and we immediately have $v(m) \overset{m\to\infty}{\longrightarrow} \infty$.

(ii) Otherwise we have a subsequence $\{p_{n_k}\}_{n_k}$ such that $n_{k+1} - n_k > 1$ and $p_{n_k} > 1/2$, for all n_k. Note that, by (6.44) and a direct computation, we have

$$(a_{n-1} - 1) = (a_n - 1)(1 - 2p_n),$$

and thus for the subsequence one has

$$(a_{n_k-1} - 1) = (a_{n_k} - 1)(1 - 2p_n).$$

So, the two subsequences $\{a_{n_k-1} - 1\}_{k\geq 1}, \{a_{n_k} - 1\}_{k\geq 1}$ have opposite signs, hence we have a subsequence of $\{a_n\}_{n\geq 1}$ such that its terms are larger than 1. Consequently, $v(m) \overset{m\to\infty}{\longrightarrow} \infty$.

Moreover, the condition that $\lim_{n\to\infty} \frac{1}{n} \sum_{i=1}^{Z(n)} a_i^2 \xi_i^2 \mathbb{1}_{\{a_i^2\xi_i^2 > n\epsilon\}} = 0$ is easy to verify, since our a_n are bounded.

In conclusion, the answers to (M) and to (INV.M) are both in the affirmative, yielding the invariance principle (6.10).

6.4.13   *Proof of Theorem 6.6*

Fix $a > 0$ and let $N = N(a)$ be such that $a/N \leq 1/2$ and that also $a/n < p_n$ holds for all $n > N$. Define $\hat{p}_n$ so that it coincides with p_n for $n \leq N$ and $\hat{p}_n = a/n$ for $n > N$ (critical). Let $\hat{S}$ denote the walk for the sequence $(\hat{p}_n)$, and note that this walk depends on the parameter $a > 0$. By the monotonicity established in the proof of Theorem 6.4,

$$\mathbb{V}\mathrm{ar}\left(\frac{S_n}{n}\right) \leq \mathbb{V}\mathrm{ar}\left(\frac{\hat{S}_n}{n}\right), \quad n \geq 1.$$

It follows from Theorem 5.1 that

$$\lim_{n\to\infty} \mathbb{V}\text{ar}\left(\frac{\hat{S}_n}{n}\right)$$

$$= \frac{1}{4(2a+1)} =: v(a) \quad \Longrightarrow \quad \limsup_{n\to\infty} \mathbb{V}\text{ar}\left(\frac{S_n}{n}\right) \leq v(a).$$

Since $a > 0$ was arbitrary,

$$\lim_{n\to\infty} \mathbb{V}\text{ar}\left(\frac{S_n}{n}\right) = 0,$$

implying WLLN. □

6.5 Some Further, Closely Related Results in the Literature

In this section, we review some interesting, relatively recent (at least at the time of writing this book) results, without proofs.

In a follow-up paper to Engländer and Volkov (2018) by Bouguet and Cloez (2018), the setting has been generalized in such a way that instead of two states (heads and tails or ± 1), one considers $D \geq 2$ states, and with probability p_n in the nth step, the state changes according to a given irreducible Markov chain.[12] (They also allow a small error term.) They assume that $\{p_n\}_{n\geq 1}$ is a *decreasing* sequence and $p := \lim_n p_n$ is not necessarily zero. This excludes the $p = 1$ case we consider, except the trivial $p_n \equiv 1$ case, and the most interesting case is $p = 0$, the one we call cooling dynamics.

Bouguet and Cloez established several interesting results, generalizing/strengthening those presented in our previous chapter on coin turning (and first appearing in Engländer and Volkov (2018)). For example, they showed that if $\sum_n p_n = \infty$, $\lim_{n\to\infty} np_n = \infty$ and

---

[12]For example, when $D = 2$, one can still consider unequal probabilities for switching between the states in different directions.

$\sum_n (p_n n^2)^{-1} < \infty$, then the empirical distribution of the states converges almost surely to the unique invariant probability distribution
of the Markov chain.

For the interested reader, we note that the relationship with
Engländer and Volkov (2018) is explained in detail in Section 4.2
in Bouguet and Cloez (2018).

The paper builds on the authors' previous results with Michel
Benaïm in Benaïm *et al.* (2017), and they point out in Bouguet and
Cloez (2018) that

> "In particular, the results we use provide functional convergence
> of the rescaled interpolating processes to the auxiliary Markov
> processes..."

at which point the authors refer to Benaïm *et al.* (2017) and another
article. Also, after their Theorem 2.8, treating the critical $p_n = c/n$
case, they note that

> "It should be noted that our approach for the study of the
> long-time behavior of ... also provides functional convergence
> for some interpolated process ... from which Theorem 2.8 is a
> straightforward consequence."

On the one hand, their Theorem 2.8 is really about the convergence
of S_n/n only, and the "interpolating process" alluded to is not the
random walk S, and it is not completely clear to the authors of this
book whether (Bouguet and Cloez, 2018) tries to say that one in fact
can obtain from Benaïm *et al.* (2017) the functional convergence for
S in the critical case as stated in our Theorem 6.4(3).

On the other hand, it seems that this derivation is after all doable,
as we explain briefly in the following. Indeed, let us suppose we
already know tightness and only want to check the convergence of
the finite dimensional distributions, that is, the existence of the limit
(in law)

$$\lim_{n\to\infty} (S_{nt_1}/n, S_{nt_2}/n, \ldots, S_{nt_k}/n), \qquad (6.45)$$

for some $0 < t_1 < t_2 < \cdots < t_k$. When $p_n = c/n$ for large n's, define
$\tau_t := \sum_1^{\lfloor t \rfloor} p_n$. Define the "pasting process" $\widehat{X}$ by

$$\widehat{X}(t) := \sum_{n=1}^{\infty} \frac{S_n}{n} 1_{\tau_n \le t < \tau_{n+1}}, \quad t \ge 0.$$

After some algebra, the limit in (6.45) is equivalent to the following one:

$$\lim_{t\to\infty} \left(\widehat{X}(\tau_{tt_1})t_1, \widehat{X}(\tau_{tt_2})t_2, \ldots, \widehat{X}(\tau_{tt_k})t_k \right).$$

Using the fact that $\lim_{t\to\infty}(\tau_t - \log t)$ is a constant, we can rewrite this as

$$\lim_{t\to\infty} (\widehat{X}^{(\tau_{tt_1})}(0)\, t_1, \ \widehat{X}^{(\tau_{tt_1})}(\log(t_2/t_1))t_2, \ldots, \ \widehat{X}^{(\tau_{tt_k})}(\log(t_k/t_{k-1}))\, t_k),$$

where $\widehat{X}^{(z)}(t) := \widehat{X}(t+z)$, $z > 0$. Now, if the limit (in law)

$$\lim_{z\to\infty} (\widehat{X}^{(z)}(s_1), \ \widehat{X}^{(z)}(s_2), \ldots, \widehat{X}^{(z)}(s_k))$$

is known, we are done, and this latter type of limit of "pseudo-trajectories" is what has been derived in Benaïm *et al.* (2017) under some suitable assumptions.

In summary, Bouguet and Cloez (2018) provides a very valuable complement to Engländer and Volkov (2018). Moreover, with some further efforts, our result on the zigzag process in this chapter can apparently be recovered from the results presented in the sequence (Benaïm *et al.*, 2017; Bouguet and Cloez, 2018), and vice versa.

6.6   Exercises

(1) Consider the filtration generated by $Y_0, W_1, W_2, \ldots$. Which one of the following events is a tail event: $A := \{S_n \to \infty\}$, $B := \{|S_n| \to \infty\}$?

(2) Suppose that $q_n = 1/n$. By Corollary 6.2, S is recurrent. Is it true that its scaling limit is a time-changed Brownian motion with an appropriate scaling?

(3) When monotonicity is dropped, part (b) of Corollary 6.2 is no longer applicable. What can we say about the recurrence of S when, for example, $q_{2n} = 1/(2n)$ and $q_{2n+1} = 1/(2n+1)^2$?

(4) Assume that $p_n \le c/n$ for all large n's and $\sum_n p_n = \infty$. We already know from Corollary 6.2 that the walk is recurrent. But now we provide another proof and the reader is asked to give a similar proof for the recurrence of the scaling limit (zigzag process) when $p_n = c/n$ for large n's.

The alternative proof is as follows. Let $\tau_0 := 0$ and

$$\tau_n := \inf\{m > 2\tau_{n-1} : Y_m = -1\}, \quad n = 1, 2, \dots.$$

Since $\sum p_n = \infty$, by the Borel–Cantelli lemma, there are infinitely many turns. As a result, with probability 1, all τ_n are well defined and finite. Clearly, $\tau_n \to \infty$, as $n \to \infty$.

Let

$$A_n := \{Y_i = -1, \text{ for all } i \in [\tau_n, 2\tau_n]\} \in \mathcal{F}_{2\tau_n} =: \mathcal{G}_n,$$

and note that $A_n \subseteq \{S_{2\tau_n} \leq 0\} =: B_n$. Therefore, if we show that $\sum_n \mathbb{P}(A_n \mid \mathcal{G}_{n-1}) = \infty$, then by the extended Borel–Cantelli lemma, it follows that $\mathbb{P}(A_n \text{ i.o.}) = 1$; hence $\mathbb{P}(B_n \text{ i.o.}) = 1$, and so we obtain that $\mathbb{P}(S_n \leq 0 \text{ i.o.}) = 1$.

For $n \geq n_0$, we have[13]

$$\mathbb{P}(A_n \mid \mathcal{G}_{n-1}, \tau_n = k) = \left(1 - \frac{c}{k+1}\right)\left(1 - \frac{c}{k+2}\right)\cdots\left(1 - \frac{c}{2k}\right).$$

The product on the right-hand side tends to 2^{-c}, as $k \to \infty$. As a result, $\mathbb{P}(A_n \mid \mathcal{G}_{n-1}) \geq \frac{1}{2} \cdot 2^{-c}$ holds for all large enough n, and we are done. A completely symmetric argument shows that also $\mathbb{P}(S_n \geq 0 \text{ i.o.}) = 1$, thus proving the recurrence of the walk S.

---

[13]Of course, the following equality makes sense only for those k for which $\mathbb{P}(\tau_n = k) > 0$; in particular, we must have $k \geq n$.

Chapter 7

Urn Models and Time-Inhomogeneous Markov Processes

As mentioned in Section 2.6, the Pólya (more precisely, Pólya–Eggenberger) urn and its generalizations have been studied intensely in the last hundred years since their discovery in 1923, see, e.g., Mahmoud (2009). Probabilistic studies of urns go back in history way beyond the Pólya urn though; in fact, they were already proposed[1] in Jacob Bernoulli's *Ars Conjectandi*, published in 1713, which is often considered the first scientific discussion of a theory of probability. Thus, not surprisingly, the theory of urn models is much more standard and available than the general theory of inhomogeneous Markov chains. It is therefore beneficial to find a connection between the two. This short chapter is devoted to some initial exploration of this connection. Our two keywords will be "mimicking" and "Markovization."

We recall the (classical) Pólya urn model: one puts a white and b blue balls into an urn. At each step, one ball is drawn uniformly at random from the urn, and its color observed; it is then returned in the urn, and an additional ball of the same color is added to the urn. Let X_i^{urn} denote the indicator of the ith ball drawn being white; they form the discrete-time process X^{urn}.

---

[1]Bernoulli considered the problem of determining, given a number of pebbles drawn from an urn, the proportions of different colored pebbles within the urn.

7.1   Mimicking Another Process

We start with some definitions that play fundamental roles.

Let $X = \{X_i\}_{i \geq 1}$ be a sequence of integer-valued random variables under the law P. We consider T defined by $T_n := \sum_1^n X_i$ $n \geq 1$ the corresponding walk. Let $X' = \{X_i'\}_{i \geq 1}$ be another sequence of integer-valued random variables under the law $\mathbf{P}$ and let $T_n' := \sum_1^n X_i'$ be the corresponding walk. Our basic definition is as follows.

Definition 7.1 (Mimicking). We say that X' *mimics* X (and vice versa) if the following holds:

(M1) $P(X_i = u) = \mathbf{P}(X_i' = u)$, $i \geq 1$,
(M2) $P(T_{i+1} = v \mid T_i = u) = \mathbf{P}(T_{i+1}' = v \mid T_i' = u)$, $i \geq 1$.

That is, X and X' have the same one-dimensional marginals and the two walks have the same one-step transition probabilities.

Note that (M2) also gives

$$P(T_i = v) = \mathbf{P}(T_i' = v), \tag{7.1}$$

provided $T_1 = T_1'$ (and this latter of course follows from (M1)). Furthermore, one has $P(T_{i+1} = v, T_i = u) = \mathbf{P}(T_{i+1}' = v, T_i' = u)$, $i \geq 1$, however this may fail for any three times (see Example 7.1).

Definition 7.2 (Markovization). Suppose that T' is a Markov chain (possibly time-inhomogeneous) with respect to its natural filtration, with transition probabilities $p_{i,u,v}$ for which condition (M2) (but not necessarily (M1)) holds. Then we call T *Markovizable* and say that T' is a *Markovization* of T.

For given Y_i, $i \geq 1$, when we can "Markovize" this sequence in the sense of Definition 7.2, we achieve it by defining the one-step transitions via the matrix $\mathcal{P}$ with entries $p_{u,v} := P(Y_{i+1} = v \mid Y_i = u)$ and the n-step transitions by defining $\mathcal{P}^{(n)} := \mathcal{P}^n$. Intuitively, the process "forgets" how it got to the state where it is at n before stepping into the next state. The following lemma shows that X and T play a symmetric role.

Lemma 7.1. X' *mimics* X *if and only if*

$$\mathsf{Law}(T_i, X_i) = \mathsf{Law}(T_i', X_i'), \quad \forall i. \tag{7.2}$$

Proof. We prove this in two parts.

$(7.2) \Rightarrow (M1, M2)$:

Obviously, (7.2) implies $(M1)$. For $(M2)$, write

$$P(T_{i+1} = v \mid T_i = u) = P(T_{i+1} = v, T_i = u)/P(T_i = u)$$
$$= P(T_{i+1} = v, X_{i+1} = v - u)/P(T_i = u)$$
$$= \mathbf{P}(T'_{i+1} = v, X'_{i+1} = v - u)/P(T'_i = u)$$
$$= \mathbf{P}(T'_{i+1} = v, T'_i = v - (v - u))/P(T'_i = u)$$
$$= \mathbf{P}(T'_{i+1} = v \mid T'_i = u),$$

finishing this part.

$(M1, M2) \Rightarrow (7.2)$:

From $(M1, M2)$, we have

$$\mathsf{Law}(T_i) = \mathsf{Law}(T'_i); \quad (M2) \Leftrightarrow \mathsf{Law}(X_{i+1} \mid T_i) = \mathsf{Law}(X'_{i+1} \mid T'_i),$$

hence $\mathsf{Law}(X_{i+1}, T_i) = \mathsf{Law}(X'_{i+1}, T'_i)$, which gives

$$\mathsf{Law}(X_{i+1}, T_i + X_{i+1}) = \mathsf{Law}(X'_{i+1}, T'_i + X'_{i+1}),$$

that is, $\mathsf{Law}(X_{i+1}, T_{i+1}) = \mathsf{Law}(X'_{i+1}, T'_{i+1})$, which is (7.2). $\square$

We close this section by noting that mimicking is clearly an equivalence relation.

7.2 Urn Models Versus Inhomogeneous Markov Chains

Recall that X_i^{urn} denotes the indicator of the ith ball drawn being white in the Pólya urn model, and let

$$T_n^{\mathrm{urn}} := X_1^{\mathrm{urn}} + \cdots + X_n^{\mathrm{urn}},$$

that is, T_n is the number of white balls by time n. Importantly, T^{urn} is a Markov chain but X^{urn} is not.

The coin-turning walk, introduced in Chapter 6, has the opposite behavior, namely, the steps are Markovian but the walk is not. (In fact, the walk is a so-called "second-order Markov chain" because we need to know the last *two* steps to obtain full information about the future.)

It is also worth mentioning that while X^{urn} is an exchangeable sequence, X is not. This is summarized in the following table.

	Steps	Walk
Coin turning	Markovian, non-exchangeable	non-Markovian
Pólya urn	non-Markovian, exchangeable	Markovian

In the sequel, we show that some inhomogeneous Markov chains (certain "critical" coin turning processes) are mimicking certain urn models. In particular, this means that the corresponding urn walks are Markovizations of the walks obtained from those Markov chains as steps. *Inter alia*, these provide new proofs for certain results of the previous chapters.

The class of inhomogeneous chains for which a similar mimicking works is very poorly understood at the time of writing this book.

7.3   Coin Turning and Urn Models: Simplest Case

Consider again the random walk S obtained from "coin turning" with a given sequence $\{p_n\}_{n\geq 2}$, as introduced in Chapter 6. It turns out that for some sequences of the p_n's, there is a close connection with urn models. To understand this, we first note that, somewhat surprisingly, when the sequence is precisely[2] $(p_1 = 1/2), p_2 = 1/3$, $p_3 = 1/4, p_4 = 1/5, \ldots$, that is, when we have $X_1 = 0, 1$ with probability $\frac{1}{2}$ each and

$$X_n = \begin{cases} X_{n-1} & \text{w.p. } n/(n+1); \\ 1 - X_{n-1} & \text{w.p. } 1/(n+1), \end{cases}$$

the random variable $\frac{S_N}{N}$ has precisely *discrete uniform law* for each N. Hence, the limiting distribution is uniform on the unit interval, in accordance with Theorem 5.1.

The emergence of the discrete uniform distribution can be related[3] to Pólya urns. In fact, the number of heads (not Markovian) "mimics"

---

[2]The initial symmetrization, that is, flipping the coin, is equivalent to starting with a fixed side and defining $p_1 := 1/2$.

[3]This was first observed by and pointed out to us by Márton Balázs.

the number of white balls drawn for the Pólya urn (a Markov chain), in the sense of Definition 7.1.

To verify the above statements, let T be the coin-turning walk with $T_n := X_1 + \cdots + X_n$. First note that by symmetry, one has $P(T_n = k, X_n = 1) = P(T_n = n - k, X_n = 0)$.

Claim 7.1. $P(T_n = k, X_n = 1) = \frac{k}{n(n+1)}$, $k = 1, 2, \ldots, n$.

Proof. We prove the claim by induction on n. The $n = 1$ case is clear. If it is true for $n \geq 1$, then

$$P(T_{n+1} = k, X_{n+1} = 1) = P(T_n = k - 1, X_{n+1} = 1)$$
$$= P(T_n = k - 1, X_n = 1, X_{n+1} = 1)$$
$$\quad + P(T_n = k - 1, X_n = 0, X_{n+1} = 1)$$
$$= P(T_n = k - 1, X_n = 1) \times \left[1 - \frac{1}{n+2}\right]$$
$$\quad + P(T_n = k - 1, X_n = 0) \times \frac{1}{n+2}$$
$$= \frac{k - 1}{n(n+1)} \times \frac{n+1}{n+2} + \frac{n - (k-1)}{n(n+1)} \times \frac{1}{n+2} = \frac{k}{(n+1)(n+2)},$$

completing the proof. □

Using Claim 7.1, the uniformity result is immediate as

$$P(T_n = k) = P(T_n = k, X_n = 1) + P(T_n = k, X_n = 0)$$
$$= \frac{k}{n(n+1)} + \frac{n - k}{n(n+1)} = \frac{1}{n+1},$$

but it also follows from Theorem 7.1, once we recall Proposition 2.7.

Theorem 7.1. *The coin-turning process X mimics the urn process X^{urn}.*

Proof. We have a Pólya urn process, starting with a single white and a single blue ball at time zero. By Lemma 7.1, we need to verify that $\mathsf{Law}(T_n, X_n) = \mathsf{Law}(T_n^{urn}, X_n^{urn})$ $\forall n$. When $a = b = 1$, de Finetti's measure in Theorem 2.4 is uniform on the unit interval (see Remark 2.7). If $\mathcal{P}$ is uniformly distributed on $[0, 1]$, then conditionally on $\mathcal{P} = p$, the indicators of drawing white balls are i.i.d. Bernoulli(p); hence $T_{n-1}^{\mathrm{urn}} \sim \mathsf{Bin}(n-1, p)$. Thus,

$$P(T_n^{\mathrm{urn}} = k, X_n^{\mathrm{urn}} = 1) = P(T_{n-1}^{\mathrm{urn}} = k-1, X_n^{\mathrm{urn}} = 1)$$

$$= \int_0^1 \binom{n-1}{k-1} p^{k-1}(1-p)^{(n-1)-(k-1)} p\, dp = \frac{k}{n(n+1)},$$

using an identity for the beta function. Hence, by Claim 7.1,

$$P(T_n^{\mathrm{urn}} = k, X_n^{\mathrm{urn}} = 1) = P(T_n = k, X_n = 1),$$

and so indeed $\mathsf{Law}(T_n, X_n) = \mathsf{Law}(T_n^{\mathrm{urn}}, X_n^{\mathrm{urn}})\ \forall n.$ □

Remark 7.1. The equation

$$P(X_{i+1} = v \mid X_i = u) = \mathbf{P}(X_{i+1}^{\mathrm{urn}} = v \mid X_i^{\mathrm{urn}} = u),\ i \geq 1$$

is, unfortunately, too much to hope for, and this is clear from the fact that the Pólya urn defines an exchangeable process. Indeed, one can easily see that $P(X_2^{\mathrm{urn}} = 1 \mid X_1^{\mathrm{urn}} = 1) = \frac{1}{2} \cdot \frac{2}{3} = \frac{1}{3}$, and by the exchangeable property, one also has $P(X_{i+1}^{\mathrm{urn}} = 1 \mid X_i^{\mathrm{urn}} = 1) = \frac{1}{3}$ for all $i \geq 1$, contradicting the equation above, as the left-hand side depends on i. Thus, an attempt to mimic the transition probabilities $P(X_{i+1} \mid X_i)$ too, by the urn model,[4] fails, and we must be content with having only the identity $\mathbf{P}(T_{i+1}' \mid T_i') = P(T_{i+1} \mid T_i)$ for the walks. ◇

Example 7.1 (Triples). Even though the joint distributions for two consecutive steps match, this property breaks down when one considers three consecutive ones. For example, the probability of seeing the sequence H-T-H at the beginning of the coin turning model is $\frac{1}{2} \cdot \frac{1}{3} \cdot \frac{1}{4} = \frac{1}{24}$. For the Pólya urn, however, the probability of drawing white-blue-white balls is twice that $\frac{1}{2} \cdot \frac{1}{3} \cdot \frac{1}{2} = \frac{1}{12}$. ◇

---

[4]One can still talk about transition probabilities regarding the non-Markovian urn process and say that the mismatch with coin turning is caused by their time homogeneity.

7.4 Coin Turning and Urn Models: General Case

More generally, and also recalling Theorem 5.1, let $a \in \mathbb{N}_+$ and consider the Pólya urn process, starting with a white and a blue balls, and the coin-turning process with the sequence

$$p_n = p_n^{(a)} := \frac{a}{2a + n - 1}, \quad n = 1, 2, \ldots.$$

(Here we assume that X_0 is deterministic and note that $p_1 = 1/2$.)

Remark 7.2. As we see in the following, in this case again, it turns out[5] that X mimics X^{urn}, hence, of course, the limiting distribution for $\frac{S_N}{N}$ is $\mathsf{Beta}(a, a)$, by (7.1) and Theorem 2.4. The reader should compare this argument with Theorem 5.1 and its proof; the common feature is that in both cases, $p_n \sim a/n$.

When $a = 1$, one gets back the case discussed at the beginning of Section 7.3, where the limit of S_N/N is uniform on the unit interval, that is, $\mathsf{Beta}(1, 1)$-distributed. $\diamond$

We now state and proof the relationship between the two processes.

Theorem 7.2. *The coin-turning process X mimics the urn process X^{urn}.*

Proof. We now have a Pólya urn process, starting with a white and a blue balls at time zero. By Lemma 7.1, we need to verify that $\mathsf{Law}(T_n, X_n) = \mathsf{Law}(T_n^{\mathrm{urn}}, X_n^{\mathrm{urn}}) \ \forall n$. We note again that de Finetti's measure for the Pólya urn process is the symmetric beta distribution and so it has density

$$f_a(x) = \frac{x^{a-1}(1-x)^{a-1}}{B(a, a)}, \quad 0 < x < 1.$$

One has

$$q_{n,k;1} := P(T_n^{\mathrm{urn}} = k, X_n^{\mathrm{urn}} = 1) = P(T_{n-1}^{\mathrm{urn}} = k - 1, X_n^{\mathrm{urn}} = 1)$$

$$= \int_0^1 \binom{n-1}{k-1} p^{k-1}(1-p)^{(n-1)-(k-1)} \cdot p \cdot \frac{p^{a-1}(1-p)^{a-1}}{B(a, a)} \, dp$$

---

[5]This was first observed by and pointed out to us by Edward Crane.

$$= \binom{n-1}{k-1} \frac{B(k+a, n-k+a)}{B(a,a)}$$

$$= \frac{(n-1)!(2a-1)!(n-k+a-1)!(k+a-1)!}{(n-k)!(k-1)!(n-1+2a)!\left((a-1)!\right)^2}, \quad k \geq 1,$$

using the definition of the beta function. As before, by symmetry,

$$q_{n,k;0} := P(T_n^{\mathrm{urn}} = k, X_n^{\mathrm{urn}} = 0) = q_{n,n-k;1}.$$

Let us now show, by induction on n, that $P(T_n = k, X_n = 1)$ equals the same quantity. Indeed, for $n = 1$,

$$q_{1,1;1} = 1/2, \quad q_{1,0;0} = 1/2.$$

Assume that $P(T_n = k, X_n = 1) = q_{n,k;1}$ holds up to n, and let us compute it for $n+1$. We have

$$P(T_{n+1} = k, X_{n+1} = 1)$$
$$= P(T_n = k-1, X_n = 0, X_{n+1} = 1)$$
$$\quad + P(T_n = k-1, X_n = 1, X_{n+1} = 1)$$
$$= P(T_n = k-1, X_n = 0)p_{n+1} + P(T_n = k-1, X_n = 1)(1 - p_{n+1})$$
$$= q_{n,k-1;0}\, p_{n+1} + q_{n,k-1;1}\left(1 - p_{n+1}\right) = q_{n+1,k;1}$$

using elementary algebra. Hence,

$$P(T_n^{\mathrm{urn}} = k, X_n^{\mathrm{urn}} = 1) = P(T_n = k, X_n = 1),$$

yielding $\mathsf{Law}(T_n, X_n) = \mathsf{Law}(T_n^{\mathrm{urn}}, X_n^{\mathrm{urn}}) \; \forall n.$ $\qquad\square$

In summary, mimicking the coin-turning process with an urn model, one obtains an "elementary" proof for Theorem 5.1 in the particular case when $a \in \mathbb{N}_+$ and the sequence is *precisely* the one given above. One can drop this latter restriction however, as the following theorem shows.

Theorem 7.3. *Let $a \in \mathbb{N}_+$ and*

$$p_n = p_n^{(a)} := \frac{a}{2a + n - 1}, \quad n = M+1, M+2, \dots, \tag{7.3}$$

for some $M \geq 0$. Then for the coin-turning process corresponding to (7.3), the limiting distribution of $\frac{S_N}{N}$ is $\mathsf{Beta}(a,a)$.

Proof. First, symmetrize the setting by assuming that $p_1 = 1/2$ and furthermore let

$$p_n = p_n^{(a)} = \frac{a}{2a + n - 1}, \quad n = M + 1, M + 2, \ldots,$$

with some $M \geq 1$. Clearly (by Slutsky's theorem), it is enough to analyze the limit (in law) as $N \to \infty$ of

$$\frac{Y_{M+1} + Y_{M+2} + \cdots + Y_N}{N}.$$

By the Markov property, the distribution of $\frac{Y_{M+1} + Y_{M+2} + \cdots + Y_N}{N}$, given Y_M, does not depend on $\{Y_i, \, i < M\}$. When $M = 1$, we know that the limit is the symmetric beta distribution with parameters $\alpha = \beta = a$. In the general case, Y_M has the same distribution because $|Y_M| = 1$ and Y_M is symmetric. Hence, the variable $\frac{Y_{M+1} + Y_{M+2} + \cdots + Y_N}{N}$ has exactly the same distribution as before, and the same limit statement holds.

Finally, let us check that we may also drop the assumption that $p_1 = 1/2$. First note that the mixing condition is satisfied and thus there exists (in law) $\lim_M Y_M =: Y_\infty \in \{-1, +1\}$, a symmetric variable. Therefore, if $M < N$ and $I \subset [-1, 1]$ is an interval of the real line, then

$$P\left(\frac{Y_{M+1} + \cdots + Y_N}{N} \in I\right)$$

$$= P\left(\frac{Y_{M+1} + \cdots + Y_N}{N} \in I \mid Y_M = 1\right) P(Y_M = 1)$$

$$+ P\left(\frac{Y_{M+1} + \cdots + Y_N}{N} \in I \mid Y_M = -1\right) P(Y_M = -1)$$

$$= P\left(\frac{Y_{M+1} + \cdots + Y_N}{N} \in I \mid Y_M = 1\right) \left(\frac{1}{2} + \epsilon_M\right)$$

$$+ P\left(\frac{Y_{M+1} + \cdots + Y_N}{N} \in I \mid Y_M = -1\right) \left(\frac{1}{2} - \epsilon_M\right),$$

where $\epsilon_M \to 0$. Hence,

$$P\left(\frac{Y_{M+1} + \cdots + Y_N}{N} \in I\right)$$

$$= \left[\frac{1}{2}P\left(\frac{Y_{M+1} + \cdots + Y_N}{N} \in I \mid Y_M = 1\right)\right.$$

$$\left. + \frac{1}{2}P\left(\frac{Y_{M+1} + \cdots + Y_N}{N} \in I \mid Y_M = -1\right)\right] + O(\epsilon_M).$$

When M is fixed and $N \to \infty$, the expression in the square brackets converges to $Q_{\text{Beta}}(I)$, where Q_{Beta} denotes the above symmetric beta-distribution. When $M \to \infty$, $O(\epsilon_M) \to 0$ (uniformly in N). Consequently, we may safely drop the assumption that $p_1 = 1/2$. $\quad \square$

Chapter 8

The Rademacher Walk

8.1 Introduction

A natural counterpart of one-dimensional coin-turning walk from Definition 6.1 is a process we call "Rademacher walk," which can be described as follows.

Definition 8.1 (Rademacher distribution). We say that a random variable X has a Rademacher distribution and write $X \sim$ Rademacher if $\mathbb{P}(X = +1) = \mathbb{P}(X = -1) = \frac{1}{2}$.

Let $X_i \sim$ Rademacher, $i = 1, 2, \ldots$, be i.i.d., and $\mathcal{F}_n = \sigma(X_1, X_2, \ldots, X_n)$ be the sigma algebra generated by the first n members of this sequence. Let $\mathbf{a} = (a_1, a_2, \ldots)$ be a non-random sequence of positive numbers. Define the **a**-walk as

$$S(n) = a_1 X_1 + a_2 X_2 + \cdots + a_n X_n = \sum_{k=1}^{n} a_k X_k$$

with the convention $S(0) = 0$.

Indeed, one can see the coin-turning walk as a Rademacher walk where the step sizes $a_i, i \geq 1$ are random.

Definition 8.2 (C-recurrent walk). Let $C \geq 0$. We call the **a**-walk S defined above C-*recurrent*, if the event $\{|S(n)| \leq C\}$ occurs for infinitely many n. (In case when $C = 0$, this is equivalent to the usual recurrence, i.e., $S(n) = 0$ for infinitely many n, so we call the walk just *recurrent*.)

We call the **a**-walk *transient* if it is not C-recurrent for any $C \geq 0$.

Our aim is to determine the probability that the **a**-walk for given **a** and C is recurrent; in principle, this probability may be different from 0 and 1 (for example, if $\mathbf{a} = (1, 1, 3, 3, 3, 3, \ldots)$, then the **a**-walk is recurrent with probability $1/2$). A simplest example of an **a**-walk is when all $a_i \equiv a \in \mathbb{R}_+$. Such a random walk is obviously a.s. recurrent since it is equivalent to the one-dimensional simple random walk.

The question of recurrence is naturally related to the Littlewood–Offord problem which deals with the maximization of probability $\mathbb{P}(S(n) = v)$ over all v, subject to various hypotheses on **a**. In particular, in Tao and Vu (2009), the authors develop the so-called *inverse Littlewood–Offord theory*, using which they show that this probability is large only when the elements of **a** are contained in a generalized arithmetic progression; see also Nguyen (2012).

The study of **a**-walk is also somewhat relevant to the conjecture by Tomaszewski (1986), which says that

$$\mathbb{P}\left(|S(n)| \leq \sqrt{a_1^2 + \cdots + a_n^2}\right) \geq \frac{1}{2},$$

for all sequences **a** and all $n \geq 1$. The conjecture was recently proved by N. Keller and O. Klein in Keller and Klein (2022).

Let us first start with some general statements. First, we show that the choice of $C > 0$ is sometimes unimportant for the definition of C-recurrence.

Theorem 8.1. *Suppose that $a_n \to \infty$ and at the same time $|a_n - a_{n-1}| \to 0$ as $n \to \infty$. Then if an **a**-walk is C-recurrent with a positive probability for some $C > 0$, then it is $\tilde{C}$-recurrent with a positive probability for all $\tilde{C} > 0$.*

Proof. Since the notion of C-recurrence is monotone in C, i.e., if an **a**-walk is C_1-recurrent for $C_1 > 0$, then it is C_2-recurrent for all $C_2 \geq C_1$, it suffices to prove that C-recurrence implies $\frac{2C}{3}$-recurrence.

Indeed, suppose the **a**-walk is C-recurrent; formally, if we define the events

$$E = \{S(n) \in [-C, C] \text{ for infinitely many } n\},$$

$$\tilde{E} = \{S(n) \in [-2C/3, 2C/3] \text{ for infinitely many } n\},$$

then $\mathbb{P}(E) > 0$. We want to show that $\mathbb{P}(\tilde{E}) > 0$ as well.

Let n_1 be so large that $|a_i - a_{i-1}| < C/6$ for all $i \geq n_1$. Define the sequence n_k, $k \geq 2$, by setting

$$n_k = \min\{i \geq n_{k-1} + 1 \colon\ a_i \geq a_{n_{k-1}} + C/6\}$$

(which is well defined since $a_i \to \infty$), then trivially

$$\frac{C}{6} \leq a_{n_{k+1}} - a_{n_k} \leq \frac{C}{3} \quad \text{for each } k = 1, 2, \ldots. \tag{8.1}$$

Fix a positive integer K and for $y = (y_1, y_2, \ldots, y_K) \in \Omega_K := \{-1, +1\}^K$ define

$$\bar{X}_K = \{X_1, X_2, \ldots, X_K\},$$

$$s_y = a_1 y_1 + a_2 y_2 + \cdots + a_k y_K.$$

Let $y \in \Omega_K$ be such that $\mathbb{P}(\{\bar{X}_K = y\} \cap E) > 0$. Observe that

$$\{\bar{X}_K = y\} \cap E = \{\bar{X}_K = y\} \cap B_K(s_y),$$

where

$$B_K^+(u) = \left\{ \text{there exist } m_1 < m_2 < \cdots \text{ such that } u \right.$$

$$\left. + \sum_{i=K+1}^{m_j} a_i X_i \in [0, C] \right\},$$

$$B_K^-(u) = \left\{ \text{there exist } m_1' < m_2' < \cdots \text{ such that } u \right.$$

$$\left. + \sum_{i=K+1}^{m_j'} a_i X_i \in [-C, 0] \right\},$$

$$B_K(u) = B_K(u)^+ \cup B_K(u)^-.$$

Since $\{\bar{X}_K = y\}$ and $B_K(u)$ are independent, we have

$$\mathbb{P}(\{\bar{X}_K = y\} \cap B_K(s_y)) = \mathbb{P}(\bar{X}_K = y)\,\mathbb{P}(B_K(s_y)).$$

Consequently, $\mathbb{P}(B_K(s_y)) > 0$, and as a result, $\mathbb{P}(B_K^+(s_y)) > 0$ or $\mathbb{P}(B_K^-(s_y)) > 0$ (or both).

Let $\Omega_K^* \subseteq \Omega_K$ contain those ys for which there is an index k such that $n_{k+2} \leq K$ and $y_{n_k} = -1$, $y_{n_{k+1}} = +1$, $y_{n_{k+2}} = -1$; let k be the smallest such index. For $y \in \Omega_K^*$, define the mappings $\sigma^+, \sigma^- : \Omega_K^* \to \Omega_K$ by

$$\sigma^+(y) = \begin{cases} -y_i, & \text{if } i = n_k \text{ or } i = n_{k+1}; \\ y_i, & \text{otherwise}; \end{cases}$$

$$\sigma^-(y) = \begin{cases} -y_i, & \text{if } i = n_{k+1} \text{ or } i = n_{k+2}; \\ y_i, & \text{otherwise}. \end{cases}$$

Then for $y \in \Omega_K^*$

$$s_{\sigma^+(y)} = s_y + 2a_{n_k} - 2a_{n_k+1} \in [s_y - 2C/3, s_y - C/3],$$
$$s_{\sigma^-(y)} = s_y - 2a_{n_k} + 2a_{n_k+1} \in [s_y + C/3, s_y + 2C/3].$$

As a result, it is not hard to see that

$$\{\bar{X}_K = \sigma^+(y)\} \cap B_K^+(s_y) \subseteq \tilde{E},$$
$$\{\bar{X}_K = \sigma^-(y)\} \cap B_K^-(s_y) \subseteq \tilde{E},$$

where

$$\tilde{E} = \left\{ \sum_{i=1}^m a_i X_i \in \left[-\frac{2C}{3}, \frac{2C}{3} \right] \text{ for infinitely many } ms \right\}.$$

Since at least one of $B_K^+(s_y)$ and $B_K^-(s_y)$ has a positive probability, $\mathbb{P}(\bar{X}_K = \sigma^\pm(y)) = 2^{-K}$ and the events on the left-hand side are independent, we conclude that $\mathbb{P}(\tilde{E}) > 0$.

Now, it only remains to show that there exists $y \in \Omega_K^*$ such that $\mathbb{P}(\{\bar{X}_K = y\} \cap E) > 0$. Let $\kappa := \kappa(K) = \max\{k \in \mathbb{Z}_+ : n_k \leq K\}$; obviously, $\kappa(K) \to \infty$ as $K \to \infty$. If we choose y from Ω_K uniformly, we can trivially bound the probability that $y \notin \Omega_K^*$ by[1]

$$\left(1 - \frac{1}{8}\right)^{\lfloor \kappa/3 \rfloor} \to 0 \quad \text{as } \kappa \to \infty$$

---

[1]Exact: See the sequence A005251 in the Online Encyclopedia of Integer Sequences (https://oeis.org/A005251), $\mathbb{P}(\bar{X}_K \notin \Omega_K^*) \approx \lambda^\kappa$, $\lambda = \frac{\sqrt[3]{100+12\sqrt{69}}}{6} + \frac{2}{3\sqrt[3]{100+12\sqrt{69}}} + \frac{2}{3} = 0.877...$, $\kappa = \kappa(K)$.

by grouping together triples $(X_{n_1}, X_{n_2}, X_{n_3})$, $(X_{n_4}, X_{n_5}, X_{n_6})$, etc.; in each such triple,

$$\mathbb{P}\left((X_{n_k}, X_{n_{k+1}}, X_{n_{k+2}}) = (-1, +1, -1)\right) = 1/8.$$

Hence,

$$\mathbb{P}(E) = \sum_{y \in \Omega_K} \mathbb{P}(\{\bar{X}_K = y\} \cap E) = \sum_{y \in \Omega_K^*} \mathbb{P}(\{\bar{X}_K = y\} \cap E) \quad (8.2)$$

$$+ \mathbb{P}\left(\{\bar{X}_K \in \Omega_K \setminus \Omega_K^*\} \cap E\right).$$

Since $\mathbb{P}\left(\{\bar{X}_K \in \Omega_K \setminus \Omega_K^*\} \cap E\right) \leq \mathbb{P}\left(\bar{X}_K \in \Omega_K \setminus \Omega_K^*\right)$, by making K sufficiently large, we can ensure that the second term on the right-hand side of (8.2) is less than $\mathbb{P}(E)$, implying that there exist some $y \in \Omega_K^*$ such that $\mathbb{P}(\{\bar{X}_K = y\} \cap E) > 0$, as required. $\qquad \square$

Our next result shows that if the sequence **a** is non-decreasing, then the walk will always "jump" over 0 infinitely many times, even if the walk is not C-recurrent.

Theorem 8.2. *Suppose that a_i is a non-decreasing positive sequence. Then the event $\{S(n) > 0\}$ holds for infinitely many n a.s. The same is true for the event $\{S(n) < 0\}$.*

The theorem immediately follows from symmetry and the following more general:

Proposition 8.1. *Suppose that a_i is a non-decreasing sequence, m is an integer such that $a_{m+1} > 0$ and $S(m) = A > 0$. Define*

$$\tau = \inf\{k \geq 0: \ S(m + k) \leq 0\}.$$

Let $Y_j \sim$ Rademacher be i.i.d., and

$$\tilde{\tau} = \inf\{k \geq 0: \ Y_1 + Y_2 + \cdots + Y_k \leq -r\},$$

where $r = \lceil A/a_{m+1} \rceil$; note that $\tilde{\tau} < \infty$ a.s. and that, in fact, $Y_1 + \cdots + Y_{\tilde{\tau}} = -r$. Then τ is stochastically smaller than $\tilde{\tau}$, that is,

$$\mathbb{P}(\tau > m) \leq \mathbb{P}(\tilde{\tau} > m), \quad m = 0, 1, 2, \ldots.$$

Proof. We use coupling. Indeed, we can write

$$S(m+j) = A + a_{m+1}Y_1 + a_{m+2}Y_2 + \cdots + a_{m+j}Y_j, \quad j = 1, 2, \ldots.$$

Suppose that $\tilde{\tau} = k$, that is,

$$Y_1 > -r, \; Y_1 + Y_2 > -r, \; \ldots, \; Y_1 + Y_2 + \cdots + Y_{k-1} > -r,$$

$$Y_1 + Y_2 + \cdots + Y_{k-1} + Y_k = -r.$$

Then, recalling that a_i is a non-decreasing sequence,

$$\begin{aligned}
S(m+k) &= A + a_{m+1}Y_1 + \cdots + a_{m+k-1}Y_{k-1} + a_{m+k}Y_k \\
&\leq A + a_{m+1}Y_1 + \cdots + a_{m+k-2}Y_{k-2} + a_{m+k-1}Y_{k-1} \\
&\quad + a_{m+k-1}Y_k (\text{since } Y_k = -1) \\
&= A + a_{m+1}Y_1 + \cdots + a_{m+k-2}Y_{k-2} + a_{m+k-1}[Y_{k-1} + Y_k] \\
&\leq A + a_{m+1}Y_1 + \cdots + a_{m+k-2}Y_{k-2} + a_{m+k-2}[Y_{k-1} + Y_k] \\
&= A + a_{m+1}Y_1 + \cdots + a_{m+k-2}[Y_{k-2} + Y_{k-1} + Y_k] \\
&\leq \cdots \leq A + a_{m+1}[Y_1 + \cdots + Y_k] = A - r a_{m+1} \leq 0,
\end{aligned}$$

since $Y_k, Y_{k-1}+Y_k, Y_{k-2}+Y_{k-1}+Y_k, \ldots, Y_1+\cdots+Y_k$ are all negative. Therefore, $\tau \leq \tilde{\tau}$. $\qquad\square$

We will need a version of the Azuma–Hoeffding inequality; compare with the results of Montgomery-Smith (1990).

Lemma 8.1 (Azuma–Hoeffding inequality). *Suppose that $b_1, b_2, \ldots, b_m$ is a sequence of non-negative numbers and $S = b_1 Y_1 + b_2 Y_2 + \cdots + b_m Y_m$, where $Y_j \sim$ Rademacher are i.i.d. Then*

$$\mathbb{P}\left(|S| \geq A\right) \leq 2 \exp\left(-\frac{A^2}{2(b_1^2 + \cdots + b_m^2)}\right) \quad \text{for all } A > 0. \qquad (8.3)$$

We also state the following fairly standard result.

Lemma 8.2 (CLT for SRW). *Let $T_i = Y_1 + \cdots + Y_i$ be a simple random walk. Suppose that L_k and y_k, $k = 1, 2, \ldots$, are two sequences such that $L_k \to \infty$, $y_k \to \infty$ and $y_k/\sqrt{L_k} \to r > 0$ as $k \to \infty$. Then*

$$\lim_{k \to \infty} \mathbb{P}\left(\max_{1 \leq i \leq L_k} T_i \geq y_k\right) = 2\,\mathbb{P}(\eta \geq r) = 2 - 2\,\Phi(r),$$

where $\eta \sim \mathcal{N}(0, 1)$ and $\Phi(\cdot)$ is its CDF.

Proof. Let $\tilde{y}_k = \lceil y_k \rceil \in \mathbb{Z}_+$. By the reflection principle,

$$\mathbb{P}\left(\max_{1 \le i \le L_k} T_i \ge y_k\right)$$

$$= \mathbb{P}\left(\max_{1 \le i \le L_k} T_i \ge \tilde{y}_k\right) = 2\,\mathbb{P}(T_{L_k} \ge \tilde{y}_k) - \mathbb{P}(T_{L_k} = \tilde{y}_k)$$

$$= 2\,\mathbb{P}\left(\frac{T_{L_k}}{\sqrt{L_k}} \ge \frac{\tilde{y}_k}{\sqrt{L_k}}\right) + O\left(\frac{1}{\sqrt{L_k}}\right) \to 2\mathbb{P}(\eta \ge r),$$

where we used the central limit theorem, along with the fact that $\tilde{y}_k/y_k \to 1$ as $k \to \infty$. $\qquad\square$

8.2 Integer-Valued A-Walks

Suppose that the sequence **a** contains only integers.

Proposition 8.2. *Let $z \in \mathbb{Z}$. Assume that the sequence*

$$\int_0^\pi \cos(tz) \prod_{k=1}^n \cos(ta_k)\,dt, \quad n = 1, 2, \ldots$$

is summable. Then the events $\{S(n) = z\}$ occur for finitely many n a.s.

Proof. The result follows from standard Fourier analysis. Indeed,

$$\mathbb{E}e^{itS(n)} = \sum_{k \in \mathbb{Z}} e^{itk}\mathbb{P}(S(n) = k),$$

where the sum above goes, in fact, effectively over a finite number of ks (as $|S(n)| \le a_1 + \cdots + a_n$). At the same time,

$$\int_{-\pi}^\pi e^{it(k-z)}\,dt = \begin{cases} 2\pi, & \text{if } k = z; \\ 0, & \text{if } k \in \mathbb{Z} \setminus \{z\}. \end{cases}$$

By changing the order of summation and integration, we obtain

$$\frac{1}{2\pi}\int_{-\pi}^\pi \mathbb{E}e^{it(S(n)-z)}\,dt$$

$$= \frac{1}{2\pi}\sum_{k \in \mathbb{Z}}\int_{-\pi}^\pi e^{it(k-z)}\,\mathbb{P}(S(n) = k)\,dt = \mathbb{P}(S(n) = z).$$

On the other hand,

$$\frac{1}{2\pi}\int_{-\pi}^{\pi}\mathbb{E}e^{it(S(n)-z)}\,\mathrm{d}t = \frac{1}{2\pi}\int_{-\pi}^{\pi}e^{-itz}\prod_{k=1}^{n}\mathbb{E}e^{ita_kX_k}\mathrm{d}t$$

$$= \frac{1}{\pi}\int_{0}^{\pi}\cos(tz)\prod_{k=1}^{n}\cos(ta_k)\,\mathrm{d}t$$

by the symmetry of cos and the fact that the imaginary part must equal zero. Now, the result follows from the Borel–Cantelli lemma, since $\sum_{n}\mathbb{P}(S(n)=z)<\infty$. $\square$

Corollary 8.1 (Condition for transience). *Assume that the sequence*

$$\int_{0}^{\pi}\left|\prod_{k=1}^{n}\cos(ta_k)\right|\mathrm{d}t, \quad n=1,2,\dots$$

*is summable. Then the **a**-walk is transient a.s.*

Proof. From Proposition 8.2, we know that for each $z \in \mathbb{Z}$

$$\pi\mathbb{P}(S(n)=z) = \int_{0}^{\pi}\cos(tz)\prod_{k=1}^{n}\cos(ta_k)\,\mathrm{d}t$$

$$\leq \int_{0}^{\pi}\left|\cos(tz)\prod_{k=1}^{n}\cos(ta_k)\right|\mathrm{d}t \leq \int_{0}^{\pi}\left|\prod_{k=1}^{n}\cos(ta_k)\right|\mathrm{d}t.$$

Hence, the event $\{S(n)=z\}$ occurs finitely often a.s. for each z. Since for each $C>0$ there are only finitely many integers in $[-C,C]$, we conclude that the walk is not C-recurrent a.s. for every C. $\square$

An interesting and quite natural example is when $\mathbf{a}=(1,2,3,\dots)$, i.e., $a_i=i$. It was previously published in the IMS Bulletin, in the Student Puzzle Corner no. 37.

Theorem 8.3. *The **a**-walk with $\mathbf{a}=(1,2,3,\dots)$ is a.s. transient.*

This statement follows from a much stronger Theorem 8.4, but for the sake of completeness, we present its self-contained proof.

Proof of Theorem 8.3. Let $A_n = \{S(n) = 0\} = \{X_1 + 2X_2 + \cdots + nX_n = 0\}$. Then $\mathbb{P}(A_n) = Q_n/2^n$, where Q_n denotes the number of ways to put $\pm$ in the sequence $*1 * 2 * 3 * \cdots * n$ such that the sum equals 0. For example, $Q_1 = Q_2 = 0$, $Q_3 = Q_4 = 2$, $Q_5 = Q_6 = 0$, $Q_7 = 8$ and $Q_8 = 14$. It was essentially shown in Sullivan (2013) that

$$Q_n \sim \sqrt{\frac{6}{\pi}} \frac{2^n}{n^{3/2}} \qquad \text{when } n \bmod 4 \in \{0, 3\}$$

(and zero otherwise) as $n \to \infty$, meaning that the ratio of the right-hand side and the left-hand side converges to one. Consequently, $\sum_n \mathbb{P}(A_n) \sim \sum_n \frac{\text{const}}{n^{3/2}} < \infty$ and the events A_n occur a.s. finitely often by the Borel–Cantelli lemma. Hence, the walk is a.s. not recurrent.

Moreover, since for any $m \in \mathbb{Z}$

$$\mathbb{P}(S(n + 2|m|) = S(n) - m \mid \mathcal{F}_n) \geq \frac{1}{2^{2m}}$$

(by making the signs of $X_{n+1}, X_{n+2}, \ldots, X_{n+2|m|}$ alternate), we conclude that if the event $\{S(n) = m\}$ occurs infinitely often, then A_n shall also occur infinitely often a.s., leading to contradiction. As a result, $\mathbb{P}(\{S(n) = m\} \text{ i.o.}) = 0$ for all integer ms, and thus the walk is a.s. not C-recurrent for any non-negative C. □

Remark 8.1. Though the $(1, 2, 3, \ldots)$-walk is transient, it still can jump over zero infinitely many times, as it was shown in Theorem 8.2.

In fact, Theorem 8.3 can be generalized greatly, using the result from Sárközi and Szemerédi (1965), or even a weaker result of Erdős (1965), which provide the estimates for the maximum number of solutions of the equation $\sum_{i=1}^{n} \epsilon_i a_i = t$, where $\epsilon_i \in \{0, 1\}$, while a_i's and t are all integers.

Theorem 8.4 (Transience for distinct a_i's). *Let $\mathbf{a}$ be such that all a_i's are distinct integers. Then $\mathbf{a}$-walk is a.s. transient.*

Proof. The main result of Sárközi and Szemerédi (1965) implies that for any $\epsilon > 0$

$$\text{card}(\{(x_1, x_2, \ldots, x_n) : \text{ all } x_i = \pm 1, \ a_1 x_1 + \cdots + a_n x_n = m\})$$

$$\leq \frac{(1 + \epsilon)2^{n+3}}{n^{3/2}\sqrt{\pi}}$$

for all $n \geq n_0(\epsilon)$ and all m. Setting $\epsilon = 1$, and fixing $m \in \mathbb{Z}$, we obtain that

$$\sum_{n=n_0(1)}^{\infty} \mathbb{P}(S(n) = m) \leq \sum_{n=n_0(1)}^{\infty} \frac{2 \cdot 2^{n+3}}{n^{3/2}\sqrt{\pi}} \times \frac{1}{2^n} = \frac{16}{\sqrt{\pi}} \sum_{n=n_0(1)}^{\infty} \frac{1}{n^{3/2}} < \infty.$$

Therefore, by the Borel–Cantelli lemma, only finitely many events $\{S(n) = m\}$ occur a.s. Since $S(n)$ takes only integer values, this implies that $\{|S(n)| \leq C\}$ happens finitely often a.s. for any $C > 0$.

$\square$

Remark 8.2.

(a) It is not difficult to see that under the condition of Theorem 8.4 it suffices that all a_k's are distinct only starting from some $k_0 \geq 1$.

(b) If $a_k = \lfloor k^\beta \rfloor$ with $\beta \geq 1$, then we immediately have a.s. transience by Theorem 8.4.

(c) In the proof of Theorem 8.4, we use the result of A. Sárközy and E. Szemerédi from 1965. The constant in their bound can, in fact, be replaced by the constant $\sqrt{6/\pi}$ from Sullivan's result. Even though the value of the constant does not matter for out proof, it is worth mentioning the (1980) result of Stanley (1980) that the set $\{1, 2, 3, \ldots, n\}$ is extremal among sets of n distinct integers for maximizing the maximum concentration probability of its Rademacher sum. This fact was proved by Stanley using some high-powered algebraic geometry but was then proved again soon afterward in a simpler way using Lie algebras by Proctor (1982).

8.3   A Non-Trivial Recurrent Example

We assume here that $\mathbf{a} = (B_1, B_2, B_3, \ldots)$, where each B_k is a consecutive block of k's of length precisely $L_k \geq 1$. Denote also by $i_k = 1 + L_1 + L_2 + \cdots + L_{k-1}$ the index of the first element of the k-th block. For example, if $L_k = 2^k$, then $i_k = 2^k - 1$ and

$$\mathbf{a} = \left(1, 1, \underbrace{2, 2, 2, 2}_{L_2 \text{ times}}, \underbrace{3, 3, 3, 3, 3, 3, 3, 3}_{L_3 \text{ times}}, \underbrace{4, \ldots, 4}_{L_4 \text{ times}}, \ldots \right),$$

one can also note that $a_i = \lfloor \log_2(i+1) \rfloor = \lceil \log_2(i+2) \rceil - 1$.

Theorem 8.5. *Suppose that for some $\varepsilon > 0$, $r > 0$, and k_0 we have*

$$\frac{L_k}{L_1 + L_2 + \cdots + L_{k'}} \geq (2 + \varepsilon) \ln k,$$

$$\frac{L_k}{L_{k'+1} + L_{k'+2} + \cdots + L_{k-1}} \geq 2r, \qquad (8.4)$$

$$L_k \geq k^4,$$

*whenever $k - k' \geq \frac{k}{\ln k} - 2$ and $k, k' \geq k_0$. Then the **a**-walk described above is a.s. recurrent.*

Remark 8.3. One can easily check that the conditions of the theorem are satisfied if $a_k = \lfloor (\log_\gamma k)^\beta \rfloor$, where $\gamma > 1$ and $\beta \in (0, 1]$.

Proof of Theorem 8.5. We proceed in five steps.

Step 1: Preliminaries

We first need the following lemma, which is probably known, but we could not locate it in the literature.

Lemma 8.3. *Let $m \in \mathbb{Z}_+$ and T_m be a simple symmetric random walk on $\mathbb{Z}^1$, that is, $T_m = Y_1 + \cdots + Y_m$, where $Y_i \sim$ Rademacher are i.i.d. There exists a universal constant $c_1 > 0$ such that for all integers z such that $|z| \leq 2\sqrt{m}$, assuming that m is sufficiently large and $m + z$ is even,*

$$\mathbb{P}(T_m = z) \geq \frac{c_1}{\sqrt{m}}.$$

Proof. W.l.o.g. assume $z \geq 0$. We have

$$\mathbb{P}(T_m = z) = \mathbb{P}\left(\frac{T_m + m}{2} = \frac{z + m}{2}\right) = \mathbb{P}(\tilde{T} = w),$$

where $\tilde{T} \sim \text{Bin}(m, 1/2)$ and $w = \frac{z+m}{2} \in \mathbb{Z}_+$. Note that $\tilde{m} \leq w \leq \tilde{m} + \sqrt{m}$, where $\tilde{m} = m/2$. So

$$\mathbb{P}(\tilde{T} = w) = \binom{m}{w} 2^{-m} = \binom{2\tilde{m}}{\tilde{m}} \frac{1}{2^{2\tilde{m}}} \frac{\tilde{m}! \, \tilde{m}!}{w!(m-w)!}$$

$$= \frac{1 + o(1)}{\sqrt{\pi \tilde{m}}} \frac{(2\tilde{m} - w + 1)(2\tilde{m} - w + 2) \cdots \tilde{m}}{(\tilde{m} + 1)(\tilde{m} + 2) \cdots w}$$

$$= \frac{1 + o(1)}{\sqrt{\pi \tilde{m}}} \left(1 - \frac{w - \tilde{m}}{\tilde{m} + 1}\right) \left(1 - \frac{w - \tilde{m}}{\tilde{m} + 2}\right) \cdots \left(1 - \frac{w - \tilde{m}}{w}\right)$$

$$\geq \frac{1 + o(1)}{\sqrt{\pi \tilde{m}}} \left(1 - \frac{\sqrt{m}}{\tilde{m} + 1}\right)^{w - \tilde{m}}$$

$$\geq \frac{1 + o(1)}{\sqrt{\pi \tilde{m}}} \left(1 - \frac{\sqrt{2} + o(1)}{\sqrt{\tilde{m}}}\right)^{\sqrt{2\tilde{m}}}$$

$$= \frac{e^{-2} + o(1)}{\sqrt{\pi m / 2}} \geq \frac{0.1}{\sqrt{m}}$$

for large enough m.   □

Corollary 8.2. *Let* T_m, $m = 0, 1, 2, \ldots$, *be as simple symmetric random walk as in Lemma 8.3. Assume that m and k are positive integers such that $k^2 \leq m$. Let $u \in \mathbb{Z}$, and either k is odd, or both k and $m - u$ are even. Then for large ks*

$$\mathbb{P}(T_m - u \bmod k = 0) \geq \frac{c_1}{2k},$$

where c_1 is the constant from Lemma 8.3.

Proof of Corollary 8.2. First, assume that m, and hence u, are both even. Since $(T_m - u) \bmod k = 0 \iff T_m = \tilde{u} \bmod k$, where $\tilde{u} = (u \bmod k) \in \{0, 1, 2, \ldots, k - 1\}$, it suffices to show the statement for $\tilde{u}$.

Let $M = \lfloor 2\sqrt{m} \rfloor \in (2\sqrt{m} - 1, 2\sqrt{m}]$ and define

$$\mathbb{I} = [-M, -M + 1, \ldots, -1, 0, 1, \ldots, M] = \mathbb{I}_0 \cup \mathbb{I}_1,$$

$$\mathbb{I}_0 = \{z \in \mathbb{I} : z \text{ is even}\}; \quad \mathbb{I}_1 = \{z \in \mathbb{I} : z \text{ is odd}\}.$$

There are at least M elements in each $\mathbb{I}_0$ and $\mathbb{I}_1$.

If k is odd, then each of these two sets contains at least $\lfloor \frac{M}{k} \rfloor$ elements z such that $z = \tilde{u} \bmod k$. If m is even (odd, resp.) for all z either in $\mathbb{I}_0$ (in $\mathbb{I}_1$, resp.) by Lemma 8.3 for large ks (and hence large m), we have $\mathbb{P}(T_m = z) \geq c_1/\sqrt{m}$. Consequently,

$$\mathbb{P}(T_m = \tilde{u} \bmod k)$$

$$\geq \sum_{z \in \mathbb{I},\, z=\tilde{u} \bmod k} \mathbb{P}(T_m = z) \geq \left\lfloor \frac{M}{k} \right\rfloor \frac{c_1}{\sqrt{m}} \geq \left(\frac{M}{k} - 1 \right) \frac{c_1}{\sqrt{m}}$$

$$\geq \left(\frac{2\sqrt{m} - 1}{k} - 1 \right) \frac{c_1}{\sqrt{m}} \geq \left(1 - \frac{1}{k} \right) \frac{c_1}{\sqrt{m}} = \frac{c_1}{k} - O(k^{-2})$$

since $m \geq k^2$.

If k is even, then if m is even (and thus a is also even), then $\mathbb{I}_0$ contains at least $\lfloor \frac{M}{k} \rfloor$ elements z such that $z = \tilde{u} \bmod k$ and at the same time Lemma 8.3 is applicable for $z \in \mathbb{I}_0$. On the other hand, if m (and so u) is odd, then $\mathbb{I}_1$ contains at least $\lfloor \frac{M}{k} \rfloor$ elements z such that $z = \tilde{u} \bmod k$ and Lemma 8.3 is applicable for $z \in \mathbb{I}_1$. The rest of the proof is the same as for the case when k is odd. □

Step 2: Splitting $S(n)$

Recall that i_k denotes the first index of block k and note that the sum of all the steps within block k can be represented as

$$S(i_{k+1} - 1) - S(i_k - 1) = k \cdot T_k, \quad T_k = X_1^{(k)} + \cdots + X_{L_k}^{(k)},$$

where the $X_j^{(k)}$ are i.i.d. random variables and $X_j^{(k)} \sim$ Rademacher. For $m = 2, \ldots$, let

$$k_m = \begin{cases} \lfloor m \ln m \rfloor & \text{if } \lfloor m \ln m \rfloor \text{ is odd;} \\ \lfloor m \ln m \rfloor + 1 & \text{if } \lfloor m \ln m \rfloor \text{ is even.} \end{cases} \tag{8.5}$$

Thus, k_m is *always* odd; k_m, $m = 2, 3, \ldots$ equal $1, 3, 5, 7, 9, 13, 15, 19, 23$, etc. Define also

$$A_m = \{S(j) = 0 \text{ for some } i_{k_m} \leq j < i_{k_m+1}\},$$

the event that $S(j)$ hits zero for the steps within block B_{k_m}, and the sequence of sigma algebras

$$\mathcal{G}_m = \mathcal{F}_{i_{k_m+1}-1} = \sigma \left(\bigcup_{\ell=1}^{k_m} \sigma \left(X_1^{(\ell)}, X_2^{(\ell)}, \ldots, X_{L_\ell}^{(\ell)} \right) \right).$$

Intuitively, $\mathcal{G}_m$ contains all the information about the walk during its steps corresponding to the first k_m blocks.

To simplify notations, let us now write $k = k_m$ and $k' = k_{m-1}$ and observe that

$$k - k' = k_m - k_{m-1} \geq m \ln m - (m - 1) \ln(m - 1) - 2$$

$$= \ln m - 1 + O\left(\frac{1}{m}\right) \geq \ln m - 2 \qquad (8.6)$$

for large m.

Let us split $S(j)$, where $j \in [i_k, i_{k+1})$, as follows:

$$S(j) = S(i_{k'}) + \sum_{n=k'+1}^{k-1} \left(X_1^{(n)} + \cdots + X_{L_n}^{(n)}\right)$$

$$+ k \cdot \left(X_1^{(k)} + \cdots + X_{j-i_k}^{(k)}\right)$$

$$= S(i_{k'}) + \left[\sum_{n=k'+1}^{k-2} \left(X_1^{(n)} + \cdots + X_{L_n}^{(n)}\right)\right.$$

$$\left. + (k - 1) \sum_{\ell=1}^{i_k - 2k^2 - 1} X_\ell^{(k-1)}\right]$$

$$+ (k - 1) \cdot \Sigma_3 + k \cdot \left(X_1^{(k)} + \cdots + X_{j-i_k}^{(k)}\right)$$

$$= \Sigma_1 + \Sigma_2 + (k - 1) \cdot \Sigma_3 + k \cdot \Sigma_4,$$

where $\Sigma_1 = S(i_{k'})$ and

$$\Sigma_2 = \sum_{n=k'+1}^{k-2} n T_n + (k - 1) T'_{k-1}, \quad T'_{k-1} = \sum_{\ell=i_{k-1}}^{i_k - 1 - 2k^2} X_\ell^{(k-1)};$$

$$\Sigma_3 = X_{i_k - 2k^2}^{(k-1)} + X_{i_k - 2k^2 + 1}^{(k-1)} + \cdots + X_{i_k - 2}^{(k-1)} + X_{i_k - 1}^{(k-1)};$$

$$\Sigma_4 = X_1^{(k)} + X_2^{(k)} + \cdots + X_{j-i_k}^{(k)}.$$

Note that Σ_i, $i = 1, 2, 3, 4$ are independent, and Σ_3 has precisely $2k^2$ terms.

Step 3: Estimating Σ_1

Recall that $k = k_m$, $k' = k_{m-1}$ and let

$$E_{m-1} = \left\{ |\Sigma_1| < k\sqrt{L_k} \right\} \in \mathcal{G}_{m-1}.$$

By Lemmas 8.1 and (8.4), assuming k is large,

$$\mathbb{P}(E_{m-1}^c) \leq \mathbb{P}(|S(k')| \geq k'\sqrt{L_k})$$

$$\leq 2 \exp\left(-\frac{k'^2 \cdot L_k}{2(L_1 + 2^2 \cdot L_2 + 3^2 \cdot L_3 + \cdots + k'^2 \cdot L_{k'})} \right)$$

$$\leq 2 \exp\left(-\frac{L_k}{2(L_1 + L_2 + L_3 + \cdots + L_{k'})} \right)$$

$$\leq 2 \exp\left(-(1 + \varepsilon/2) \ln k \right) = \frac{2}{k^{1+\varepsilon/2}} =: \varepsilon_m. \tag{8.7}$$

Step 4: Estimating Σ_2

Again, by Lemmas 8.1 and (8.4), assuming that k is sufficiently large,

$$\mathbb{P}\left(|\Sigma_2| \geq k\sqrt{\frac{L_k}{r}} \right)$$

$$\leq 2 \exp\left(-\frac{k^2 r^{-1} L_k}{\begin{array}{l} 2[(k'+1)^2 L_{k'+1} + \cdots \\ + (k-2)^2 L_{k-2} + (k-1)^2 (L_{k-1} - 2k^2)] \end{array}} \right)$$

$$\leq 2 \exp\left(-\frac{r^{-1} L_k}{2[L_{k'+1} + \cdots + L_{k-1} - 2k^2]} \right)$$

$$\leq 2 \exp(-1) = 0.7357588824\ldots.$$

Consequently,

$$\mathbb{P}\left(|\Sigma_2| < k\sqrt{L_k/r} \right) \geq 0.2 \quad \text{for large } k. \tag{8.8}$$

Step 5: Finishing the proof

We have a trivial lower bound

$$\mathbb{P}\left(A_m \mid E_{m-1}, \mathcal{G}_{m-1}\right) \geq \mathbb{P}\left(A_m \mid |\Sigma_2| < k\sqrt{\frac{L_k}{r}}, E_{m-1}, \mathcal{G}_{m-1}\right)$$

$$\times \mathbb{P}\left(|\Sigma_2| < k\sqrt{\frac{L_k}{r}} \mid E_{m-1}, \mathcal{G}_{m-1}\right)$$

$$=: (*) \times 0.2 \quad \text{for large } k \tag{8.9}$$

by (8.8), since the second multiplier equals $\mathbb{P}\left(|\Sigma_2| < k\sqrt{L_k/r}\right)$ by independence.

Let

$$\text{Div}_k = \{\Sigma_1 + \Sigma_2 + (k-1)\Sigma_3 = 0 \bmod k\}$$

$$= \{\Sigma_1 + \Sigma_2 - \Sigma_3 = 0 \bmod k\}.$$

Since only on the event Div_k, it is possible that $S(j) = 0$ for some j (since the step sizes are $\pm k$ in the block B_k), we conclude that for large k

$$(*) = \mathbb{P}\left(A_m \mid \text{Div}_k, |\Sigma_2| < k\sqrt{\frac{L_k}{r}}, E_{m-1}, \mathcal{G}_{m-1}\right)$$

$$\times \mathbb{P}\left(\text{Div}_k \mid |\Sigma_2| < k\sqrt{\frac{L_k}{r}}, E_{m-1}, \mathcal{G}_{m-1}\right)$$

$$\geq \mathbb{P}\left(A_m \mid \text{Div}_k, |\Sigma_2| < k\sqrt{\frac{L_k}{r}}, E_{m-1}, \mathcal{G}_{m-1}\right) \times \frac{c_1}{2k} \tag{8.10}$$

due to the fact that by Corollary 8.2, $\mathbb{P}\left(\text{Div}_k \mid \mathcal{F}_{i_k - 2k^2 - 1}\right) \geq c_1/(2k)$.
On the other hand,

$$\mathbb{P}\left(A_m \mid \text{Div}_k, |\Sigma_2| < k\sqrt{\frac{L_k}{r}}, E_{m-1}, \mathcal{G}_{m-1}\right)$$

$$\geq \min_{z \in Z_k} \mathbb{P}(z + T_m = 0 \text{ for some } m \in [0, L_k])$$

$$\geq \beta := 1 - \Phi\left(r^{-1/2} + 3\right) > 0, \tag{8.11}$$

where $z + T_m$ is a simple random walk starting at z (see Lemma 8.3), and

$$Z_k = \left\{ z \in \mathbb{Z} : |z| \leq (r^{-1/2} + 3) \sqrt{L_k} \right\}.$$

Indeed, using the last part of (8.4), and the conditions we imposed, we have

$$|\Sigma_1 + \Sigma_2 + (k-1)\Sigma_3| \leq k\sqrt{L_k} + k\sqrt{L_k/r} + 2(k-1)k^2$$
$$< (1 + r^{-1/2} + 2)k\sqrt{L_k}$$

for large k, $S(j) = [\Sigma_1 + \Sigma_2 + (k-1)\Sigma_3] + k \cdot \Sigma_4$, and by Lemma 8.2,

$$\liminf_{k \to \infty} \min_{z \in Z_k} \mathbb{P}(z + T_m = 0 \text{ for some } m \in [0, L_k])$$

$$\geq 2\,\mathbb{P}(\eta > r^{-1/2} + 3) = 2\beta,$$

so the minimum in (8.11) is $\geq \beta$ for all sufficiently large k.

Finally, from (8.9), (8.10) and (8.11), we get that

$$\sum_m \mathbb{P}\left(A_m \mid E_{m-1}, \mathcal{G}_{m-1}\right) \geq \sum_m \frac{0.2\,c_1\beta}{2m \log m} = +\infty$$

and $\mathbb{P}(E_m^c)$ is summable by (8.7), so we can apply Lemma 1.1 to conclude that events A_m occur infinitely often and thus our **a**-walk is recurrent. □

8.4 Continuous Example

The example of **a**-walk described in Theorem 8.5 roughly corresponds to the case $a_k = \lceil \log_\gamma k \rceil$, $k = 1, 2, \ldots$. But what if a_k's take non-integer values but, for example, equal

$$a_k = \log_\gamma k \equiv c \ln k, \quad k = 1, 2, \ldots,$$

where $\gamma = e^{1/c} > 1$? In this section, we study this example. It is unreasonable to assume that such **a**-walk is recurrent because of the irrationality of the step sizes, however, we might want to investigate if this walk is C-recurrent for *some* $C > 0$. Our main result is as follows.

Theorem 8.6 (Logarithmic growth). *Let $c > 0$ and $a_k = c \ln k$. Then the **a**-walk is a.s. C-recurrent for every $C > 0$.*

To prove this theorem, it is sufficient to show that whatever the value $c > 0$ is, $\{|S(n)| \leq 3\}$ happens i.o. almost surely. Indeed, take any $C > 0$. Then the statement that $\mathbf{a}'$-walk with $a'_k = \frac{3c}{C} \ln k$, $k = 1, 2, \ldots$, is 3-recurrent is equivalent to the statement that $\mathbf{a}$-walk with $a_k = c \ln k$ is C-recurrent.

The proof will proceed similarly to that of Theorem 8.5. Let us define k_m slightly differently from (8.5); namely, let

$$
k_m = \begin{cases} \lfloor m \ln m \rfloor & \text{if } \lfloor m \ln m \rfloor \text{ is even;} \\ \lfloor m \ln m \rfloor - 1 & \text{if } \lfloor m \ln m \rfloor \text{ is odd.} \end{cases}
$$

Thus, now k_m are always *even*. As before, set $k = k_m$, and $k' = k_{m-1}$, and define

$$
i_k = \lceil \gamma^k \rceil = \max\{i \geq 1 : a_i < k\} + 1
$$
$$
= \min\{i \geq 1 : a_i \geq k\} \in [\gamma^k, \gamma^k + 1),
$$

i.e., the first index when a_i starts exceeding k. For $i \in J_k := [i_k, i_{k+1})$, write

$$
S(i) = S(i_{k'} - 1) + [S(i_k - 1) - S(i_{k'} - 1) - \Sigma_3] + \Sigma_3 + [S(i) - S(i_k - 1)]
$$
$$
= \Sigma_1 \qquad\qquad +\Sigma_2 \qquad\qquad +\Sigma_3 \qquad +\Sigma_4(i),
$$
$$
\tag{8.12}
$$

where

$$
\Sigma_3 = \left[S(i_{k-1} + k^2 - 1) - S(i_{k-1} - 1) \right] + \left[S(i_k) - S(i_k - k^2) \right].
$$

Note that Σ_i, $i = 1, 2, 3, 4$, are independent, and Σ_3 has $2 \cdot k^2$ terms and contains the first k^2 and the last k^2 steps of the walk, when the step sizes lie in $[k, k + 1)$.

Let

$$
E_{m-1} = \left\{ |\Sigma_1| < k\sqrt{i_k} \right\} = \left\{ S(i_{k_{m-1}}) < k_m \sqrt{i_{k_m}} \right\} \tag{8.13}
$$

By Lemma 8.1, since $a_i < k' < k$ for $i < i_{k'}$,

$$
\mathbb{P}(E^c_{m-1}) = \mathbb{P}\left(|\Sigma_1| \geq k\sqrt{i_k} \right)
$$
$$
\leq 2 \exp\left(-\frac{i_k k^2}{2 \sum_{j=1}^{i_{k'}} a_j^2} \right) \leq 2 \exp\left(-\frac{i_k}{2 i_{k'}} \right)
$$

$$\leq 2\exp\left(-\frac{\gamma^k - 1}{2\gamma^{k'}}\right) = 2\exp\left(-\frac{\gamma^{k-k'}(1 + o(1))}{2}\right)$$

$$= 2\exp\left(-\frac{\gamma^{\ln m - 2}}{2 + o(1)}\right) = 2\exp\left(-\frac{m^{\ln \gamma}}{2\gamma^2 + o(1)}\right) =: \varepsilon_{m-1}$$

$$(8.14)$$

using (8.6) for k sufficiently large.[2] Observe that ε_m is summable.
Similarly, by Lemma 8.1,

$$\mathbb{P}\left(|\Sigma_2| \geq 2k\sqrt{i_k}\right)$$

$$\leq 2\exp\left(-\frac{4k^2 i_k}{2k^2\left(i_k - i_{k'} - 2k^2\right)}\right) < 2\,e^{-2} = 0.27\ldots.$$

Hence,

$$\mathbb{P}(F_m) \geq 0.72, \quad \text{where } F_m = \left\{|\Sigma_2| < 2k\sqrt{i_k}\right\}. \tag{8.15}$$

Lemma 8.4. *Let $n = k^2$, where k is an even positive integer, and assume also that k is sufficiently large. Suppose that X_i, Y_i, $i = 1, 2, \ldots, n$, are i.i.d. Rademacher. Let*

$$T = (k-1)(X_1 + X_2 + \cdots + X_n) + k(Y_1 + Y_2 + \cdots + Y_n). \tag{8.16}$$

Then

$$\mathbb{P}(T = j) \geq \frac{c_1^2}{4n} \quad \text{for each } j = 0, \pm2, \pm4, \ldots, \pm n,$$

where c_1 is the constant from Lemma 8.3.

Proof. It follows from Corollary 8.2 that

$$\mathbb{P}(X_1 + \cdots + X_n = \ell) \geq \frac{c_1}{2k}, \quad \mathbb{P}(Y_1 + \cdots + Y_n = \ell) \geq \frac{c_1}{2k} \tag{8.17}$$

for all even ℓ such that $|\ell| \leq 2k$.

---

[2]Note that (8.6) was stated for k_m defined slightly differently, however, it holds here as well.

Let $j = 2\tilde{j} \in \{0, 2, 4, \ldots, n-2, n\}$. Consider the sequence of $k-1$ numbers

$$\tilde{j}, \; \tilde{j} - k, \; \tilde{j} - 2k, \; \tilde{j} - 3k, \ldots, \tilde{j} - (k-2)k,$$

they all give different remainders when divided by $k-1$. Hence, there must be an $m \in \{0, 1, \ldots, k-2\}$ such that $\tilde{j} - mk = b(k-1)$ and b is an integer; moreover, since $0 \le \tilde{j} \le n/2$, we have $b \in \left[-\frac{k(k-2)}{k-1}, \frac{n}{2(k-1)}\right]$. For such m and b, we have $j = 2\tilde{j} = (2m)k + (2b)(k-1)$, and, since both $|2m|$ and $|2b| \le 2k$,

$$\mathbb{P}(T = j) \ge \mathbb{P}(X_1 + \cdots + X_n = 2b) \cdot \mathbb{P}(Y_1 + \cdots + Y_n = 2m) \ge \left(\frac{c_1}{2k}\right)^2 = \frac{c_1^2}{4n}$$

by (8.17). The result for negative j follows by symmetry. □

Corollary 8.3. *Let* $\varepsilon = \frac{2ck^4}{\gamma^{k-1}}$. *Then for large even* k,

$$\mathbb{P}\left(\Sigma_3 \in [j - \varepsilon, j + \varepsilon]\right) \ge \frac{c_1^2}{4k^2} \quad \text{for each } j = 0, \pm 2, \pm 4, \ldots, \pm k^2.$$

Proof. Σ_3 has the same distribution as

$$\sum_{\ell=1}^{k^2} c \ln(i_{k-1} - 1 + \ell) X_\ell + \sum_{\ell=1}^{k^2} c \ln(i_k - \ell) Y_\ell$$

for some i.i.d. $X_\ell, Y_\ell \sim$ Rademacher. At the same time, for $\ell \ge 1$,

$$|c \ln(i_{k-1} - 1 + \ell) - (k-1)| = |c \ln(\lceil \gamma^{k-1} \rceil + \ell - 1) - (k-1)|$$
$$\le |c \ln(\gamma^{k-1} + \ell) - (k-1)|$$
$$= c \ln\left(1 + \frac{\ell}{\gamma^{k-1}}\right) \le \frac{c\ell}{\gamma^{k-1}}.$$

Similarly,

$$k - c \ln(i_k - \ell) = k - c \ln(\lceil \gamma^k \rceil - \ell) = k - c \ln(\gamma^k - \ell')$$
$$= -c \ln\left(1 - \frac{\ell'}{\gamma^k}\right) \in \left[0, \frac{c\ell}{\gamma^{k-1}}\right]$$

for some $\ell' \in [\ell - 1, \ell]$, assuming $\ell = o(\gamma^k)$. As a result, for T defined by (8.16),

$$|\Sigma_3 - T| \leq \sum_{\ell=1}^{k^2} \frac{2c\ell}{\gamma^{k-1}} = \frac{ck^2(k^2+1)}{\gamma^{k-1}} \leq \frac{2ck^4}{\gamma^{k-1}}.$$

Now, the result follows from Lemma 8.4. □

Proof of Theorem 8.6. Recall that $J_k = [i_k, i_{k+1})$ and define

$$A_m = \{S(i) = 0 \text{ for some } i \in J_{k_m}\}.$$

Then

$$\mathbb{P}(A_m \mid E_{m-1}, \mathcal{G}_{m-1}) \geq 0.72 \times \mathbb{P}(A_m \mid F_m, E_{m-1}, \mathcal{G}_{m-1}) \qquad (8.18)$$

(please see the definition of F_m in (8.15)). Recall formula (8.12) and write

$$\tilde{S}(i) = S(i) - \Sigma_3 = \Sigma_1 + \Sigma_2 + \Sigma_4(i).$$

From now on assume that $|\Sigma_1| < k\sqrt{i_k}$ and $|\Sigma_2| < 2k\sqrt{i_k}$, that is, E_{m-1} and F_m occur. Also assume w.l.o.g. that $\Sigma_1 + \Sigma_2 \geq 0$. Let

$$L_k = i_{k+1} - i_k - k^2 = (\gamma - 1)\gamma^k + o(\gamma^k).$$

Consider a simple random walk with steps $Y_i \sim$ Rademacher during its first L_k steps. The probability that its minimum will be equal to or below the level $-3\sqrt{i_k} = -\frac{3+o(1)}{\sqrt{\gamma-1}}\sqrt{L_k}$ converges by Lemma 8.2 to

$$2\,\mathbb{P}\left(\eta > \frac{3}{\sqrt{\gamma-1}}\right) = 2 - 2\,\Phi\left(\frac{3}{\sqrt{\gamma-1}}\right) =: 2c_2 \in (0,1)$$

as $k \to \infty$ (recall that $\eta \sim \mathcal{N}(0,1)$). As a result, by Proposition 8.1, the probability that for some $j_0 \in \{i_k, i_k+1, i_k+2, \ldots, i_k + L_k - 1\}$ we have the down-crossing, that is,

$$\tilde{S}(j_0 - 1) \geq 0 > \tilde{S}(j_0),$$

is bounded below by c_2 for k sufficiently large. Formally, let

$$j_0 = \inf\{j > i_k : \tilde{S}(j) < 0\},$$
$$\mathcal{C}_0 = \{i_k \leq j_0 \leq i_k + L_k - 1\},$$

so we have showed that on $E_{m-1} \cap F_m \cap \{\Sigma_1 + \Sigma_2 > 0\}$ we have $\mathbb{P}(\mathcal{C}_0) > c_2$ for large k.

Now, assume that event $\mathcal{C}_0$ occurred and define additionally

$$b_0 = \tilde{S}(j_0) \in (-k-1, 0],$$

$$\mathcal{C} = \left\{ \max_{0 \le h \le k^2} \sum_{g=1}^{h} X_{j_0+g} \ge k \right\}.$$

Again, from Lemma 8.2, as $k \to \infty$,

$$\mathbb{P}(\mathcal{C}) = 2\, \mathbb{P}(X_{j_0+1} + X_{j_0+2} + \cdots + X_{j_0+k^2} \ge k)$$
$$\to 2(1 - \Phi(1)) = 0.3173\ldots.$$

From now on assume that k is so large that $\mathbb{P}(\mathcal{C}) > 0.2$. On the event $\mathcal{C}$, there exists an increasing sequence $j_1, j_2, \ldots, j_k$ such that

$$j_0 < j_1 < j_2 < \cdots < j_k \le j_0 + k^2 < i_{k+1}$$

such that $X_{j_0+1} + X_{j_0+2} + \cdots + X_{j_\ell} = \ell$ for each $\ell = 1, 2, \ldots, k$ since the random walk must pass through each integer in $\{1, 2, \ldots, k\}$ in order to reach level k.

For $\ell = 1, 2, \ldots, k$, define

$$b_\ell := \tilde{S}(j_\ell) = b_0 + \sum_{h=j_0+1}^{j_\ell} a_h X_h = b_0 + a_{j_0} \sum_{h=j_0+1}^{j_\ell} X_h$$

$$+ \sum_{h=j_0+1}^{j_\ell} (a_h - a_{j_0})\, X_h$$

$$= b_0 + a_{j_0}\ell + O\left(\frac{k^4}{\gamma^k}\right)$$

since for $h \in [j_0, j_0 + k^2] \subseteq [i_k, i_{k+1})$ we have

$$|a_h - a_{j_0}| = c\left|\ln \frac{a_h}{a_{j_0}}\right| \le c\left|\ln \frac{i_k + k^2}{j_k}\right| = O\left(\frac{k^2}{\gamma^k}\right).$$

As a result,

$$-(k+1) < b_0 < b_1 < b_2 < \cdots < b_{k-1} < (k-1)(k+1) < k^2$$

and moreover the distance between consecutive b_g's is at least two (provided k is large). For $\ell = 0, 1, \ldots, k - 1$, define

$$\tilde{b}_\ell = \sup\left\{ x \in 2\mathbb{Z} : x + b_\ell \in \left[\frac{1}{2}, 3 - \frac{1}{2}\right) \right\} \equiv -2\left\lfloor \frac{b_\ell}{2} - \frac{1}{4} \right\rfloor.$$

Then $\tilde{b}_\ell$'s are all distinct even integers satisfying $|\tilde{b}_\ell| \leq k^2$.

As a result,

$$\mathbb{P}(A_m \mid F_m, E_{m-1}, \mathcal{G}_{m-1}) \geq \frac{c_2}{5} \times \mathbb{P}(S(i) \in [0, 3])$$

for some $i \in J_k \mid \mathcal{C}, \mathcal{C}_0)$

$$\geq \frac{c_2}{5} \times \mathbb{P}(\tilde{S}(i_\ell) + \Sigma_3 \in [0, 3] \text{ for some } \ell = 0, 1, \ldots, k - 1 \mid \mathcal{C}, \mathcal{C}_0)$$

$$= \frac{c_2}{5} \times \mathbb{P}(b_\ell + \Sigma_3 \in [0, 3] \text{ for some } \ell = 0, 1, \ldots, k - 1 \mid \mathcal{C}, \mathcal{C}_0)$$

$$\geq \frac{c_2}{5} \times \mathbb{P}(\Sigma_3 \in [\tilde{b}_\ell - \varepsilon, \tilde{b}_\ell + \varepsilon] \text{ for some } \ell = 0, 1, \ldots, k - 1 \mid \mathcal{C}, \mathcal{C}_0)$$

$$= \frac{c_2}{5} \times \sum_{\ell=0}^{k-1} \mathbb{P}(\Sigma_3 \in [\tilde{b}_\ell - \varepsilon, \tilde{b}_\ell + \varepsilon] \mid \mathcal{C}, \mathcal{C}_0) \geq \frac{c_2}{5} \times k \times \frac{c_1^2}{4k^2} = \frac{c_1^2 c_2}{20k}$$

assuming that ε in Corollary 8.3 is sufficiently small. Finally,

$$\mathbb{P}(A_m \mid E_{m-1}, \mathcal{G}_{m-1}) \geq 0.72 \times \mathbb{P}(A_m \mid F_m, E_{m-1}, \mathcal{G}_{m-1}) \geq \frac{0.72\, c_1^2 c_2}{20 k_m}$$

$$\geq \frac{0.036\, c_1^2 c_2}{m \ln m}$$

the sum of which diverges. Hence, recalling (8.14), we can again apply Lemma 1.1. □

8.5 Sublinear Growth of Step Sizes

Throughout this section, we assume

$$a_k = \lfloor k^\beta \rfloor, \quad 0 < \beta < 1.$$

Proposition 8.3. *Let $S(n) = a_1 X_1 + \cdots + a_n X_n$, where $a_k = \lfloor k^\beta \rfloor$, $0 < \beta < 1$. Then*

$$\mathbb{P}(|S(n)| = z) \leq \frac{\nu}{n^{1/2+\beta}} \quad \text{for all large } n,$$

for some $\nu > 0$.

Proof. Let $F_n(t) = \prod_{k=1}^{n} |\cos(ta_k)|$. For all $z \in \mathbb{Z}$, we have

$$\mathbb{P}(S(n) = z) = \frac{1}{2\pi} \int_{-\pi}^{\pi} e^{-itz} \mathbb{E} e^{itS(n)} dt \le \frac{1}{2\pi} \int_{-\pi}^{\pi} \left| \mathbb{E} e^{itS(n)} \right| dt$$

$$= \frac{1}{2\pi} \int_{-\pi}^{\pi} F_n(t) dt \le \frac{\nu}{n^{1/2+\beta}}$$

for some $\nu > 0$, for all large n, by Lemma 8.6 from the Appendix. □

Theorem 8.7. *Suppose that $a_k = \lfloor k^\beta \rfloor$, $0 < \beta < 1$. Then the **a**-walk is a.s. transient.*

Proof. In the case $\beta > 1/2$, the a.s. transience follows immediately from Borel–Cantelli lemma and Proposition 8.3, as $\sum_n \mathbb{P}(|S(n)| \le C) < \infty$ for each $C \ge 1$.

Assume from now on that $0 < \beta \le 1/2$. Define k_m, Δ_m, m_n as in Case 3 of the proof of Lemma 8.6. Fix a positive integer M and consider now only those n for which $m_n = M$. Let $I_M = \{k_M, k_M + 1, \ldots, k_{M+1} - 1\}$. Note that the elements of I_M are precisely those n for which $a_n = M$, and that the cardinality of I_M is of order $M^{1/\beta-1}$. Next, fix some $z \in \mathbb{Z}$ and define

$$E_M = E_M(z) = \{S(n) = z \text{ for some } n \in I_M\}\}.$$

For each z, we show that $\sum_M \mathbb{P}(E_M) < \infty$, and so by the Borel–Cantelli lemma, a.s. only finitely many events E_M occur. Since $S(n)$ takes only integer values, this will imply that the walk is not C-recurrent for any $C \ge 0$.

So, fix z from now on, write $S(n) = S(k_M) + R(n)$, where

$$R(n) = \sum_{i=k_M}^{n} a_i X_i = M \sum_{i=k_M}^{n} X_i.$$

Observe also that $S(k_M)$ and $R(n)$ are independent. In order $S(n) = z$ for some $n \in I_M$, we need that $S(k_M) = z \bmod M$. Let $Q = M^{\frac{1}{2\beta}+1-\varepsilon}$ for an $\varepsilon \in (0, 1/2)$. Assuming M is so large that $Q \geq 2|z|$,

$$\mathbb{P}(|R(n)| \geq Q - |z|) \leq 2 \exp\left(-\frac{(Q/2)^2}{2M^2 \cdot (k_{M+1} - k_M)}\right)$$

$$= 2 \exp\left(-\frac{\beta Q^2}{(8 + o(1))M^2 \cdot M^{\frac{1}{\beta}-1}}\right)$$

$$= 2 \exp\left(-\frac{\beta M^{1-2\varepsilon}}{8 + o(1)}\right) \quad \text{for all } n \in I_M$$

by Lemma 8.1; hence

$$\mathbb{P}\left(\max_{n \in I_M} |R(n)| \geq Q - |z|\right) \leq |I_M| \times 2 \exp\left(-\frac{\beta M^{1-2\varepsilon}}{8 + o(1)}\right) =: \alpha_M,$$

where $\alpha_M = O\left(M^{\frac{1}{\beta}-1} e^{-\frac{\beta M^{1-2\varepsilon}}{8+o(1)}}\right)$ is summable in M. So,

$$\mathbb{P}(E_M) \leq \mathbb{P}\left(E_M, \max_{n \in I_M} |R(n)| < Q - |z|\right)$$

$$+ \mathbb{P}\left(\max_{n \in I_M} |R(n)| \geq Q - |z|\right)$$

$$= \mathbb{P}\left(E_M, S(k_M) = z \bmod M, \max_{n \in I_M} |R(n)| < Q - |z|\right) + \alpha_M$$

$$= (*) + \alpha_M,$$

where the term α_M is summable since $1 - 2\varepsilon > 0$. Since E_M implies $-S(k_M) = R(n) - z$ for some $n \in I_M$,

$$(*) \leq \mathbb{P}(|S(k_M)| < Q, \ S(k_M) = z \bmod M)$$

$$= \sum_{j: \ |j|<Q, \ j=z \bmod M} \mathbb{P}(|S(k_M)| = j)$$

$$\leq \frac{\nu}{k_M^{1/2+\beta}} \times |\{j : \ |j| < Q, \ j = z \bmod M\}|$$

$$\leq \frac{\nu + o(1)}{M^{1+\frac{1}{2\beta}}} \times \left[\frac{2Q+1}{M} + 1\right] = \frac{(\nu + o(1))}{\pi M^{1+\varepsilon}}$$

by Proposition 8.3. The right-hand side is summable in M, which concludes the proof. $\qquad\square$

Remark 8.4. By setting $\varepsilon = 1/2 - \delta/2$, where $\delta > 0$ is very close to zero, we can ensure that

$$\mathbb{P}(|S(n)| < M^{1/2-\delta} \text{ for some } n \in I_M) \leq \sum_{z:|z|<M^{1/2-\delta}} \mathbb{P}(E_M(z))$$

$$\leq \left[\frac{\text{const}}{M^{1+\varepsilon}} + \alpha_M\right] \times 2M^{1/2-\delta}$$

$$= \frac{2\,\text{const}}{M^{1+\delta/2}} + 2M^{1/2-\delta}\alpha_M$$

is summable. Hence, a.s. eventually $|S(n)|$ will be larger that $n^{\beta/2-\delta}$ for any $\delta > 0$.

8.6   Rademacher Walk in Higher Dimensions

While we do not have clear results for the two-dimensional version of the process, we can prove transience for *any* sequence of a_n in dimensions three and above.

Formally, let $a_1, a_2, \ldots$ be a given sequence of positive integers. Consider the random walk $S(n)$ in $\mathbb{Z}^d$, $d \geq 1$, starting at the origin, which first makes a_1 steps in one of the $2d$ directions with equal probability, then again chooses one of the $2d$ directions and makes a_2 steps in that directions, and so on. Formally, $S(0) = \mathbf{0}$,

$$S(n) = a_1\mathbf{f}_1 + a_2\mathbf{f}_2 + \cdots + a_k\mathbf{f}_k,$$

where $\mathbf{f}_k$ are i.i.d. random vectors uniformly distributed over all $2d$ unit vectors $\pm e_1, \pm e_2, \ldots, \pm e_d$ in $\mathbb{Z}^d$, and for any t such that we have the following.

Lemma 8.5. *There exists a universal constant C_d, depending on d only, such that*

$$\mathbb{P}(S(n) = \mathbf{0}) \leq \frac{C_d}{n^{d/2}}.$$

Proof. For any given n, the walk $S(n)$ on average makes n/d steps in the direction $j = 1, 2, \ldots, d$. Let $A_{j;n}$ be the event that the walk makes less than $n/(d+1)$ steps in the direction j during n steps; then by the large deviation estimate (e.g., one can use Lemma 8.1),

$$\mathbb{P}(A_{j,n}) \leq \exp(-\alpha n)$$

for some $\alpha > 0$. Let $A_n = \cup_{j=1}^{d} A_{j;n}$, then $\mathbb{P}(A_n) \leq d e^{-\alpha n}$. Let $Y_{j;n}$ be the projection of $S(n)$ in the direction $j \in \{1, 2, \ldots, d\}$, then on $A_{j;n}^c$, we can write

$$Y_{j;n} = \sum_{i=1}^{m_j} a_{r_i} X'_{j;i},$$

where $m_j \geq n/(d+1)$, $1 \leq r_1 < r_2 < \cdots < r_{m_j}$, and $X'_{j;i}$ are Rademacher i.i.d. random variables. Moreover, for a given sequence of m_j, $Y_{j;n}$ are independent. Hence,

$$\mathbb{P}(S(n) = \mathbf{0})$$
$$\leq \mathbb{P}(A_n) + \mathbb{P}(Y_n = \mathbf{0}, A_n^c)$$
$$\leq d e^{-\alpha n} + \sup_{m_1,\ldots,m_d: m_j \geq \frac{n}{d+1}} \mathbb{P}(Y_{1;n} = 0, \ldots, Y_{d;n} = 0 \mid m_1, \ldots, m_d)$$
$$\leq d e^{-\alpha n} + \prod_{j=1}^{d} \sup_{m_j \geq \frac{n}{d+1}} \mathbb{P}(Y_{j;n} = 0 \mid m_j) \leq d e^{-\alpha n} + \frac{C_1(d)^d}{n^{d/2}}$$

since by Corollary 1.1,

$$\sup_{m_j \geq \frac{n}{d+1}} \mathbb{P}(Y_{j;n} = 0 \mid m_j) \leq \sup_{m_j \geq \frac{n}{d+1}} \frac{C}{\sqrt{m_j/2}} = \frac{C_1(d)}{n^{1/2}}$$

for some $C_1(d)$. $\qquad\square$

As a result of Lemma 8.5, we have

$$\sum_n \mathbb{P}(S(n) = 0) < \infty$$

for $d \geq 3$, so from the Borel–Cantelli lemma, we obtain the following result.

Theorem 8.8 (Transience for high dimensions). *For any sequence $\{a_k\}$, the walk S is transient for $d \geq 3$.*

8.7   Appendix: Generalization of Blair Sullivan's Results

Let $a_k = \lfloor k^\beta \rfloor$, where $0 < \beta < 1$.

Lemma 8.6. *Let $F_n(t) = \prod_{k=1}^n |\cos(t a_k)|$. Then*

$$\int_{-\pi}^{\pi} F_n(t) dt = \frac{\sqrt{8\pi(1+2\beta)} + o(1)}{n^{\beta+1/2}} \qquad \text{as } n \to \infty.$$

Remark 8.5. Note that for $\beta = 1$ we would have obtained the same result as in Sullivan (2013).

Proof. We proceed in the spirit of Sullivan (2013). Note that by symmetry

$$\int_{-\pi}^{\pi} F_n(t) dt = 2 \int_0^{\pi} F_n(t) dt = 2 \int_0^{\pi/2} F_n(t) dt + 2 \int_0^{\pi/2} F_n(\pi - t) dt$$

$$= 4 \int_0^{\pi/2} F_n(t) dt$$

since $|\cos((\pi - t) a_k)| = |\cos(\pi a_k - t a_k)| = |\cos(t a_k)|$ as a_k is an integer. Let $\varepsilon > 0$ be very small and define

$$I_0 = \left[0, \frac{1}{n^{\beta+1/2-\varepsilon}} \right], \qquad I_1 = \left[\frac{1}{n^{\beta+1/2-\varepsilon}}, \frac{1}{n^\beta} \right],$$

$$I_2 = \left[\frac{1}{n^\beta}, \frac{c_1}{n^\beta} \right], \qquad I_3 = \left[\frac{c_1}{n^\beta}, \frac{\pi}{2} \right],$$

for some $c_1 > 1$ to be determined later. Then

$$\int_0^{\pi/2} F_n(t)\, dt = \int_{I_0} F_n(t)\, dt + \int_{I_1} F_n(t)\, dt + \int_{I_2} F_n(t)\, dt + \int_{I_3} F_n(t)\, dt.$$

We show that the contribution of all the integrals, except the first one, is negligible and estimate the value of the first one.

First, observe that when $0 \le t a_k \le \pi/2$ for all $k \le n$, by the elementary inequality $|\cos u| \le e^{-u^2/2}$ valid for $|u| \le \pi/2$, we have

$$F_n(t) \le \prod_{k=1}^n \exp\left(-\frac{t^2 a_k^2}{2} \right) = \exp\left(-\frac{t^2}{2} \sum_{k=1}^n a_k^2 \right)$$

$$= \exp\left(-\frac{t^2\, n^{2\beta+1}(1 + o(1))}{2(1+2\beta)} \right) \qquad (8.19)$$

since $a_k^2 = k^{2\beta}(1 + o(1))$.

Case 0: $t \in I_0$

Here $ta_k \leq \frac{1}{n^{1/2-\varepsilon}} \ll 1$, hence for n large enough

$$\frac{(ta_k)^2}{2} \leq -\ln\cos(ta_k) = \frac{(ta_k)^2}{2} + O\left((ta_k)^4\right) \leq (1+o(1))\frac{(ta_k)^2}{2}$$

yielding $F_n(t) = \exp\left(-\frac{t^2\, n^{2\beta+1}\rho_{n,t}}{2(1+2\beta)}\right)$, where $\rho_{n,t} = 1+o(1)$ for large n (compare with (8.19)). Since for any $r > 0$ we have

$$\int_0^{n^{-\beta-1/2+\varepsilon}} \exp\left(-\frac{t^2\, n^{2\beta+1}r}{2(1+2\beta)}\right) dt$$

$$= \frac{1}{n^{1/2+\beta}} \int_0^{n^{\varepsilon}} \exp\left(-\frac{s^2\, r}{2(1+2\beta)}\right) ds$$

$$= \frac{1}{n^{1/2+\beta}} \left[\sqrt{\frac{\pi(1+2\beta)}{2r}} + o(1)\right]$$

where the main term is monotone in r, by substituting $r = \rho_{n,t} = 1+o(1)$ we conclude that

$$\int_{I_0} F_n(t)dt = \frac{1}{n^{1/2+\beta}}\left[\sqrt{\pi(1/2+\beta)} + o(1)\right].$$

Case 1: $t \in I_1$

Since $ta_k \leq 1 < \pi/2$, by (8.19) for some $C_2 > 0$ we have $F_n(t) \leq \exp\left(-\frac{n^{2\varepsilon}(1+o(1))}{2(1+2\beta)}\right) \leq e^{-C_2 n^{2\varepsilon}}$, so $\int_{I_1} F_n(t)dt \leq e^{-C_2 n^{2\varepsilon}}$, which decays faster than polynomially.

Case 2: $t \in I_2$

As in Case 2 in Sullivan (2013), we use monotonicity of $F_n(t)$ in n. Let $r = c_1^{-1/\beta} \in (0,1)$ then $\lfloor rn \rfloor^\beta \leq (rn)^\beta = \frac{n^\beta}{c_1}$, consequently by (8.19), since $t \leq \frac{c_1}{n^\beta}$,

$$F_{\lfloor rn \rfloor}(t) \leq \exp\left(-\frac{t^2\,(rn)^{2\beta+1}(1+o(1))}{2(1+2\beta)}\right)$$

and since $F_n(t) \leq F_{\lfloor rn \rfloor}(t)$, we get a similar bound as in Case 1.

Case 3: $t \in I_3$

Let

$$k_m = \inf\{k \in \mathbb{Z}_+ : k^\beta \geq m\} = \lceil m^{1/\beta} \rceil, \quad m = 1, 2, \ldots,$$

$$\Delta_m = k_{m+1} - k_m = \beta^{-1} m^\gamma + O(m^{1/\beta - 2}) + \rho_0, \quad \gamma := \frac{1-\beta}{\beta},$$

where $|\rho_0| \leq 1$. Then

$$a_k = m \quad \text{if and only if} \quad k \in \{k_m, k_m + 1, \ldots, k_m + \Delta_m - 1 \, (\equiv k_{m+1} - 1)\}.$$

For $n \in \mathbb{Z}_+$, let

$$m_n = \max\{m : k_m \leq n\} = n^\beta (1 + o(1)), \quad n \in [k_{m_n}, k_{m_n} + \Delta_{m_n} - 1].$$

By the inequality between the mean geometric and the mean arithmetic,

$$F_n(t) = \sqrt{\prod_{k=1}^{n} \cos^2(ta_k)} = \left(\sqrt[n]{\prod_{k=1}^{n} \cos^2(ta_k)} \right)^{n/2}$$

$$\leq \left(\frac{\sum_{k=1}^{n} \cos^2(ta_k)}{n} \right)^{n/2}$$

$$= \left(\frac{1}{2} + \frac{U_n(t)}{2n} \right)^{n/2} \quad \text{where } U_n(t) = \sum_{k=1}^{n} \cos(2ta_k).$$

We show that if t is not too small, then for some $0 \leq c < 1$ we have $U_n(t) \leq cn$ and hence $F_n(t) \leq \left(\frac{1+c}{2} \right)^{n/2}$. In order to do that, first note that

$$U_n(t) \leq \sum_{k=1}^{k_{m_n}} \cos(2ta_k) + (n - k_{m_n}) = \sum_{m=1}^{m_n} \Delta_m \cos(2tm) + (n - k_{m_n}).$$

Let $r \in (0, 1)$ and assume w.l.o.g. that rm_n is an integer. For $m \in [rm_n + 1, m_n]$, we have

$$A \leq \Delta_m \leq \bar{A}, \quad \text{where } A = \beta^{-1}(rm_n)^\gamma + O(1), \quad \bar{A} = \beta^{-1} m_n^\gamma + O(1).$$

Consequently,

$$\sum_{m=rm_n+1}^{m_n} \Delta_m \cos(2tm)$$

$$\leq \sum_{m=rm_n+1}^{m_n} \left[\bar{A}\cdot\mathbf{1}_{\cos(2tm)\geq 0} + A\cdot\mathbf{1}_{\cos(2tm)<0}\right]\cos(2tm)$$

$$= \sum_{m=rm_n+1}^{m_n} \left[\bar{A}-A\right]\cos(2tm)\,\mathbf{1}_{\cos(2tm)\geq 0} + \sum_{m=rm_n+1}^{m_n} A\cos(2tm)$$

$$\leq (1-r)m_n\left(\bar{A}-A\right) + A\sum_{m=rm_n+1}^{m_n}\cos(2tm) = (1-r)m_n\left(\bar{A}-A\right)$$

$$+ A\left(\cos^2\left(rtm_n+t\right)-\cos^2(tm_n+t)\right)$$

$$+ A\frac{\cos t}{2\sin t}\left(\sin(2t(m_n+1))-\sin(2t(rm_n+1))\right)$$

$$\leq \frac{m_n^{1/\beta}}{\beta}(1-r)\left[1-r^\gamma+O(m_n^{-\gamma})\right]$$

$$+ m_n^\gamma\left(\frac{r^\gamma}{\beta}+O(m_n^{-\gamma})\right)\left(1+\frac{1}{|\sin t|}\right).$$

Hence, since $m_n^{1/\beta}=n+o(n)$,

$$U_n(t) \leq \sum_{m=1}^{rm_n}\Delta_m + \sum_{m=rm_n+1}^{m_n}\Delta_m\cos(2tm) + (n-k_{m_n})$$

$$= k_{rm_n+1} + \sum_{m=rm_n+1}^{m_n}\Delta_m\cos(2tm) + O(\Delta_{m_n})$$

$$\leq r^{1/\beta}n + \frac{m_n^{1/\beta}}{\beta}(1-r)\left[1-r^\gamma+O(m_n^{-\gamma})\right]$$

$$+ m_n^\gamma\left(\frac{r^\gamma}{\beta}+O(m_n^{-\gamma})\right)\left(1+\frac{1}{|\sin t|}\right) + O(m_n^{\gamma\beta})$$

$$\leq n\left(r^{1/\beta}+\frac{(1-r)(1-r^\gamma)}{\beta}\right) + \frac{4n^{1-\beta}\beta^{-1}r^\gamma}{|\sin t|} + o(n).$$

Consider now the function

$$h(r, \beta) := r^{1/\beta} + \frac{(1-r)(1-r^{\gamma})}{\beta} = r^{1/\beta} + \frac{(1-r)\left(1 - r^{1/\beta-1}\right)}{\beta}$$

and note that

$$h(1 - \beta, \beta) = (1 - \beta)^{1/\beta} + 1 - (1 - \beta)^{1/\beta-1} = 1 - \beta(1 - \beta)^{1/\beta-1}$$

$$\leq 1 - e^{-1}\beta < 1 - \beta/3$$

since $\sup_{\beta \in (0,1)} (1 - \beta)^{1/\beta-1} = e^{-1}$ by elementary calculus. So, if we set $r = 1 - \beta \in (0, 1)$, by noting $t \leq 2\sin t$ for $t \in [0, \pi/2]$, we conclude that $U_n(t) \leq \left(1 - \frac{\beta}{4}\right)n$ provided that $t \geq \frac{c_1}{n^\beta}$ for some $c_1 > 0$. Consequently, $\int_{I_3} F_n(t)dt \leq \left(1 - \frac{\beta}{8}\right)^{n/2}$ for large n, which converges to zero exponentially fast. $\square$

Chapter 9

Higher Dimensions: The Conservative Random Walk

The process described in this chapter is a higher-dimensional analog of the coin-turning walk, and it shall henceforth be referred to as the "conservative random walk." We opted for this term because this process demonstrates a tendency to retain its direction over long intervals. It is worth noting that analogous processes, albeit under varying nomenclature, have been studied in the literature, particularly within continuous-time frameworks.

In settings characterized by temporal homogeneity, although possibly with spatial dependencies or within the context of a spatially random environment, processes akin to our conservative walks have been denoted as "persistent random walks" or "Newtonian random walks." Among the earliest references to continuous processes with memory of this nature is the work by Goldstein (1951) and Kac (1974). The term *persistent random walk* was introduced in Cénac *et al.* (2018, 2020), where these studies delved into the general criteria distinguishing recurrence from transience.

A planar motion with just three directions was studied in Di Crescenzo (2002). Kolesnik recently published a book on *Markov random flights* (Kolesnik, 2021). The paper Orsingher and Ratanov (2002) investigated *planar random motions with drifts* involving four directions/speeds, switching at Poisson times. Applications of telegraph processes to option pricing are detailed in Ratanov and Kolesnik (2022). In Chen and Renshaw (1994), one explores characteristic functions of correlated random walks, while Szász and Tóth (1984) examined a version of this process in a random environment.

The main distinction between the conservative random walk and most of the models studied in the literature (except, perhaps, Vasdekis and Roberts (2023), which leans toward applied research) lies in the time-inhomogeneous nature of the underlying process of direction switching, thereby giving rise to various new phenomena.

9.1   Introduction

We are going to study a non-classical random walk. This kind of process has been studied in one dimension as the coin-turning walk in Chapter 6, and we now define and study the higher-dimensional analogs. To avoid ambiguity, in the sequel, by geometric distribution we mean the probability distribution of the number of Bernoulli trials (and not failures) needed to get one success, i.e., the random variable with support on $\{1, 2, \dots\}$. It is denoted by $\mathsf{Geom}(p)$. Finally, by *symmetrized geometric distribution* with parameter $p \in (0, 1)$ (or $\mathsf{Sgeom}(p)$), we mean $\{p_m\}$ with

$$p_m = \frac{1}{2}(1-p)^{m-1}p, \quad m = \pm 1, \pm 2, \pm 3, \dots. \tag{9.1}$$

9.1.1   *Generalizing the "coin-turning walk" to higher dimensions: "conservative random walk"*

We define a random walk corresponding to a given sequence p_n, $n = 2, 3, \dots$ in $[0, 1]$ for $d \geq 2$, similarly to the case $d = 1$ in Definition 6.1. Now, instead of turning a "coin," we have to roll a die that has $2d$ sides.

The steps are defined as follows. Let $Y_n \in \{\pm e_1, \dots, \pm e_d\}$, where e_i are the $2d$ unit vectors in $\mathbb{R}^d$, and let Y_1 be chosen uniformly from these vectors. Let the vectors $Y_1, Y_2, \dots$ form an inhomogeneous Markov chain with the transition matrix between times $n - 1$ and n given by

$$(1 - p_n)\mathbf{I}_{2d} + \frac{p_n}{2d}A_{2d}, \ n \geq 2,$$

where $\mathbf{I}_{2d}$ is the $2d \times 2d$ identity matrix and

$$A_{2d} := \begin{pmatrix} 1 & 1 & \cdots & 1 \\ 1 & 1 & \cdots & 1 \\ \vdots & \vdots & \ddots & \vdots \\ 1 & 1 & \cdots & 1 \end{pmatrix}$$

is the $2d \times 2d$ matrix of ones.

Now, we define the random walk S on $\mathbb{Z}^d$, starting at z. Let $S_n := z + \sum_{i=1}^{n} Y_i$ for $n \geq 0$ (with the usual convention that $\sum_{i=1}^{0} = 0$), and denote by $\mathbb{P}_z$ the law of this walk. Sometimes we simply write $\mathbb{P}$ when $z = \mathbf{0}$. Equivalently, we can define a sequence of independent Bernoulli random variables η_i, $i = 0, 1, \ldots$, such that $\mathbb{P}(\eta_i = 1) = p_i$, and the increasing sequence of stopping times τ_j, such that $\tau_0 = 0$ and

$$\tau_{j+1} = \inf\{k > \tau_j : \eta_k = 1\}, \quad j = 0, 1, 2, \ldots .$$

At times τ_j, the walk S_n behaves just like a simple symmetric random walk, while in between those times, it keeps going in the direction it was going before.

For the sake of completeness, we include the time-homogeneous case too, that is, the case when $p_n = p$ for $n \geq 2$, where $0 < p < 1$ (when $p = 0$, the walk moves in a straight line, while the case $p = 1$ corresponds to the classical simple symmetric random walk, so we do not consider these two degenerate cases).

Intuitively, the walker is more "reluctant" to change direction than an ordinary random walker, motivating the following definition.

Definition 9.1 (Conservative random walk). We dub the process S the **conservative random walk** in d dimensions, corresponding to the sequence $\{p_n\}$. (The reader should be aware that in the literature the name "persistent walk" is also used sometimes.)

Remark 9.1. Regarding the sequence of the p_n's, we note the following:

(i) In this chapter, we focus on the case when the p_n's are non-increasing ("cooling dynamics"). Nevertheless, studying growing p_n's and mixed cases also makes sense. This question we suggest as a research topic for the interested reader.

(ii) The probability of changing the direction is $p_n \cdot \frac{2d-1}{2d} \leq \frac{2d-1}{2d}$. Our setting thus rules out the kind of heating dynamics (allowed in the setup of Chapter 6) when the probability of changing the direction approaches one.   ◇

We now present a fundamental definition.

Definition 9.2 (Recurrence/transience). We call the walk S

- **recurrent** if $\mathbb{P}_0(S_n = \mathbf{0} \text{ i.o.}) = 1$,
- **weakly transient**, if it is not recurrent in the above sense,
- **strongly transient** if $\mathbb{P}_0(\lim_{n\to\infty} |S_n| = \infty) = 1$.

Remark 9.2 (Differences to traditional categorization). It is easy to see that strong transience implies weak transience and that, in fact, $\mathbb{P}_z(\lim_{n\to\infty} |S_n| = \infty) = 1$ for each $z \in \mathbb{Z}^d$. On the other hand, for recurrence, the probability might depend on the starting point as well as on the "target."

Unlike in the case of a simple random walk, it is not a priori clear whether weak transience necessarily implies strong transience. For example, the walk might come close to the origin infinitely often, without hitting it, see Figure 9.1. Also, as mentioned above, it is hypothetically possible that the walker visits the origin infinitely often yet visits some other fixed point only finitely often.

Note that both of these scenarios are possible only if the probability p_n of updating the direction is not bounded away from zero.

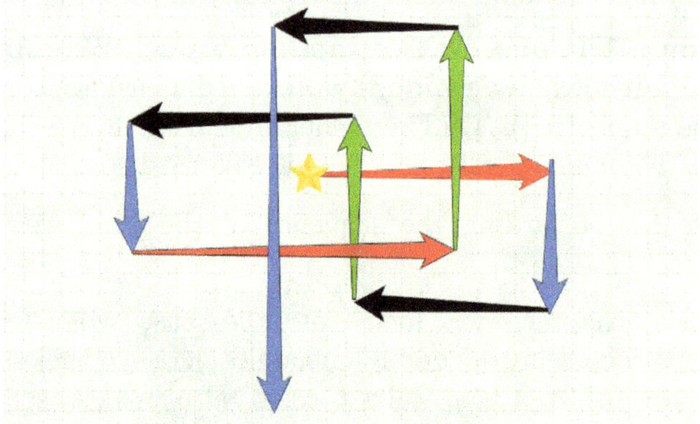

Fig. 9.1.   A sample path of the walk; the origin is denoted by star. Each arrow keeps track of the "total" horizontal/vertical relocation only.

Indeed, for a usual random walk, if it hits the origin infinitely often, every time it does, it has a fixed positive probability of, e.g., going to $(1,0)$ on the next step. Hence, by the usual arguments, it will also hit $(1,0)$ infinitely often. Our random walk, while possibly hitting zero only from a vertical direction, say, at times η_1, η_2, η_3, etc., might never change its direction at the origin and simply continue going vertically if $\sum_k p_{\eta_k} < \infty$. Thus, the previous argument will not work.

Finally, although a simple application of Kolmogorov's $0 - 1$ law shows (the η_j are independent and the directions chosen at updates too) that $\mathbb{P}_z(\lim_{n\to\infty} |S_n| = \infty) \in \{0,1\}$, we cannot rule out the possibility that, e.g., $\mathbb{P}_0(S_n = \mathbf{0} \text{ i.o.}) \in (0,1)$.   ◇

In the sequel, in Section 9.2, as a warm-up, we prove recurrence when $d = 2$ and $p_n = p \in (0,1)$. In Section 9.5, we derive the scaling limit in this case, while in Section 9.6, we do that for the critical case when the scaling limit is very different (a "zigzag process"). In Section 9.4, we consider the (recurrent) case when the sequence of the p_n's is periodic, followed by the proof of transience for the two-dimensional walk when p_n has a sufficiently strong decay, as well as that of strong transience for certain multidimensional cases. In Section 9.7, we formulate some open problems. Finally, the Appendix states and proves some technical lemmas.

9.2 Recurrence in $\mathbb{Z}^1$ and $\mathbb{Z}^2$

If the p_n are summable, then the walk will make only finitely many turns and then it will trivially drift to infinity. Hence, in the rest of this section, we may and will assume that

$$\sum_{n\geq 1} p_n = \infty, \qquad (9.2)$$

that is, the walk makes infinitely many turns a.s.

In the one-dimensional case, the result is quite easy.

Theorem 9.1. *Assuming* (9.2), *S on* $\mathbb{Z}^1$ *is always recurrent.*

Proof. The proof is identical to that of the first part in Corollary 6.2, once we observe that the conservative random walk can be viewed as a coin-turning walk with $\widehat{p}_n := p_n/2$.   □

Consider now the case $d = 2$, and define the two-dimensional random walk by $S_n = (X_n, Y_n) \in \mathbb{Z}^2$, $n \geq 1$. The walker keeps going in one of the four directions (up, left, down, or right) at time n with probability $1 - p_n$, while with probability p_n, the walker changes its direction in one of the four possible directions, all directions having equal probability.

We start with a result that is easy to prove.

Theorem 9.2 (Periodic sequence). *Let the sequence $\{p_n\}$ be periodic, that is, assume that there exists an $r \geq 1$ such that $p_{n+r} = p_n$ for all $n \geq n_0$ with some $n_0 \in \mathbb{N}$. Then for $d = 2$, the walk is not strongly transient:* $\mathbb{P}(\lim_n |S_n| = \infty) = 0$.

Proof. Let $\mathcal{T} \subset \mathbb{N}$ be the set of "update times." The proof is based on the following decomposition of the random walk. Let us define a sequence of stopping times: $\tau_0 := n_0$ and

$$\tau_{n+1} = \min\{m > n \ : \ r \mid m - n_0, \ m \in \mathcal{T}\}.$$

Define the random walk U^1 on the time interval $\{1, 2, \ldots, r\}$ as follows. Let

$$U_i^1 := S_{n_0+i} - S_{n_0}.$$

We possibly concatenate further copies $U^2, U^3, \ldots$ to it in such a way that the last step of the previous piece is the same as the first step of the next one. The first step in U^1 can be each one of the unit base vectors with equal probabilities, and, by symmetry, this property is inherited for further pieces. The number of the pieces (including U^1) is geometric with parameter p_{n_0}. The total length of the walk we obtained this way is exactly τ_1; this is the first time we have an "update" time which is a multiple of r. Then repeat the same with the next finite piece of random walk of length $\tau_2 - \tau_1$, which is independent of the previous piece, and continue this construction ad infinitum. It is easy to see that the random walk obtained this way is exactly S.

Define the embedded walk S^* by $S_n^* := S_{\tau_n}$. The steps of this walk are i.i.d. vectors and the length of each one is bounded by the total length of the corresponding piece of the random walk. This latter is a random, geometrically (with parameter p_{n_0}) distributed multiple of r. In particular, the step size for S^* has a finite second moment, while this walk is clearly symmetric and hence has zero drift. It follows from Section 8, T1 in Spitzer (1976) that $\mathbb{P}(\lim_n |S_n^*| = \infty) = 0$ for $d = 2$. The same must hold for S too, since $\lim_n |S_n(\omega)| = \infty$ implies $\lim_n |S_n^*(\omega)| = \infty$. □

The following result is perhaps not too surprising.

Theorem 9.3 (Recurrence on $\mathbb{Z}^2$ when p_n is uniformly bounded). *Let $d = 2$ and assume that $\liminf_{n\to\infty} p_n > 0$. Then the random walk S is recurrent, and in fact, the event $\{S_n = v\}$ occurs infinitely often a.s. for all $v \in \mathbb{Z}^2$.*

Proof. Let $\mathcal{A}_v := \{S_n = v \text{ for infinitely many } n\text{'s}\}$ and $\mathcal{A} := \bigcap_{v\in\mathbb{Z}^2} \mathcal{A}_v$. Our goal is to show that $\mathcal{A}$ occurs almost surely. We may and will assume, without loss of generality, that $p_n \geq \varepsilon$ for some $\varepsilon > 0$ for all n because if $\mathcal{A}$ occurs almost surely for this case, then in the general case we may simply ignore a finite number of initial times and for the rest, all p_n are bounded away from zero and $\mathcal{A}$ occurs almost surely, using a (random) translation.

Next, using that $p_n \geq \varepsilon$, it is elementary to show that $\mathbb{P}(\mathcal{A}_v \triangle \mathcal{A}_w) = 0$ for all $v, w \in \mathbb{Z}^2$. This fact implies that we have the usual dichotomy: with probability 1, either the walk visits all vertices infinitely often or it visits all vertices finitely often; in our definitions, this means that weak transience implies strong transience.

Recall that $\tau_1, \tau_2, \ldots$ are the consecutive times when the random walk S updates its direction, let $\tau_0 = 0$ and introduce the *embedded walk* $\bar{S} = (\bar{S}_k)_{k\geq 0}$, where $\bar{S}_k := S_{\tau_k}$, $k = 0, 1, 2, \ldots$. This process is a two-dimensional long-range random walk with increments ξ_l such that

$$\bar{S}_{l+1} - \bar{S}_l = (\kappa_l \xi_l, (1 - \kappa_l)\xi_l),$$

where ξ_l has a symmetric distribution on $\{\pm 1, \pm 2, \ldots\}$ satisfying

$$\mathbb{P}(|\xi_l| > k \mid \mathcal{F}_{\tau_l}) = (1 - p_{\tau_l+1})(1 - p_{\tau_l+2})\cdots(1 - p_{\tau_l+k})$$
$$\leq (1 - \varepsilon)^k, \quad k = 1, 2, \ldots, \tag{9.3}$$

and κ_l is Bernoulli$(1/2)$; moreover, all κ_l are independent of each other and the sequence $\{\xi_l\}$. Equivalently, we can describe $\bar{S}$ as a process with increments, conditional on $\mathcal{F}_{\tau_l}$, distributed as

$$\bar{S}_{l+1} - \bar{S}_l = \begin{cases} (|\xi_l|, 0), & \text{with probability } 1/4; \\ (-|\xi_l|, 0), & \text{with probability } 1/4; \\ (0, |\xi_l|), & \text{with probability } 1/4; \\ (0, -|\xi_l|), & \text{with probability } 1/4. \end{cases}$$

We show that, in fact, even the embedded process $\bar{S}$ is recurrent, and, as a result, so is S. If ξ_l were i.i.d., one could directly use Proposition 4.2.4 from Lawler and Limic (2010) which says that any time-homogeneous random walk on $\mathbb{Z}^2$ with zero drift and a finite second moment is recurrent, but because of the inhomogeneity, we need a proof based on Lyapunov functions.

Recall that $|(x, y)| = \sqrt{x^2 + y^2}$. We use the same idea as in the proof of Theorem 2.2.1 in Fayolle *et al.* (1995) (see also Theorem 2.5.2 in Menshikov *et al.* (2017)) which says that to establish recurrence, it is sufficient to find some function $f : \mathbb{Z}^2 \to \mathbb{R}$ and $A > 0$ such that

- $f(z) \geq 0$,
- $f(z) \to \infty$ as $|z| \to \infty$,
- if $M_l := f(\bar{S}_l)$ then the process $M = \{M_l\}_{l=0,1,2,\dots}$ satisfies

$$\mathbb{E}\left[M_{l+1} - M_l \mid \bar{S}_l = (x, y)\right] \leq 0, \qquad (9.4)$$

whenever $S_l \notin B_A := \{(x, y) \in \mathbb{Z}^2 : x^2 + y^2 \leq A^2\}$. So, informally, M is "a supermartingale, outside of some disc."

Once such function f is found, it will follow that the embedded walk visits the disc B_A infinitely often a.s. Indeed, if $\tau_A^{(n)} := \inf\{l \geq n : |S_l| \leq A\}$ is the first time after $n \geq 0$ when S_l enters the disc B_A, then $M_{\text{stopped}}^{(n)}(l) := M_{l \wedge \tau_A^{(n)}}$ is a non-negative supermartingale with respect to the canonical filtration of the embedded walk by (9.4), converging almost surely to some limit $\tilde{M}^{(n)} \geq 0$. By Fatou's lemma and the supermartingale property,

$$
\mathbb{E}\left[\tilde{M}^{(n)} \mid \bar{S}_n\right] = \mathbb{E}\left[\lim_{l\to\infty} M_{\text{stopped}}^{(n)}(l) \mid \bar{S}_n\right]
$$

$$
\leq \liminf_{l\to\infty} \mathbb{E}\left[M_{\text{stopped}}^{(n)}(l) \mid \bar{S}_n\right]
$$

$$
\leq M_{\text{stopped}}^{(n)}(n) = M_n = f(\bar{S}_n). \qquad (9.5)
$$

On the event $\{\tau^{(n)} = \infty\}$ (i.e., when the walk never hits B_A after time n), the dichotomy discussed above implies that the walk is a.s. transient, and since $f(z) \to \infty$ as $|z| \to \infty$, if $\mathbb{P}(\tau^{(n)} = \infty) > 0$, then $\mathbb{E}\left[\tilde{M}^{(n)}\right] = \infty$ which contradicts (9.5). Hence, $\tau_A^{(n)}$ is finite a.s. for all $n \geq 0$, and thus the walk visits B_A, and hence all vertices of $\mathbb{Z}^2$, infinitely often a.s.

It only remains to find an appropriate function, and we claim that $f : \mathbb{Z}^2 \to \mathbb{R}_+$ defined as

$$
f(x,y) := \begin{cases} \ln(x^2 + y^2 - a) = \ln(r^2 - a), & \text{if } r \geq \sqrt{a+1}; \\ 0, & \text{otherwise,} \end{cases}
$$

works, with a suitable constant $a > 0$ to be chosen later, and $r = r(x,y) = \sqrt{x^2 + y^2}$. Denote $x_l := x + \kappa \xi_l$ and $y_l := y + (1 - \kappa_l)\xi_l$, and assume that $r \geq 3$. Define the event

$$
\mathcal{E}_l := \left\{ |\xi_l| \leq r - \sqrt{a+1} \right\}.
$$

By the triangle inequality, on $\mathcal{E}_l$ we have $\sqrt{x_l^2 + y_l^2} \geq \sqrt{a+1}$ and thus $f(x_l, y_l) = \ln(x_l^2 + y_l^2 - a)$.

Also define the random variable $\psi_l = \psi_l(x,y)$ for $r \geq \sqrt{a+1}$ as

$$
\psi := \frac{2\xi_l \zeta_l + \xi_l^2}{r^2 - a},
$$

where

$$
\zeta_l := x_l\kappa_l + y_l(1 - \kappa_l) = \begin{cases} x_l, & \text{with probability } 1/2; \\ y_l, & \text{with probability } 1/2, \end{cases}
$$

and ζ_l is independent of ξ_l. Letting $S_l = (x,y)$ and $S_{l+1} = (x_l, y_l)$, a straightforward computation yields that if both r^2 and $x_l^2 + y_l^2 \geq a+1$, then

$$\Delta_l := f(\bar{S}_{l+1}) - f(\bar{S}_l) = \ln(x_l^2 + y_l^2 - a) - \ln(r^2 - a) = \ln(1 + \psi_l).$$

Consequently, if $r > \sqrt{a+1}$, then

$$\mathbb{E}\left[\Delta_l \mid \bar{S}_l = (x, y)\right] = \mathbb{E}\left[\Delta_l \mathbb{1}_{\mathcal{E}_l} \mid \bar{S}_l = (x, y)\right] + \mathbb{E}\left[\Delta_l \mathbb{1}_{\mathcal{E}_l^c} \mid \bar{S}_l = (x, y)\right]$$

$$= \mathbb{E}\left[\ln(1 + \psi_l)\mathbb{1}_{\mathcal{E}_l}\right] + \mathbb{E}\left[\Delta_l \mathbb{1}_{\mathcal{E}_l^c} \mid \bar{S}_l = (x, y)\right]$$

$$\leq \mathbb{E}\left[\ln(1 + \psi_l)\mathbb{1}_{\mathcal{E}_l}\right] + \mathbb{E}\left[\ln(x_l^2 + y_l^2)\mathbb{1}_{\mathcal{E}_l^c}\right]$$

$$=: \text{(I)} + \text{(II)}. \tag{9.6}$$

Another simple computation, using the independence of ζ_l and ξ_l, along with the fact that ξ_l has vanishing odd moments, gives

$$\mathbb{E}\psi_l = \frac{\mathbb{E}\xi_l^2}{r^2 - a},$$

$$\mathbb{E}\psi_l^2 = \frac{\mathbb{E}\xi_l^4 + 2(x^2 + y^2)\mathbb{E}\xi_l^2}{(r^2 - a)^2},$$

$$\mathbb{E}\psi_l^3 = \frac{6(x^2 + y^2)\mathbb{E}\xi_l^4}{(r^2 - a)^3} + \mathcal{O}(r^{-6})\mathbb{E}\xi_l^6. \tag{9.7}$$

Next, we use the elementary inequality

$$\ln(1 + u) \leq u - \frac{u^2}{2} + \frac{u^3}{3} \quad \text{for } u > -1 \tag{9.8}$$

and observe that (as a brief computation reveals) for fixed (x, y) satisfying $r > \sqrt{a+1}$ we have $\psi_l > -1$ almost everywhere on the event $\mathcal{E}_l$. Hence, by (9.7), and (9.8), we have that

$$\text{(I)} \leq \mathbb{E}\left[\left(\psi_l - \frac{\psi_l^2}{2} + \frac{\psi_l^3}{3}\right)\mathbb{1}_{\mathcal{E}_l}\right] = \left(\mathbb{E}\psi_l - \frac{\mathbb{E}(\psi_l^2)}{2} + \frac{\mathbb{E}(\psi_l^3)}{3}\right) + \text{(III)}$$

$$= -\frac{6a\,\mathbb{E}\xi_l^2 - 9\,\mathbb{E}\xi_l^4 + o(1)}{r^4} + \text{(III)} \leq -\frac{1}{r^4} + \text{(III)} \tag{9.9}$$

for all large r, where

$$\text{(III)} := -\mathbb{E}\left[\left(\psi_l - \frac{\psi_l^2}{2} + \frac{\psi_l^3}{3}\right)\mathbb{1}_{\mathcal{E}_l^c}\right],$$

provided $a > 0$ is large enough, since from (9.3) we easily have $\mathbb{E}\xi_l^2 \geq 1$ and $\mathbb{E}\xi_l^4 \leq \frac{24}{e^4}$. Fix this a from now on.

From the inequalities

$$\ln\left(x_l^2 + y_l^2\right) \leq \ln\left[(2x^2 + 2\kappa_l^2\xi_l^2) + (2y^2 + 2(1 - \kappa_l)^2\xi_l^2)\right]$$
$$= \ln\left(2x^2 + 2y^2 + 2\xi_l^2\right) \leq \ln(2r^2) + \ln(2\xi_l^2)$$

(which use the fact that $\ln(b + c) \leq \ln(b) + \ln(c)$ whenever $\min(b, c) \geq 2$), we conclude that the second term in (9.6) satisfies

$$(\mathrm{II}) \leq \mathbb{E}\left[(\ln(2r^2) + 2\ln|\xi_l|)\, \mathbb{1}_{\mathcal{E}_l^c}\right]$$
$$= \ln(2r^2)\,\mathbb{P}\left(|\xi_l| > r - \sqrt{a + 1}\right) + 2\,\mathbb{E}\left[\ln\left(2\xi_l^2\right)\mathbb{1}_{|\xi_l| > r - \sqrt{a+1}}\right]$$
$$\leq C_1\ln(r)\, e^{-C_2 r} \tag{9.10}$$

for some $C_1, C_2 > 0$ since $|\xi_l|$ is stochastically smaller than a geometric random variable with parameter $1 - \varepsilon$ (see (9.3)) and then using the properties of the geometric distribution.

We also have

$$|(\mathrm{III})| \leq \mathbb{E}\left[\left(|\psi_l| + \frac{\psi_l^2}{2} + \frac{|\psi_l|^3}{3}\right)\mathbb{1}_{\mathcal{E}_l^c}\right] \leq C_3 e^{-C_2 r}, \tag{9.11}$$

for some constants $C_2, C_3 > 0$ not depending on l. Indeed, assuming that r, and hence $\max(|x|, |y|)$, are sufficiently large, we get

$$x^2 + y^2 - a \geq 4\max(|x|, |y|) \geq 2,$$

yielding $|\psi_l| \leq \frac{1}{2}(|\xi_l| + \xi_l^2) \leq \xi_l^2$ and hence for every positive integer m, one has

$$\mathbb{E}\left(|\psi_l|^m \mathbb{1}_{\mathcal{E}_l^c}\right) = \mathbb{E}\left(|\psi_l|^m \mathbb{1}_{|\xi_l| > r - \sqrt{a+1}}\right)$$
$$< \mathbb{E}\left(\xi_l^{2m}\mathbb{1}_{|\xi_l| > r - \sqrt{a+1}}\right) < C_3'(m)e^{-C_2 r},$$

where $C_3'(m) > 0$, $C_2 > 0$, exploiting again well-known properties of geometric distribution.

Finally, from (9.6), (9.9), (9.11) and (9.10), we conclude that

$$\mathbb{E}\left[\Delta_l \mid \bar{S}_l = (x, y)\right] \leq -\frac{1 + o(1)}{r^4}, \quad \text{as } r \to \infty,$$

which is negative for $r = |(x, y)|$ sufficiently large. We thus have established (9.4), completing the proof. $\qquad\square$

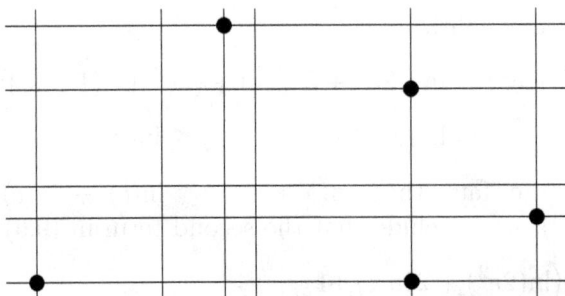

Fig. 9.2. Random grid. The points on the lines are recurrent. The dark circles are infinite update points.

9.3   The Structure of Recurrent Points in Dimensions $d = 2, 3$

Definition 9.3 (Recurrent and infinite update points). Points visited infinitely often by the walk are called *recurrent points*. Points, where the walk's direction is updated infinitely often, are called *infinite update points*. A priori, these sets can be random, as they depend on the realization of the walk. (See Figure 9.2.)

Theorem 9.4. *Assume that* $\limsup p_n < 1$ *and that* $d = 2, 3$. *Then we have the following:*

(1) *The set of recurrent points is a (random) union of straight lines parallel to the axes, which we dub a "random grid." To be more precise, it consists of H horizontal and V vertical lines, where $H, V \in \{0, 1, 2, \dots, \} \cup \{\infty\}$.*
(2) *For $d = 3$, there are no infinite update points a.s.*
(3) *For $d = 2$ and $z \in \mathbb{Z}^2$, if the probability of z being an infinite update point is positive, then, conditionally on z being an infinite update point, it a.s. holds that the points on both the horizontal and the vertical lines passing through z are recurrent points.*

Proof. If z is a recurrent point, then there is a direction from which it is visited infinitely often, say, from the left. This implies that $z - e_1$ is also visited infinitely often. Since $\limsup p_n < 1$, there exists an $\epsilon > 0$ such that $p_n \le \epsilon$ for all sufficiently large ns. Therefore, every time the walk visits z from the left, with probability at least ϵ, it will continue going the same direction thus visiting $z + e_1$. Hence, $z + e_1$ is also a recurrent point visited infinitely often from the left (formally, it follows from Lévy's Borel–Cantelli lemma). Now, by induction, we get that all the points on the right are recurrent points.

To show that all the points on the left of z are recurrent, observe that $z - e_1$ is a recurrent point as we mentioned above, and if $z - e_1$ is an infinite update point, then this claim follows from the second part of the statement (proven in the following). Otherwise, $z - e_1$ is eventually only visited from the left and we can apply induction.

In dimension $d = 2$, unlike when $d = 3$, infinite update points might exist. On the event $z \in \mathbb{Z}^2$ is an infinite update point, when the walk comes to z, it can choose any direction with probability $1/4$. Let k be any positive integer. For large ns $p_n \leq 1 - \epsilon$, therefore with probability at least $\frac{1}{4}\epsilon^k$ the walk visits $z + ke_1$ after k steps. Hence, by Lévy's Borel–Cantelli lemma, $z + ke_1$ will be visited infinitely often a.s. The same statement holds for the other three directions. □

Remark 9.3. To show that this grid indeed can be random, consider the following sequence of p_i's: $1/2, 1, 0, 0, 1, 0, 0, 1, 0, 0, 1, \ldots$ (with period three). This leads to two possible grids both congruent to the shifted and stretched $\mathbb{Z}^2$.

In dimension $d = 3$, Lemma 8.5 and the Borel–Cantelli lemma imply that the set of the points where the conservative random walk updates its direction infinitely often is empty a.s. Suppose that some point x is a recurrent point. Then there is at least one direction from which it is visited infinitely often, say e_j. Then by the same argument as in the two-dimensional case, we see that all the points on the ray $x + ke_j$, $k \geq 0$, are also recurrent points. At the same time, there is no point $x - ke_j$, $k \geq 0$ where the walk updates its direction infinitely many times. This is only possible if there are infinitely many distinct points on $x - ke_j$, $k \geq 0$ where the walk updates its direction. Hence, the whole line (in both directions) passing through x is visited infinitely often. So, the recurrent set consists of a collection of lines parallel to one of the axes.

We conclude this subsection with two proposed problems.

Problem 9.1 (Connectedness; $d = 3$). Is the recurrent set connected in $d = 3$?

Problem 9.2 (The asymptotic mesh of the two-dimensional grid). Let $d = 2$ and consider the recurrence grid. Let L_i be the

x-coordinate of the ith vertical line to the right of the y-axis. Are there infinitely many such lines a.s.? If so, does the limit $\lim_{i\to\infty} L_i/i$ exists a.s.? If so, is that limit deterministic?

9.4   Transience in Higher Dimensions

Theorem 9.5 (Strong transience in two dimensions). *When* $d \geq 2$ *and* $p_n < n^{-1/2-\varepsilon}$, $n \geq n_0$, *for some* n_0 *and* $\varepsilon > 0$, *the walk is strongly transient, i.e.,* $|S_n| \to \infty$ *a.s.*

Proof. We present the proof only for the case $d = 2$; the extension of the proof for $d \geq 3$ is straightforward.

First, we prove that S is weakly transient, namely, that with probability one, it hits $(0,0)$ only finitely often. This statement is obviously true if $\varepsilon > 1/2$, as $\sum p_n < \infty$, so without loss of generality from now on we assume that $\varepsilon \leq 1/2$. Moreover, $p_n < n^{-1/2-\varepsilon'}$ implies $p_n < n^{-1/2-\varepsilon}$ for $0 < \varepsilon < \varepsilon'$; hence, it suffices to prove the theorem only for small positive ε.

As before, let τ_n be the times when the direction of the walk might change (the nth update time); hence, for a fixed $m \geq 1$,

$$\mathbb{P}(\tau_{n+1} > m + k \mid \tau_n = m, \mathcal{F}_m) = (1 - p_{m+1})(1 - p_{m+2})$$

$$\cdots (1 - p_{m+k}) \overset{k\to\infty}{\to} 0$$

as $\sum p_n = \infty$. Let us define a subsequence of these stopping times, by choosing only those at which the walk switches the direction from horizontal to vertical or vice versa, see Figure 9.3. To do so formally, let $\kappa_n \in \{\pm e_1, \pm e_2\}$ be the random direction the walk chooses at time τ_n, set $\eta_0 := 0$ and

$$\eta_{j+1} = \inf\{\mathbb{N} \ni n > \eta_j : \kappa_{\tau_n} \perp \kappa_{\tau_{n-1}}\}.$$

Also observe that $\eta_{j+1} - \eta_j$ are i.i.d. $\mathsf{Geom}(1/2)$.

Define the events A_j $(j \geq 1)$ as

$$A_j := \{\exists a \in \mathbb{Z} \setminus \{0\} : X_{\tau_{\eta_j}} = (0, a) \text{ or } X_{\tau_{\eta_j}} = (a, 0)\},$$

that is, A_j is the event that at time τ_{η_j} the walker is located either on the x- or on the y-axis. The crucial observation is that in order to

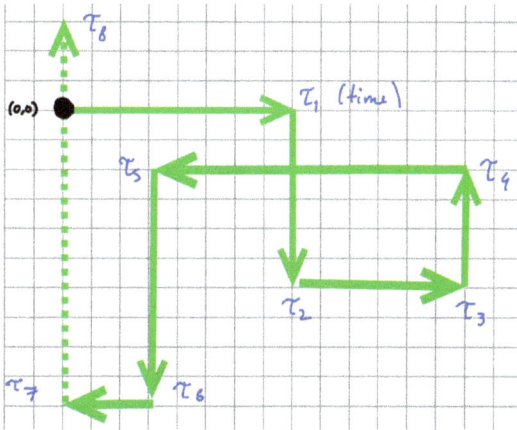

Fig. 9.3. A sample path of S_n assuming the walk always turns by 90 degrees.

hit the origin $(0,0)$ between τ_{η_j} and $\tau_{\eta_{j+1}}$, the event A_j must occur. Indeed, between the times τ_{η_j} and $\tau_{\eta_{j+1}}$ the walk moves only along the same line, either (x,t) or (t,y), $t \in \mathbb{Z}$, and unless $x = 0$ (resp. $y = 0$) the origin cannot be hit. Furthermore, if the walk is already on the horizontal or vertical axis, then it can hit zero only finitely many times[1] a.s. before leaving this axis. We show that almost surely, only a finite number of the A_j occur, hence the origin is only visited finitely many times, thus proving non-recurrence.

Fix j, denote $\eta_j =: \ell$, and without the loss of generality, suppose that at time τ_ℓ the walker starts moving horizontally along the line $(t,y)_{t \in \mathbb{Z}}$ for some $y \neq 0$. Let $m \geq 0$ and

$$
\begin{aligned}
B_{m+1} &:= \{\eta_{j+1} = \ell + m + 1\} \\
&= \{\kappa_{\ell+1}, \kappa_{\ell+2}, \ldots, \kappa_{\ell+m} \in \{-e_1, e_1\}, \kappa_{\ell+m+1} \in \{-e_2, e_2\}\}
\end{aligned}
$$

and note that κ's are i.i.d. uniform on $\{\pm e_1, \pm e_2\}$ and are independent of all η's. We have for all $x \in \mathbb{Z} \setminus \{0\}$ that almost surely

$$
\begin{aligned}
&\mathbb{P}\left(A_{j+1} \mid \mathcal{F}_{\tau_\ell}, B_{m+1}, S_{\tau_{\ell+m}} = (x, y)\right) \\
&= \frac{1}{2}(1 - p_{\tau_{\ell+m}+1})(1 - p_{\tau_{\ell+m}+2}) \cdots (1 - p_{\tau_{\ell+m}+|x|-1}) p_{\tau_{\ell+m}+|x|}
\end{aligned}
$$

---

[1]In fact, bounded by a Geometric(1/4) random variable.

$$< p_{\tau_{\ell+m}+|x|}$$

$$\leq \frac{1}{(\tau_{\ell+m}+|x|)^{1/2+\varepsilon}} \leq \frac{1}{(\tau_{\ell+m})^{1/2+\varepsilon}} \leq \frac{1}{(\tau_{\eta_j})^{1/2+\varepsilon}} \tag{9.12}$$

(if $x = 0$, then the probability on the left-hand side is zero) where the factor $1/2$ comes from the fact that at time $\tau_{\ell+m}$ the walk has an option of going toward or away from zero with equal probabilities; in the final two inequalities, we assumed that j is sufficiently large (i.e., $j \geq n_0$, where n_0 is as in the statement) and monotonicity of τ_k's in k.

Since the right-hand side of (9.12) does not depend on x, y and m, we conclude that

$$\mathbb{P}(A_{j+1} \mid \mathcal{F}_{\tau_{\eta_j}}) \leq \frac{1}{\tau_{\eta_j}^{1/2+\varepsilon}}, \quad \text{a.s.} \tag{9.13}$$

for large js.

Our next goal is to show that the τ_j are "rare" in the sense that for $n_j := (j/8)^{\frac{1}{1/2-\varepsilon}}$,

$$\tau_j \geq n_j, \quad \text{with finitely many exceptions, a.s.} \tag{9.14}$$

Then (9.14) implies $\sum_j \mathbb{P}(A_{j+1} \mid \mathcal{F}_{\tau_{\eta_j}}) < \infty$ a.s., and thus (by the conditional Borel–Cantelli lemma) that only a finite number of A_js can occur a.s. Indeed, by (9.13),

$$\sum_{j=n_0+1}^{\infty} \mathbb{P}(A_{j+1} \mid \mathcal{F}_{\tau_{\eta_j}}) \leq \sum_{j=n_0+1}^{\infty} \frac{1}{\tau_j^{1/2+\varepsilon}}, \quad \text{a.s.}$$

and, using (9.14), the sum on the right-hand side is a.s. finite, as

$$\sum_j \frac{1}{n_j^{1/2+\varepsilon}} = \sum_j \frac{1}{(j/8)^{\frac{1+2\varepsilon}{1-2\varepsilon}}} < \infty.$$

It remains to verify (9.14), and as we discussed at the beginning, we may (and will) assume that $\varepsilon \in (0, 1/4)$. For this, note that

$$\mathbb{P}(\tau_j < n) = \mathbb{P}(\xi_1 + \cdots + \xi_n > j), \tag{9.15}$$

where the ξ_js are independent Bernoulli random variables with $\mathbb{P}(\xi_k = 1) = p_k$. Using that $1 + x < e^x$, $x > 0$ along with Markov's inequality, we obtain that for some positive constant C_1,

$$\mathbb{P}(\xi_1 + \cdots + \xi_n > j)$$

$$= \mathbb{P}(2^{\xi_1 + \cdots + \xi_n} > 2^j) \leq 2^{-j} \prod_{i=1}^{n} \mathbb{E} 2^{\xi_i} = 2^{-j} \prod_{i=1}^{n} (1 + p_i)$$

$$< C_1 2^{-j} \prod_{i=1}^{n} \left(1 + \frac{1}{i^{\frac{1}{2}+\varepsilon}} \right) < C_1 2^{-j} \exp \left\{ \sum_{i=1}^{n} i^{-\frac{1}{2}-\varepsilon} \right\}$$

$$< C_1 2^{-j} \exp \left\{ \int_{i=0}^{n} x^{-\frac{1}{2}-\varepsilon} \, dx \right\} = C_1 2^{-j} \exp \left(\frac{n^{\frac{1}{2}-\varepsilon}}{\frac{1}{2} - \varepsilon} \right)$$

$$\leq C_1 \exp \left(4 n^{\frac{1}{2}-\varepsilon} - j \ln 2 \right),$$

which, along with (9.15) and plugging in $n_j = (j/8)^{\frac{1}{1/2-\varepsilon}}$, leads to the estimate

$$\mathbb{P}(\tau_j < n_j) < C_1 e^{(1/2 - \ln 2)j} < C_1 e^{-0.193j}.$$

Thus, (9.14) follows by the Borel–Cantelli lemma. This completes the proof of non-recurrence.

To upgrade the proof to strong transience is straightforward. Fix any point $(x_0, y_0) \in \mathbb{Z}^2$ and redefine the events A_j as

$$A_j := \{ \exists a \in \mathbb{Z} \setminus \{0\} : X_{\tau_{n_j}} = (x_0, a) \text{ or } X_{\tau_{n_j}} = (a, y_0) \}.$$

Following the weak transience proof verbatim, we obtain that a.s. the point (x_0, y_0) is visited only finitely many times. Hence, the same is true for *all* the points (x_0, y_0) such that $|(x_0, y_0)| \leq r$ for a fixed $r \geq 0$. We conclude that the walk leaves eventually each given bounded set a.s., which completes the proof of the theorem. $\square$

The following statement is more general than the two-dimensional one in Theorem 9.5, as it works for all $d > 1$, however, it requires quite strong regularity conditions.

Theorem 9.6 (Strong transience for $d \geq 2$). *Let $d \in \{2, 3, 4, \ldots\}$ and consider the d-dimensional conservative random walk S described in Section 9.1.1 with $S_n = Y_1 + \cdots + Y_n$. Assume that for* *some $\varepsilon > \varepsilon' > 0$ and $r > 1$, the sequence of p_n's satisfies the following conditions:*

$$\limsup_{n\to\infty} \frac{\max\limits_{k\in[0,n^{1-\varepsilon'}]} p_{n-k}}{\min\limits_{k\in[0,n^{1-\varepsilon'}]} p_{n-k}} < r, \tag{9.16}$$

$$\lim_{n\to\infty} \frac{p_n n^{1-\varepsilon}}{\ln n} = \infty, \tag{9.17}$$

$$\sum_{n=1}^{\infty} \left(\frac{p_n}{n^{1-\varepsilon}}\right)^{d/2} < \infty. \tag{9.18}$$

Then $\sum_{n=1}^{\infty} \mathbb{P}(S_n = w) < \infty$ for any $w \in \mathbb{Z}^d$, and the walk S is thus strongly transient.

Example 9.1 (Inverse sublinear decay). Assume that $\gamma \in (0,1)$, and $p_n = (c+o(1))/n^{\gamma}$. Under this assumption (9.16) is automatically satisfied for any $\varepsilon' > 0$, and assumptions (9.17) and (9.18) hold too (pick $\varepsilon < \min\{\gamma, 1-\gamma\}$). In this case, therefore, the walk exhibits strong transience. Finally, note that, as we have seen in Chapter 6, the $\gamma = 1$ assumption (critical case) produces a behavior that is dramatically different from that in the $\gamma \in (0,1)$ case. Nevertheless, by Theorem 9.5, strong transience still holds.

Remark 9.4. Concerning the assumptions in Theorem 9.6, note the following:

(i) Assumption 9.16 implies that $p_n > 0$ for all sufficiently large n.

(ii) Assumptions 9.16 and (9.18) for $d = 2$ imply that $p_n \to 0$ as $n \to \infty$. Indeed, if along a subsequence, $p_{n_k} \geq c > 0$ for $\forall k \geq 1$, then $\sum_{n=1}^{\infty} \frac{p_n}{n^{1-\varepsilon}} \geq \frac{c}{r} \sum_{k=1}^{\infty} \frac{n_k^{1-\varepsilon'}}{n_k^{1-\varepsilon}} = \infty$.

(iii) Assumptions 9.16, (9.17) and (9.18) always hold if $d \geq 3$ and $\liminf_{n\to\infty} p_n > 0$ (by (ii), this is ruled out when $d = 2$), although, then strong transience is anything but surprising.

(iv) For $d \geq 3$, Assumption (9.18) is automatically satisfied when $0 < \varepsilon' < \varepsilon < 1 - 2/d$.   ◇

Proof. Our goal is to show that for any given lattice point $w \in \mathbb{Z}^d$, one has

$$\sum_{n \geq 1} \mathbb{P}(S_n = w) < \infty. \tag{9.19}$$

The proof proceeds in three steps. First, we introduce the embedded process, considering the walk only at the times when it updates its direction. In the second step, we show that the probability that the walk makes too many updates during a certain time interval is negligible. Finally, we estimate the probability of hitting a specific point using the inversion formula for lattice distributions.

Step 1: Recall that $\eta_j \in \{0, 1\}$ was the indicator function of the update occurring at time j; the η_j's are independent with $\mathbb{P}(\eta_j = 1) = 1 - \mathbb{P}(\eta_j = 0) = p_j$. We now consider the walk S_k for times $k = n, n-1, n-2, \ldots$ backward. Let $\tau_0 := n$ and let τ_k's be the decreasing sequence of update times; formally,

$$\tau_k = \max\{j < \tau_{k-1} : \eta_j = 1\}, \quad k = 1, 2, \ldots.$$

For definiteness, if for some k we have $\eta_j = 0$ for $j = 0, 1, 2, \ldots, \tau_k - 1$, then we set $\tau_{k+1} = \tau_{k+2} = \cdots = 0$. We estimate the summands in (9.19) as follows. Let $m = m(n) := \lfloor n^{1-\varepsilon} p_n \rfloor$ and $V_n := S_n - S_{\tau_m}$. Clearly, a.s.

$$\mathbb{P}(S_n = w) = \mathbb{P}(S_n - S_{\tau_m} = w - S_{\tau_m}), \tag{9.20}$$

and note also that S_{τ_m} and $S_n - S_{\tau_m}$ are independent. Using the bound

$$\leq \sup_{z \in \mathbb{Z}^d} \mathbb{P}(S_n - S_{\tau_m} = z) = \sup_{z \in \mathbb{Z}^d} \mathbb{P}(V_n = z), \tag{9.21}$$

it is enough to find numbers γ_n such that

$$\sup_{z \in \mathbb{Z}^d} \mathbb{P}(V_n = z) < \gamma_n \quad \text{a.s. and} \quad \sum_{n \geq 1} \gamma_n < \infty. \tag{9.22}$$

For that, the distribution of V_n is handled by inverting its characteristic function, after which some elementary but tedious computations are carried out to bound the multiple integrals involved. In fact, for simplicity (and without loss of rigor), we assume that $n^{1-\varepsilon} p_n$ is an integer.

We assume that n is so large that the ratio in the lim sup in (9.16) does not exceed r and show in the following that, loosely speaking,

- the probability of the update "does not change significantly" over the time segment $[n - n^{1-\varepsilon}, n]$,
- the variables $l_k = \tau_{k-1} - \tau_k$, $k = 1, 2, \ldots, m$ are "nearly" i.i.d. $\mathsf{Geom}(p_n)$,
- $\mathbb{P}(l_1 + \cdots + l_m > n^{1-\varepsilon'}/2)$ is "very small,"

where the l_k are defined via

$$Z_k := S_{\tau_{k-1}} - S_{\tau_k} \in \begin{cases} \{(\pm l_k, 0), (0, \pm l_k)\}, \ d = 2, \\ \{(\pm l_k, 0, 0), (0, \pm l_k, 0), (0, 0, \pm l_k)\}, \ d = 3; \\ \ldots \end{cases}$$

Note that

$$V_n = Z_1 + \cdots + Z_m$$

(m depends on n). To be more precise, first note that, given $l_1, l_2, \ldots, l_{k-1}, l_k$, Z_k has the following distribution:

$$Z_k = \begin{cases} (l_k, 0) & \text{with probability } 1/4; \\ (-l_k, 0) & \text{with probability } 1/4; \\ (0, l_k) & \text{with probability } 1/4; \\ (0, -l_k) & \text{with probability } 1/4, \end{cases}$$

if $d = 2$, with similar formulae for $d \geq 3$.

Step 2: Denoting

$$A := \{l_1 + \cdots + l_m = n - \tau_m > n^{1-\varepsilon'}/2\}, \tag{9.23}$$

we now show that

$$\mathbb{P}(A) = o\left(e^{-n^{\varepsilon - \varepsilon'}}\right), \quad \text{as } n \to \infty. \tag{9.24}$$

The event A is the same as having less than $m = n^{1-\varepsilon}p_n$ "updates" in the time segment $[n - \frac{1}{2}n^{1-\varepsilon'}, n]$. The probability of each update at those time points is no less than p_n/r by (9.16), independent of the others. Hence, $\mathbb{P}(A) \leq \mathbb{P}(W < m)$, where $W \sim \mathsf{Bin}(N, q)$ with $N = \lfloor n^{1-\varepsilon'}/2 \rfloor$ and $q = p_n/r \in (0, 1/r]$. Since $(Nq)^i/i!$ is an increasing function in i for $i < Nq$ and $m \le Nq$ as well as $m = o(N)$ as $n \to \infty$, we have

$$
\mathbb{P}(W < m) = \sum_{i=0}^{m-1} \binom{N}{i} q^i (1-q)^{N-i} < \sum_{i=0}^{m-1} \frac{(Nq)^i}{i!} e^{-qN(1+o(1))}
$$

$$
\le m \cdot \frac{(Nq)^m}{m!} e^{-qN(1+o(1))} \le e^{m \log N - qN(1+o(1))}
$$

$$
= e^{-qN(1+o(1))}.
$$

Since $\ln n = o(m)$ by (9.17), we conclude that $\ln n \cdot n^{\varepsilon - \varepsilon'} = o(qN)$, proving (9.24).

Step 3: Recall that our goal is to find a sequence $(\gamma_n)_{n \ge 1}$ satisfying (9.22). In order to obtain the distribution of $V_n = S_n - S_{\tau_m}$, we invert its characteristic function and discuss the cases $d = 2$ and $d \ge 3$ separately. The following $\bullet$ denotes the usual dot product in $\mathbb{R}^d$.

First consider the case $d = 2$. Let $h(t_x, t_y) := \left| \mathbb{E}e^{it \bullet V_n} \right|$, $t = (t_x, t_y)$. Since V_n has a lattice distribution, the inversion formula is particularly simple (see, e.g., Chapter 15, Problem 26 in Fristedt and Gray (1997) or Dynkin and Yushkevich (1969)): for any $z = (z_x, z_y) \in \mathbb{Z}^2$, we have

$$
(*) := \mathbb{P}(V_n = z) = \frac{1}{(2\pi)^2} \int_0^{2\pi} \int_0^{2\pi} e^{-i(t \bullet z)} \mathbb{E}e^{it \bullet V_n} \, dt_x \, dt_y
$$

$$
\le \frac{1}{(2\pi)^2} \int_0^{2\pi} \int_0^{2\pi} h(t_x, t_y) \, dt_x \, dt_y
$$

$$
= \frac{1}{\pi^2} \int_0^{\pi} \int_0^{\pi} h(t_x, t_y) \, dt_x \, dt_y. \tag{9.25}
$$

To verify the ultimate equality, note that $V_n \in \mathbb{Z}^2$, and its distribution is symmetric in both coordinate variables. Hence, $h(u, v) = h(-u, v) = h(u, -v) = h(-u, -v)$ and the integrals of h on the squares $[0, \pi] \times [0, \pi]$, $[0, \pi] \times [\pi, 2\pi]$, $[\pi, 2\pi] \times [0, \pi]$ and $[\pi, 2\pi] \times [\pi, 2\pi]$ agree.

Since, given l_k, $k = 1, 2, \ldots, m$, Z_k is equally likely to be $(0, \pm l_k)$, $(\pm l_k, 0)$ independently of everything else, we have

$$\mathbb{E}[e^{it \bullet V_n} \mid l_1, \ldots, l_m] = \prod_{k=1}^{m} \mathbb{E}[e^{it \bullet Z_k} \mid l_k] = \prod_{k=1}^{m} \frac{\cos(t_x l_k) + \cos(t_y l_k)}{2}.$$

Consequently,

$$h(t_x, t_y) = \left| \mathbb{E} \left(\mathbb{E}[e^{it \bullet V_n} \mid l_1, \ldots, l_m] \right) \right| \leq \mathbb{E} \left(\left| \mathbb{E}[e^{it \bullet V_n} \mid l_1, \ldots, l_m] \right| \right)$$

$$\leq \mathbb{E} \left[\prod_{k=1}^{m} \phi(t; l_k) \right], \tag{9.26}$$

where

$$\phi(t; l_k) = \phi(t_x, t_y; l_k) = \frac{|\cos(t_x l_k)| + |\cos(t_y l_k)|}{2}.$$

Since l_k is an integer,

$$\phi(\pi - t_x, t_y; l_k) = \phi(t_x, \pi - t_y; l_k) = \phi(\pi - t_x, \pi - t_y; l_k) = \phi(t; l_k).$$

Consequently, from (9.25),

$$(*) \leq \frac{1}{\pi^2} \int_0^\pi \int_0^\pi \mathbb{E} \left[\prod_{k=1}^{m} \phi(t, l_k) \right] dt_x \, dt_y$$

$$= \frac{4}{\pi^2} \int_0^{\pi/2} \int_0^{\pi/2} \mathbb{E} \left[\prod_{k=1}^{m} \phi(t; l_k) \right] dt_x \, dt_y. \tag{9.27}$$

To proceed with the estimation, we now need a lemma. (Recall that the lengths l_k are defined for a given n.)

Lemma 9.1. *Given $(p_k)_{k \geq 1}$ satisfying our assumptions, there exists an $N \in \mathbb{N}$ with the following property. Let $n \geq N$. Let $2 \leq k \leq m = m(n)$. Then, we have ω-wise that*

$$\mathbb{E}(|\cos(s l_k)| \mid l_1, \ldots, l_{k-1}) \leq \psi(s) \mathbb{1}_{A_{k-1}^c} + \mathbb{1}_{A_{k-1}}, \quad \forall s \in [0, \pi/2], \tag{9.28}$$

where

$$A_{k-1} = \left\{ l_1 + \cdots + l_{k-1} > \frac{1}{2} n^{1-\varepsilon'} \right\} \overset{(m \geq k)}{\subseteq} A, \qquad (9.29)$$

$$\psi(s) = \psi_n(s) = \max \left(1 - \frac{c_1 s^2}{p_n^2}, 1 - c_2 \right)$$

and $c_1 > 0$, $c_2 \in (0,1)$ are constants that depend on r only.

Proof. Since the inequality is trivial when A_{k-1} occurs, from now on we assume that we are on A_{k-1}^c. For a fixed $k \in \{2, \ldots, m\}$ and $j \geq 1$, let $q_j := \mathbb{P}(l_k = j \mid l_1, \ldots, l_{k-1})$. Then

$$q_j = p_{n-l_1-\cdots-l_{k-1}-j} \times \prod_{i=1}^{j-1} (1 - p_{n-l_1-\cdots-l_{k-1}-i}), \qquad (9.30)$$

as the product of the probabilities to turn first, and then to keep the same direction for $j - 1$ steps (where the product equals 1 by definition for $j = 1$).

As long as $l_1 + \cdots + l_{k-1} + i < n^{1-\varepsilon'}$ by (9.16) we have that the above p's lie in $[p_n/r, r p_n]$. Consider two cases:

$$\text{(I) } p_n \geq \frac{1}{2r} \quad \text{and} \quad \text{(II) } p_n < \frac{1}{2r}.$$

In case (I), we apply Lemma 9.4 with $M = 1$ and thus consider only q_1, which $\geq \frac{1}{2r} \times r^{-1}$ for large n. Hence, choosing $a = a_{(I)}(r) = \frac{1}{2r^2}$ in the lemma yields that the left-hand side of (9.28) does not exceed

$$\max \left(1 - c_1' a_{(I)} s^2 M^2, 1 - c_2' a_{(I)} \right) = \max \left(1 - \frac{c_1' s^2}{2r^2}, 1 - \frac{c_2'}{2r^2} \right)$$

$$\leq \max \left(1 - \frac{c_1' s^2}{8r^4 p_n^2}, 1 - \frac{c_2'}{2r^2} \right),$$

$$(9.31)$$

where the equality holds due to the fact that $M = 1$ and the inequality holds because $p_n < 1/(2r)$.

In case (II), note that

$$q_j \geq (1 - rp_n)^{j-1} \frac{p_n}{r}$$

as long as $l_1 + \cdots + l_{k-1} + j < n^{1-\varepsilon'}$; since we are on A_{k-1}^c, this is fulfilled, provided $j < \frac{n^{1-\varepsilon'}}{2}$. Let $M := \lfloor 1/(2p_n) \rfloor$ and observe that since $p_n < 1/(2r)$,

$$\frac{1 - r^{-1}}{2p_n} \leq \frac{1 - 2p_n}{2p_n} = \frac{1}{2p_n} - 1 \leq M,$$

$$M \leq \frac{1}{2p_n} = \frac{n^{1-\varepsilon'}}{2} \cdot \frac{1}{p_n n^{1-\varepsilon'}} \overset{\text{by (9.17)}}{\leq} \frac{n^{1-\varepsilon'}}{2}, \qquad (9.32)$$

provided n is large enough. Thus, for $j = 1, 2, \ldots, M$, we have

$$q_j \geq (1 - rp_n)^{M-1} \frac{p_n}{r} \geq (1 - rp_n)^{\frac{1}{2p_n} - 1} \frac{p_n}{r}$$

$$\geq (r - 1)(1 - rp_n)^{\frac{1}{2p_n} - 1}(r^2 M)^{-1}$$

since $p_n \geq \frac{1 - r^{-1}}{M}$. An elementary calculation shows that

$$\inf_{p_n \in (0,(2r)^{-1}]} (1 - rp_n)^{\frac{1}{2p_n} - 1} \geq \inf_{p_n \in (0,(2r)^{-1}]} (1 - rp_n)^{\frac{1}{2p_n}}$$

$$= \inf_{y \in (0,1/2]} (1 - y)^{\frac{r}{2y}} = 2^{-r},$$

since $(1-y)^{1/y}$ is a decreasing function in y. Hence, $q_j \geq \frac{a_{(\text{II})}}{M}$, where $a_{(\text{II})} = \frac{r-1}{2^r r^2}$. Therefore, by Lemma 9.4 with $a = a_{(\text{II})}$, the left-hand side of (9.28) does not exceed

$$\max\left(1 - c_1' a_{(\text{II})} s^2 M^2, 1 - c_2' a_{(\text{II})}\right)$$

$$\overset{(9.32)}{\leq} \max\left(1 - \frac{c_1' a_{(\text{II})} s^2 (1 - r^{-1})^2}{p_n^2}, 1 - c_2' a_{(\text{II})}\right)$$

$$\leq \max\left(1 - \frac{c_1' s^2 (r-1)^3}{2^{r-1} r^4 p_n^2}, 1 - c_2' \frac{r-1}{2^{r-1} r^2}\right).$$

Finally, choosing

$$c_1 := c_1' \min\left(\frac{1}{8r^4}, \frac{(r-1)^3}{2^{r-1}\,r^4}\right),$$

$$c_2 := \min\left(\frac{c_2'}{2r^2}, \frac{c_2'(r-1)}{2^r\,r^2}, \frac{1}{2}\right) \leq \frac{1}{2}$$

concludes the proof. ☐

Let us now return to the proof of Theorem 9.6. Setting $k := m$, from Lemma 9.1, it follows that when $t_x, t_y \in [0, \pi/2]$,

$$\mathbb{E}\left[\prod_{u=1}^{m}\phi(t, l_u)\right] \leq \mathbb{E}\left(\mathbb{E}\left[\prod_{u=1}^{m}\phi(t, l_u) \mid l_1, \ldots, l_{m-1}\right]\right)$$

$$\leq \mathbb{E}\left(\prod_{u=1}^{m-1}\phi(t, l_u)\left[\frac{\psi(t_x)+\psi(t_y)}{2}\,\mathbb{1}_{A^c_{m-1}} + \mathbb{1}_{A_{m-1}}\right]\right)$$

$$\leq \frac{\psi(t_x)+\psi(t_y)}{2}\,\mathbb{E}\left[\prod_{u=1}^{m-1}\phi(t, l_u)\right] + \mathbb{P}(A_{m-1})$$

$$= \cdots = \left(\frac{\psi(t_x)+\psi(t_y)}{2}\right)^m + \mathbb{P}(A_{m-1}) + \mathbb{P}(A_{m-2}) + \cdots + \mathbb{P}(A_1)$$

$$\leq \left[\frac{\psi(t_x)+\psi(t_y)}{2}\right]^m + m\,\mathbb{P}(A) = \left[\frac{\psi(t_x)+\psi(t_y)}{2}\right]^m$$

$$+ o\left(n^{1-\varepsilon}e^{-n^{\varepsilon-\varepsilon'}}\right), \tag{9.33}$$

by induction over $k = m, m-1, \ldots, 3, 2$ for $\mathbb{E}\left[\prod_{u=1}^{k}\phi(t, l_u)\right]$ and using (9.24) and (9.29).

This, along with (9.27), yields

$$(*) \leq \frac{4}{\pi^2}\int_0^{\pi/2}\int_0^{\pi/2}\left[\frac{\psi(t_x)+\psi(t_y)}{2}\right]^m dt_x\,dt_y + o\left(n^{1-\varepsilon}e^{-n^{\varepsilon-\varepsilon'}}\right).$$
$$\tag{9.34}$$

The second summand in this inequality can be ignored, as it is summable. To estimate the double integral, we split the area of integration into four regions, depending on whether t_x (t_y resp.) is greater or smaller than $\bar{t} := p_n \sqrt{\frac{c_2}{c_1}} < \frac{\pi}{2}$ (this is the value of t_x (t_y resp.) which makes the candidates for the maximum in $\psi(t)$ equal.) We also assume that n is large enough (since $p_n \to 0$ when $d = 2$ by (ii) of Remark 9.4). Then

$$(*) \leq \frac{4}{\pi^2} \int_0^{\bar{t}} \int_0^{\bar{t}} \left[\frac{\psi(t_x) + \psi(t_y)}{2} \right]^m dt_x\, dt_y$$

$$+ \frac{4}{\pi^2} \iint_{t\in[0,\frac{\pi}{2}]^2\setminus[0,\bar{t}]^2} \left[\frac{\psi(t_x) + \psi(t_y)}{2} \right]^m dt_x\, dt_y =: (I) + (II).$$

Turning to scaled polar coordinates: $t_x = \sqrt{2 p_n} \rho \cos\theta$, $t_y = \sqrt{2 p_n} \rho \sin\theta$, we have

$$(I) = \frac{1}{\pi^2} \int_{-\bar{t}}^{\bar{t}} \int_{-\bar{t}}^{\bar{t}} \left[1 - \frac{c_1 t_x^2 + c_1 t_y^2}{2 p_n^2} \right]^m dt_x\, dt_y$$

$$\leq \frac{2 p_n^2}{\pi^2} \int_0^{2\pi} \int_0^{\sqrt{1/c_1}} \rho \left(1 - c_1 \rho^2 \right)^m d\rho\, d\theta$$

$$= \frac{4 p_n^2}{\pi} \cdot \frac{1}{2 c_1 (m+1)} = \frac{2 p_n^2}{\pi c_1 (m+1)} \leq \frac{2 p_n}{\pi c_1\, n^{1-\varepsilon}}.$$

On the other hand,

$$(II) \leq \frac{4}{\pi^2} \iint_{t\in[0,\frac{\pi}{2}]^2\setminus[0,\bar{t}]^2} \left[\frac{1 + (1 - c_2)}{2} \right]^m dt_x\, dt_y$$

$$\leq \frac{4}{\pi^2} \iint_{t\in[0,\frac{\pi}{2}]^2} \left[\frac{2 - c_2}{2} \right]^m dt_x\, dt_y = \left(1 - \frac{c_2}{2} \right)^m.$$

Consequently, almost surely,

$$\mathbb{P}(S_n - S_{\tau m} = z) = \mathbb{P}(V_n = z) \leq (I) + (II) = \frac{p_n}{\pi c_1\, n^{1-\varepsilon}}$$

$$+ \left(1 - \frac{c_2}{2} \right)^{p_n n^{1-\varepsilon}} =: \gamma_n.$$

Since the bound is uniform in z, we actually obtained that

$$\sup_{z \in \mathbb{Z}^2} \mathbb{P}(V_n = z) \leq \gamma_n, \tag{9.35}$$

which is summable in n because of (9.17) and (9.18). Thus, (9.22) is achieved.

Consider now the case $d \geq 3$. The proof of the $d = 2$ case carries through up to formula (9.25), but now instead of double integrals, one has to deal with multiple ones. Namely, analogously to (9.34), one now obtains that

$$
\mathbb{P}(V_n = z) \leq \left(\frac{2}{\pi}\right)^d \int_{[0,\pi/2]^d} \cdots \int \left(\frac{\psi(t_1) + \psi(t_2) + \cdots + \psi(t_d)}{d}\right)^m
$$

$$
\times \, dt_1 \, dt_2 \ldots dt_d + o\left(n^{1-\varepsilon} e^{-n^{\varepsilon - \varepsilon'}}\right),
$$

and we split the area of integration in the same way as in the two-dimensional case, using the same $\bar{t}$ as in the two-dimensional case. However, since $d \geq 3$, we can no longer assume that $p_n \to 0$, and it might happen that $\bar{t} \geq \pi/2$; in this case, a multiple integral analogous to (II) is no longer present in the computation, making it easier. On the other hand, when that term is present (i.e., when $\bar{t} < \pi/2$), its estimate is similar to the two-dimensional case:

$$
(II) = \left(\frac{2}{\pi}\right)^d \int_{t \in [0,\pi/2]^d \setminus [0,\bar{t}]^d} \cdots \int \left(\frac{\psi(t_1) + \psi(t_2) + \cdots + \psi(t_d)}{d}\right)^m
$$

$$
\times \, dt_1 \, dt_2 \ldots dt_d
$$

$$
\leq \left(\frac{2}{\pi}\right)^d \int_{t \in [0,\pi/2]^d} \cdots \int \left(\frac{1 + \cdots + 1 + (1 - c_2)}{d}\right)^m dt_1 \, dt_2 \ldots dt_d
$$

$$
= \left(1 - \frac{c_2}{d}\right)^m.
$$

The estimate of the multiple integral defined analogously to (I) however is different: using scaled d-dimensional spherical coordinates $t_1 = p_n \cdot \rho \cos \theta_1$, $t_2 = p_n \cdot \rho \sin \theta_1 \cos \theta_2$, $t_3 = p_n \cdot \rho \sin \theta_1 \sin \theta_2 \cos \theta_3$, $\ldots$, $t_d = p_n \cdot \rho \sin \theta_1 \sin \theta_2 \ldots \sin \theta_{d-1}$ and exploiting the facts that the expression in the parentheses in the integrand below is positive in the cube $[-\bar{t}, \bar{t}]^d$, which lies inside the ball centered at the origin with radius $\sqrt{d/c_1}$ (note that $c_2 < 1$), and the Jacobian J of the transform satisfies $|\det(J)| \le \rho^{d-1}$, one arrives at[2]

$$(I) = \frac{1}{\pi^d} \int \cdots \int_{[-\bar{t},\bar{t}]^d} \left(1 - \frac{c_1 t_1^2 + c_1 t_2^2 + \cdots + c_1 t_d^2}{p_n^2 \, d}\right)^m dt_1 \, dt_2 \ldots dt_d$$

so

$$(I) \le \frac{p_n^d}{\pi^d} \int \cdots \int_{\rho \in \left[0, \sqrt{\frac{d}{c_1}}\right], \theta_1, \ldots, \theta_{d-2} \in [0,\pi], \theta_{d-1} \in [0,2\pi]} \rho^{d-1} \left(1 - \frac{c_1 \rho^2}{d}\right)^m d\rho$$

$$\times \, d\theta_1 \ldots d\theta_{d-1}$$

$$= \frac{2 p_n^d}{\pi} \int_0^{\sqrt{d/c_1}} \rho^{d-1} \left[1 - \frac{c_1 \rho^2}{d}\right]^m d\rho = \frac{p_n^d}{\pi} \left(\frac{d}{c_1}\right)^{d/2}$$

$$\int_0^1 u^{\frac{d}{2}-1} (1-u)^m du$$

$$= \text{const } p_n^d \cdot \text{B}\left(\frac{d}{2}, m+1\right) \sim \text{const } \Gamma(d/2) p_n^d \cdot m^{-d/2}$$

$$\sim \text{const } \Gamma(d/2) \left(\frac{p_n}{n^{1-\varepsilon}}\right)^{d/2},$$

where B denotes the Beta function, and we exploited its well-known asymptotics, while we also recalled that $m = \lfloor n^{1-\varepsilon} p_n \rfloor$. By (9.18), the last expression is summable in n. As a result, the bound (I) + (II) is uniform in z and it is summable in n, just as in the $d = 2$ case. Again, (9.22) is confirmed. □

Luckily, we have a much stronger result for the case $d \ge 4$.

Theorem 9.7 (Strong transience; $d \ge 4$). *The conservative random walk is strongly transient for any sequence $\{p_n\}$ as long as $d \ge 4$.*

Proof. Let $S(n)$ denote the embedded process corresponding to the conservative random walk (previously denoted by $\tilde{S}_n$), which is the

---

[2]Note: One must replace the limits in the integrals by $\pm \pi/2$ if $\bar{t} \ge \pi/2$.

Rademacher walk introduced in Section 8.6. Then

$$S_t = S(K(t)) + (t - a_1 - \cdots - a_{K(t)})\mathbf{f_k},$$

where

$$K(s) = \min\{k \in \mathbb{Z}_+ : a_1 + \cdots + a_k \geq s\}.$$

Let also $T_k := (a_1 + \cdots + a_{k-1}, a_1 + \cdots + a_k] \cap \mathbb{Z}$.

Note that in order that $S_n = \mathbf{0}$, assuming that $k = K(t)$ and $\mathbf{f_k} = \pm e_j$, where e_j is the unit vector in the j-th direction, we necessary have that all but the jth coordinate of $S(k)$ are equal to zero. Moreover, only for at most one $t \in T_k$ we might have $S_t = \mathbf{0}$. Let

$$B_k = \{S_t = \mathbf{0} \text{ for some } t \in T_k).$$

Then

$$\mathbb{P}(B_k) \leq \mathbb{P}\left(\cup_{j=1}^d \{\text{all but the } j\text{-th coordinate of } S(k) \text{ are zeros}\}\right)$$

$$= d \cdot \mathbb{P}\left(\text{all but the } d\text{-th coordinate of } S(k) \text{ are zeros}\right)$$

$$\leq d \cdot \frac{C_{d-1}}{k^{(d-1)/2}} \tag{9.36}$$

by arguments identical to those in the proof of Lemma 8.5. Note also that the right-hand side does not depend on values of $a_1, a_2, \ldots$. Now, since $\sum_n \mathbb{P}(B_n) < \infty$ for $d \geq 4$, by the Borel–Cantelli lemma, we get the result. □

Unfortunately, the previous theorem does not cover the case $d = 3$, even though we believe its statement is still true.

Problem 9.3. Prove that the conservative random walk is transient for $d = 3$ for any sequence $\{p_n\}$.

9.5 Scaling Limit of the Conservative Walk in the Homogeneous Case

We exploit the following lemma later, but we think it is also of independent interest. Let L_n be a one-dimensional conservative walk, where $p_n = p \in (0, 1)$.

Proposition 9.1 (Large deviation-type bound; $d = 1$). *There exists an $N_0 \in \mathbb{N}$ such that if $n \geq N_0$, then for all $a \geq 1$,*

$$\mathbb{P}(|L_n| > a\sqrt{n}) \leq 2f(p, a), \quad \text{where } f(p, a) := \exp\left\{-p^2 a/5\right\}.$$
$$(9.37)$$

Before presenting the proof of Proposition 9.1, we state and prove a lemma which is a consequence of this proposition.

Lemma 9.2 (Upper bound on distance; $d \geq 2$). *If $d \geq 2$, then there is an $N_0 \in \mathbb{N}$ such that for all $a \geq \sqrt{d}$ and $n \geq N_0$,*

$$\mathbb{P}(|S_n| > a\sqrt{n}) \leq d f\left(p, a/\sqrt{d}\right),$$

where f is given in (9.37).

Proof. Let $S^{(j)}$, $j = 1, 2, \ldots, d$ be the jth coordinate of S. Since after an update, with probability $1/d$ the walk will be moving along the same axis as before, and with probability $1 - 1/d$, it will start moving in a perpendicular direction, we can write

$$S_n^{(j)} = L_\kappa^{(j)}$$

for some $\kappa = \kappa_n^{(j)} \in \{1, 2, \ldots, n\}$, and $L^{(j)}$ has the distribution of the one-dimensional walk as in Proposition 9.1. By Proposition 9.1, we have

$$\mathbb{P}\left(\left|S_n^{(j)}\right| > a\sqrt{\frac{n}{d}}\right) = \mathbb{P}\left(\left|L_\kappa^{(j)}\right| > \left(a\sqrt{\frac{n}{d\kappa}}\right)\sqrt{\kappa}\right) \leq 2f\left(p, a\sqrt{\frac{n}{d\kappa}}\right)$$

$$\leq 2f\left(p, a/\sqrt{d}\right)$$

since $a \mapsto f(p, a)$ is decreasing, $n/\kappa \geq 1$, and $a/\sqrt{d} \geq 1$. Finally,

$$\mathbb{P}(|S_n| \geq a\sqrt{n}) = \mathbb{P}\left(\sqrt{\sum_{j=1}^{d}\left(S_n^{(j)}\right)^2} \geq a\sqrt{n}\right)$$

$$\leq \mathbb{P}\left(|S_n^{(j)}| \geq a\sqrt{\frac{n}{d}} \text{ for some } j \in \{1, 2, \ldots, d\}\right)$$

$$\leq d\mathbb{P}\left(|L_n| \geq a\sqrt{\frac{n}{d}}\right) \leq d f\left(p, a/\sqrt{d}\right),$$

where Proposition 9.1 is used in the last inequality. $\qquad\square$

Proof of Proposition 9.1. Define the strictly increasing integer sequence of stopping times

$$0 = \tau_0 < \tau_1 < \tau_2 < \cdots,$$

when the walk updates its direction; it keeps going in the same direction between times τ_i and τ_{i+1}. Then the $\tau_k - \tau_{k-1} \sim \mathsf{Geom}(p)$, $k = 1, 2, \ldots$, are i.i.d. Moreover, $\tilde{L}_k = L_{\tau_k}$ defines the embedded walk, where

$$\tilde{L}_k = \xi_1 + \cdots + \xi_k,$$

with the i.i.d. variables $\xi_i \sim \mathsf{Sgeom}(p)$, while trivially, $|\xi_i| = \tau_i - \tau_{i-1}$. Let

$$\nu(n) := \min\{j \in \mathbb{Z}_+ : \tau_j \geq n\} \quad \text{so that } \tau_{\nu(n)-1} < n \leq \tau_{\nu(n)}$$

$$\text{and } \nu(n) \leq n. \tag{9.38}$$

Since the walk moves in the same direction between $\tau_{\nu(n)-1}$ and $\tau_{\nu(n)}$,

$$|L_n| \leq \max\left(|\tilde{L}_{\nu(n)-1}|, |\tilde{L}_{\nu(n)}|\right),$$

hence

$$\mathbb{P}\left(|L_n| \geq a\sqrt{n}\right) \leq \mathbb{P}\left(\left|\tilde{L}_{\nu(n)-1}\right| \geq a\sqrt{n}\right) + \mathbb{P}\left(\left|\tilde{L}_{\nu(n)}\right| \geq a\sqrt{n}\right). \tag{9.39}$$

Using Markov's inequality, we want to bound $\tilde{L}_m = \xi_1 + \cdots + \xi_m$. Indeed, for any positive integers n, $m \leq n$ (see (9.38)), and any $t \in (-p, p)$,

$$\mathbb{P}\left(\tilde{L}_m > a\sqrt{n}\right) = \mathbb{P}\left(\prod_{i=1}^m e^{t\xi_i} > e^{at\sqrt{n}}\right) \leq \frac{\left(\mathbb{E}e^{t\xi_1}\right)^m}{e^{at\sqrt{n}}} = e^{-\Lambda(a,t,m,n)}, \tag{9.40}$$

where $\Lambda(a, t, m, n) := at\sqrt{n} - m \log\left(\mathbb{E}e^{t\xi_1}.\right)$. Since $\mathbb{E}e^{t\xi_1} = \frac{1}{2}\mathbb{E}\left(e^{t|\xi_1|} + e^{-t|\xi_1|}\right) = \mathbb{E}\cosh(t|\xi_1|) \geq 1$, the function Λ is monotone decreasing in m, hence, for given p and t, $\Lambda(a,t,m,n)$ reaches its minimum over $m \leq n$ at $m = n$. Now, set $t(n) := \frac{p^2}{(2-p)\sqrt{n}}$ $(< p)$. Using that

$$\mathbb{E}e^{t(n)\xi_1} = \frac{p}{2}\left(\frac{1}{e^{-t(n)} - (1-p)} + \frac{1}{e^{t(n)} - (1-p)}\right),$$

it follows via Taylor expansion that

$$\Lambda(a,t(n),n,n) = \frac{p^2(2a-1)}{4-2p} + O(n^{-1}) \geq \frac{p^2 a}{5} \qquad (9.41)$$

for sufficiently large n, since $a \geq 1$ and $p \geq 0$. (Here the term $O(n^{-1})$ is uniform in a.) The result now follows from (9.39), (9.40) and (9.41).

Our next result shows that diffusive scaling leads to Brownian motion, up to a constant scaling factor.

Theorem 9.8 (Scaling limit in the homogeneous case). *Let $d \geq 1$ and $p \in (0,1)$. Extend the walk S to all non-negative times using linear interpolation, and for $n \geq 1$, define the rescaled walk S^n by*

$$S^n(t) := \sqrt{\frac{p}{2-p}} \cdot \frac{S_{nt}}{\sqrt{n}}, \quad t \geq 0,$$

and finally, let $\mathcal{W}^{(d)}$ denote the d-dimensional Wiener measure. Then $\lim_{n\to\infty} \mathsf{Law}(S^n) = \mathcal{W}^{(d)}$ on $C([0,\infty), \mathbb{R}^d)$.

Remark 9.5. (i) Informally, $\frac{S_n}{\sqrt{n}} \approx \sqrt{\frac{2-p}{p}} \cdot B_\cdot$, for large n, where B is a standard d-dimensional Brownian motion. Since $\sqrt{\frac{2-p}{p}} \in (1,\infty)$ for $p \in (0,1)$, the Brownian motion is "sped up." The intuition is that the updates are less frequent compared to a simple random walk, there is thus less cancellation in the steps.

(ii) Note that, e.g., for $d = 2$, the horizontal and vertical components of the walk are not independent because, for example, the horizontal component is idle (stays at one location) for the duration of a vertical "run." $\diamond$

Proof. While using the notation of the previous section, we also take the liberty of using the notation X_t as well as $X(t)$ for a stochastic process X, whichever is more convenient at the given instance.

We follow the standard route and prove the result by checking the convergence of the finite-dimensional distributions (fidis) along with tightness.

(a) **Convergence of fidi's:** We argue that the convergence of the fidi's is easy to check for an embedded walk, and the original random walk must have the same limiting fidi's.

To carry out this plan, recall from the proof of Theorem 9.3 the long-range embedded random walk, $\bar{S} = (\bar{S}_k)_{k\geq 0}$, where $\bar{S}_k := S_{T_k}$, $k \geq 0$. In this d-dimensional setting, its increment vectors are $\bar{S}_{l+1} - \bar{S}_l = \xi_l \sum_{i=1}^{d} \mathbb{1}_{\{U_l=i\}} \mathbf{e}_i$, $l = 0, 1, \ldots$, where $\xi_l \sim \mathsf{Sgeom}(p)$ and U_l is uniform on $1, 2, \ldots, d$ (and the system $\{\xi_l, U_m\}_{l,m\geq 0}$ is independent), while $\{\mathbf{e}_i\}_{1\leq i\leq d}$ are the unit basis vectors. Using that

$$\mathsf{Var}(\bar{S}_{l+1} - \bar{S}_l) = \mathbb{E}(\bar{S}_{l+1} - \bar{S}_l)^2 = \mathbb{E}\xi_l^2 \cdot \mathbb{E}\left|\sum_{i=1}^{d} \mathbb{1}_{\{U_l=i\}} \mathbf{e}_i\right|^2$$

$$= \mathbb{E}\xi_l^2 \cdot \sum_{i=1}^{d} \mathbb{E}\mathbb{1}_{\{U_l=i\}} = \frac{2-p}{p^2},$$

it follows that the increment vectors have mean value $\mathbf{0}$ and covariance matrix $\frac{2-p}{p^2} I_d$, where I_d is the unit matrix. Therefore, denoting $\gamma_p := \frac{p}{\sqrt{2-p}}$, we may apply the multidimensional Donsker invariance principle (see, e.g., Theorem 9.3.1 in Stroock (2011)) to the process $\hat{S} := \gamma_p \bar{S}$. We obtain that the rescaled walk $\hat{S}^n$ defined by

$$\hat{S}^n(t) := \frac{\hat{S}_{nt}}{\sqrt{n}} = \gamma_p \frac{\bar{S}_{nt}}{\sqrt{n}}, \quad t \geq 0, \tag{9.42}$$

satisfies

$$\lim_{n\to\infty} \mathsf{Law}(\hat{S}^n) = \mathcal{W}^{(d)} \text{ on } C([0,\infty), \mathbb{R}^d). \tag{9.43}$$

In other words,

$$\lim_{n\to\infty} \mathsf{Law}\left(t \mapsto \gamma_p \frac{S_{T(nt)}}{\sqrt{n}}\right) = \mathcal{W}^{(d)} \text{ on } C([0,\infty), \mathbb{R}^d), \tag{9.44}$$

where T is a random time change such that for integers $t = l \geq 0$, $T(l) := \tau_l$ and for $t = l+s$, $s \in (0,1)$, $T(t) := T(l)+(T(l+1)-T(l))s$.

Next, given that the waiting times for the updates are $\mathsf{Geom}(p)$, the strong law of large numbers implies that

$$\lim_{s\to\infty} T(s)/s \to 1/p, \quad a.s. \tag{9.45}$$

Let B be a standard d-dimensional Brownian motion. We know that for $0 \leq t_1 < t_2 < \cdots < t_k$,

$$\lim_{n\to\infty} \mathsf{Law}\left(\frac{\gamma_p}{\sqrt{n}} \left(S_{T(nt_1)}, S_{T(nt_2)}, \ldots, S_{T(nt_k)} \right) \right)$$

$$= \mathsf{Law}(B_{t_1}, B_{t_2}, \ldots, B_{t_k}), \tag{9.46}$$

and what we want to see next is that this implies that

$$\lim_{n\to\infty} \mathsf{Law}\left(\frac{\gamma_p}{\sqrt{n}} \left(S_{\frac{1}{p}nt_1}, S_{\frac{1}{p}nt_2}, \ldots, S_{\frac{1}{p}nt_k} \right) \right) = \mathsf{Law}(B_{t_1}, B_{t_2}, \ldots, B_{t_k}). \tag{9.47}$$

It is enough to show (by Slutsky's theorem) that the difference vector between the vectors on the left-hand sides of (9.46) and (9.47) converges in probability to $\mathbf{0}$ as $n \to \infty$. (These vectors are such that each of their components are in $\mathbb{R}^d$.) We check this componentwise.

Denote by T^{-1} the inverse of the (strictly increasing) map T. Fix $\epsilon > 0$ and define the random variables

$$t_{i,n}^* := \frac{1}{n} T^{-1}\left(\frac{1}{p}nt_i \right).$$

Then almost surely,

$$\lim_n t_{i,n}^* = \lim_n \frac{1}{n} T^{-1}\left(\frac{1}{p}nt_i \right) \overset{(9.45)}{=} t_i.$$

Fix $1 \leq i \leq k$. The ith component of the difference vector alluded to above satisfies

$$\mathbb{P}\left(\frac{\gamma_p}{\sqrt{n}} \left| S_{\frac{1}{p}nt_i} - S_{T(nt_i)} \right| > \epsilon \right) = \mathbb{P}\left(\frac{\gamma_p}{\sqrt{n}} \left| S_{T(nt_{i,n}^*)} - S_{T(nt_i)} \right| > \epsilon \right),$$

and we now verify that this converges to zero. Given that $\lim_n t_{i,n}^* = t_i$, a.s., it is enough to check that

$$\lim_{\delta\to 0} \limsup_{n\to\infty} \mathbb{P}\left(\frac{\gamma_p}{\sqrt{n}} \left| S_{T(nt_{i,n}^*)} - S_{T(nt_i)} \right| > \epsilon \mid E_n^{i,\delta} \right) = 0,$$

where

$$E_n^{i,\delta} := \{t_{i,n}^* \in (t_i - \delta, t_i + \delta)\}.$$

Since, for any fix $\delta > 0$, $\lim_n \mathbb{P}(E_n^{i,\delta}) = 1$, there is an $N_\delta \in \mathbb{N}$ such that $\mathbb{P}(E_n^{i,\delta}) \geq 1/2$ for $n > N_\delta$, hence

$$\limsup_{n \to \infty} \mathbb{P}\left(\frac{\gamma_p}{\sqrt{n}}\left|S_{T(nt_{i,n}^*)} - S_{T(nt_i)}\right| > \epsilon \mid E_n^{i,\delta}\right)$$

$$\leq 2 \limsup_{n \to \infty} \mathbb{P}\left(\sup_{s \in (t_i - \delta, t_i + \delta)} \frac{\gamma_p}{\sqrt{n}}\left|S_{T(ns)} - S_{T(nt_i)}\right| > \epsilon\right).$$

As $\delta \to 0$, the right-hand side tends to zero since (as a consequence of the convergence in law to $\mathcal{W}^{(d)}$)

$$\lim_n \mathbb{P}\left(\sup_{s \in (t_i - \delta, t_i + \delta)} \frac{\gamma_p}{\sqrt{n}}\left|S_{T(ns)} - S_{T(nt_i)}\right| > \epsilon\right)$$

$$= \mathcal{W}^{(d)}\left(\sup_{s \in (t_i - \delta, t_i + \delta)} |B_s - B_{t_i}| > \epsilon\right).$$

We now have verified (9.47), that is, that

$$\lim_{n \to \infty} \mathsf{Law}\left(t \mapsto \frac{p}{\sqrt{2 - p}}\frac{S_{n\tilde{t}}}{\sqrt{n}}\right) = \mathcal{W}^{(d)},$$

where $\tilde{t} := \frac{t}{p}$. Finally, use Brownian scaling: $\tilde{t}$ can be replaced with t, leading to the equivalent limit

$$\lim_{n \to \infty} \mathsf{Law}\left(t \mapsto \sqrt{\frac{p}{2 - p}}\frac{S_{nt}}{\sqrt{n}}\right) = \mathcal{W}^{(d)}.$$

This completes the proof of the convergence of fidi's.

(b) Tightness:

Exploiting Lemma 9.2, we are going to check the fourth moments and use Kolmogorov's tightness condition (see, e.g., Chapter 3 in

Billingsley (1999)). To achieve our goal, we fix an $\varepsilon \in (0, p)$ and note that by Lemma 9.2, and using the fact that $|S_n| \leq n$, it follows that

$$\mathbb{E}|S_n|^4 = \sum_{i=0}^{n^4-1} \mathbb{P}\left(|S_n|^4 > i\right) \leq d^2 n^2 + \sum_{i=d^2 n^2}^{n^4-1} \mathbb{P}\left(|S_n|^4 > i\right)$$

$$= d^2 n^2 + \sum_{i=d^2 n^2}^{n^4-1} \mathbb{P}\left(|S_n| > a_i \sqrt{n}\right)$$

$$\leq d^2 n^2 + d \sum_{i=d^2 n^2}^{n^4-1} f\left(p, a_i/\sqrt{d}\right),$$

for all large n's, where $a_i := \frac{i^{1/4}}{\sqrt{n}} \geq \sqrt{d}$, as $i \geq d^2 n^2$. Since

$$\sum_{i=d^2 n^2}^{n^4-1} f\left(p, a_i/\sqrt{d}\right) = \sum_{i=d^2 n^2}^{n^4-1} e^{-\frac{p^2 a_i}{5\sqrt{d}}} \leq \sum_{i=0}^{\infty} \exp\left\{-\frac{p^2 i^{1/4}}{5\sqrt{d} n}\right\},$$

by comparing the sum on the right-hand side with the corresponding integral

$$\int_0^\infty \exp\left\{-\frac{p^2 x^{1/4}}{5\sqrt{d} n}\right\} dx \overset{x=\frac{n^2 d^2 (5u)^4}{p^8}}{=} \frac{2500\, d^2\, n^2}{p^8} \int_0^\infty u^3 e^{-u} du$$

$$= \frac{15000\, d^2\, n^2}{p^8},$$

we conclude that there exists a $C_p > 0$ such that $\mathbb{E}|S_n|^4 \leq C_p n^2$, $n \geq 1$. For the rescaled process S^n this yields

$$\mathbb{E}|S^n(t)|^4 = \frac{p^2}{2-p^2} \mathbb{E}\left|\frac{S_{nt}}{\sqrt{n}}\right|^4 \leq C_p \frac{n^2 t^2}{n^2} = C_p t^2, \; n \geq 1$$

(since $\frac{p}{2-p} < 1$), provided nt is an integer. If nt is not an integer, recall that S^n is defined by linear interpolation and use Jensen's inequality for $y = x^4$ to get the same bound with some C'_p replacing C_p. Finally, by the stationary increments property,

$$\mathbb{E}|S^n(t) - S^n(s)|^4 \leq C'_p (t-s)^2, \; 0 \leq s < t, \; n \geq 1.$$

Kolmogorov's condition for tightness is thus satisfied. $\square$

Remark 9.6 (A direct bound for $\mathbb{E}|S^4|$, establishing tightness). An alternative way of establishing tightness is via computing a bound in a more elementary (although somewhat tedious) way for $\mathbb{E}|S_n|^4$. For simplicity, we illustrate this in the $d = 2$ case.

Let $L_n := S_n^{(x)} - S_n^{(y)}$ and $R_n := S_n^{(x)} + S_n^{(y)}$. Then L_n is really a one-dimensional conservative walk with parameter p. (See the proof of Lemma 9.2.) Next, observe that $|S_n|^2 = (S_n^{(x)})^2 + (S_n^{(y)})^2 = \frac{1}{2}(L_n^2 + R_n^2)$, yielding that $|S_n|^4 \le \frac{1}{2}(L_n^4 + R_n^4)$. Since L_n and R_n have the same distribution, this implies $\mathbb{E}|S_n|^4 \le \mathbb{E}L_n^4$. The latter expectation can be computed directly, albeit that requires a bit of algebra. In this time-homogeneous case, (5.2) and (5.4) reduce to

$$e_{i,j} = \mathsf{Cov}(Y_i, Y_j) = \mathbb{E}(Y_i Y_j) = q^{j-i}, \qquad (9.48)$$

$$\mathbb{E}(Y_j \mid Y_i) = Y_i \mathbb{E}(-1)^{\sum_{i+1}^{j} W_k} = e_{i,j} Y_i,$$

for $j \ge i$, where $q := 1 - p$. We thus have

$$\mathbb{E}L_n^4 = \mathbb{E}\left(\sum_1^n Y_i\right)^4$$

$$= \mathbb{E}\left(6\sum_{i=1}^{n-1}\sum_{j=i+1}^{n} Y_i^2 Y_j^2 + 4\sum_{i=1}^{n-1}\sum_{j=i+1}^{n} Y_i^3 Y_j + 4\sum_{i=1}^{n-1}\sum_{j=i+1}^{n} Y_i Y_j^3 \right.$$

$$+ 24\sum_{i=1}^{n-3}\sum_{j=i+1}^{n-2}\sum_{k=j+1}^{n-1}\sum_{l=k+1}^{n} Y_i Y_j Y_k Y_l + \sum_{i=1}^{n} Y_i^4$$

$$+ 12\sum_{i=1}^{n} Y_i^2 \left[\sum_{k=1}^{i-2}\sum_{l=k+1}^{i-1} Y_k Y_l + \sum_{k=i+1}^{n-1}\sum_{l=k+1}^{n} Y_k Y_l \right.$$

$$\left.\left. + \sum_{k=1}^{i-1}\sum_{l=i+1}^{n} Y_k Y_l \right]\right)$$

hence

$$\mathbb{E}L_n^4 = n + 3n(n-1) + 8\sum_{i=1}^{n-1}\sum_{j=i+1}^{n} \mathbb{E}[Y_i Y_j]$$

$$+ 24\sum_{i=1}^{n-3}\sum_{j=i+1}^{n-2}\sum_{k=j+1}^{n-1}\sum_{l=k+1}^{n} \mathbb{E}[Y_i Y_j Y_k Y_l]$$

$$+ 12 \sum_{i=1}^{n} \left[\sum_{k=1}^{i-2} \sum_{l=k+1}^{i-1} \mathbb{E}\left[Y_k Y_l\right] + \sum_{k=i+1}^{n-1} \sum_{l=k+1}^{n} \mathbb{E}\left[Y_k Y_l\right] \right.$$

$$\left. + \sum_{k=1}^{i-1} \sum_{l=i+1}^{n} \mathbb{E}[Y_k Y_l] \right]. \tag{9.49}$$

From (9.48), we obtain that if $i < j < k < l$, then

$$\mathbb{E}\left[Y_i Y_j Y_k Y_l\right] = \mathbb{E}\left(Y_i Y_j Y_k \, \mathbb{E}\left[Y_l | Y_i, Y_j, Y_k\right]\right) = \mathbb{E}\left(Y_i Y_j Y_k^2 \, q^{l-k}\right)$$

$$= q^{l-k} \, \mathbb{E}\left[Y_i Y_j\right] = q^{j-i} \times q^{l-k}.$$

Substituting this into (9.49) gives

$$\mathbb{E}L_n^4 = n + 3n(n-1) + 8 \left[\frac{nq}{p} - \frac{q}{p^2} + O(q^n) \right]$$

$$+ 24 \left[\frac{n^2 q^2}{2p^2} - \frac{n(5-q)q^2}{(2p^3)} + \frac{3q^2}{p^4} + +O(nq^n) \right]$$

$$+ 12 \left[\frac{n^2 q}{p} + \frac{nq(2q-3)}{p^2} + \frac{2q}{p^2} + O(nq^n) \right]$$

$$= \frac{3n^2(2-p)^2}{p^2} - \frac{2n(2-p)(p^2 + 12(1-p))}{p^3}$$

$$+ \frac{8(1-p)(3-2p)(3-p)}{p^4}$$

$$+ O\left(nq^n\right) = O\left(n^2\right),$$

hence, Kolmogorov's tightness condition holds.   ◇

9.6   The Scaling Limit in Critical Regime

Next, we turn our attention to the case when $p_n = a/n$ for all large n's where $a > 0$. Just like in the study of coin turning, we call this case the "critical regime." We now need the definition of the "zigzag process" in higher dimensions.

9.6.1 Preparation: the zigzag process in higher dimensions

For simplicity, we start with the two-dimensional case. We describe informally a stochastic process in continuous time, moving in $\mathbb{R}^2$ and starting at the origin. The process is piecewise linear and always moves either horizontally or vertically.

First, note that if $p_n^* := \frac{3}{4}p_n$, then the direction is changed with probability p_n^* at time n. In our case, $p_n = a/n$ and thus $p_n^* = \frac{3a}{4n} =: b/n$ for large n's.

One then takes a realization of the "scale-free" Poisson point process, just like it was done for coin-turning when $d = 1$. This process is defined on $(0, \infty)$ with intensity measure $\frac{b}{x}\,dx$. For the given realization, we construct a trajectory of the zigzag process as follows. Let t_* and t^* in the point process be the left, resp., right neighbors of $t = 1$. Toss two independent fair coins and assign one of the labels "N,W,S,E" according to the outcome (that is, each has probability $1/4$) to the time interval $(t_*, t^*]$. Going backward in time, label each interval in a way so that the next interval can be labeled in three different ways, each with probability $1/3$, and the label must differ from that of the previous interval (if the interval containing 1 was, say, labeled "N," then, going backward, the next label should be W, S or E with equal probabilities, etc.) Do the same for the intervals between the points forward in time. This way, each interval between two consecutive points of the PPP is labeled. All the coin tossing is independent. These four labels indicate the direction the process is moving in the time intervals.

Let the union of intervals labeled N be $U^{(N)}$ and

$$\ell_N(t) := \mathrm{Leb}\left(U^{(N)} \cap [0, t]\right).$$

Define similarly $\ell_S(t), \ell_E(t), \ell_W(t)$.

The trajectory of the zigzag process Z will then[3] be defined by

$$Z_t := (\ell_E(t) - \ell_W(t),\ \ell_N(t) - \ell_S(t)), \quad t > 0.$$

Clearly, $\lim_{t\to 0} Z_t = (0,0)$ almost surely, even though there are infinitely many points of the PPP in any neighborhood of the origin.

---

[3]Conditionally on the realization of the PPP and the labeling.

When $d > 2$, the construction is analogous. The difference is that in general $p_n^* := \frac{2d-1}{2d} p_n$ and $p_n = a/n$ yields $p_n^* = \frac{(2d-1)a}{2dn} =: b/n$ for large n's, and, furthermore, one needs to work with $2d$ labels. The constraint is that between two consecutive time intervals the process "must change the label."

9.6.2 *Scaling limit for the multidimensional case*

In the critical case, just like in one dimension, proper scaling leads to the zigzag process.

Theorem 9.9 (Scaling limit in the critical case). *Let $d \geq 2$ and $p_n = a/n$ for $n \geq n_0$. Extend the walk S to all non-negative times using linear interpolation and for $n \geq 1$, define the rescaled walk S^n by*

$$S^n(t) := \frac{S_{nt}}{n}, \ t \geq 0,$$

and finally, let $\mathcal{Z}^{(d)}$ denote the law of the d-dimensional zigzag process, with parameter $b = \frac{(2d-1)a}{2d}$. Then $\lim_{n \to \infty} \mathsf{Law}(S^n) = \mathcal{Z}^{(d)}$ on $C([0, \infty), \mathbb{R}^d)$.

Remark 9.7. The result is still valid for $d = 1$. Note, however, that the definition of p_n for the coin-turning walk in Chapter 5 differs by a factor 2, yielding unit parameter instead of $1/2$.

Proof. The proof is very similar to that of the one-dimensional analog, the coin-turning walk.

The tightness part works similarly, namely, just like in the case of the coin-turning walk, one simply uses the Lipschitz-1 property of the paths that holds for S^n for each $n \geq 1$. (This is an advantage compared to the time homogeneous case, and it comes from the fact that the scaling is n and not n^2 in this case.)

For the convergence of the finite dimensional distributions, it will be enough to show that weak convergence holds for the processes on $[0, T]$ for any $T > 1$.

To keep the notation easier, in the rest of the proof, we work with the $d = 2$ case, however we note that the general case is completely analogous, by considering $2d$ labels instead of just four.

Consider now the space $\mathfrak{S}$ of all double infinite sequences $c_{-2}, c_{-1}, c_0, c_1, c_2, \ldots$, where $c_i \in \{N, W, S, E\}$. When assigning a unique path on $[0, T]$ to a realization of the turning points, the situation is a bit more complicated than for $d = 1$. Namely, one has to use a rule, described in Definition 9.4 with some fixed $s \in \mathfrak{S}$. (In one dimension, there are only two options to assign a path; see Chapter 6.) Informally, for the segment containing 1, we assign the label of c_0, for the segment to the left and to the right we assign the label of c_{-1} and the label of c_1, respectively, etc.

More precisely, fix $T > 1$, denote by $\mathcal{M}_T$ the set of all locally finite point measures on the interval $(0, T]$ and denote by $N^{(n)} = N^{(n,T)}$ the laws of the point processes induced by the changes of direction of the walk $S^{(n)}$ on the time interval $(0, T]$.

Let $s \in \mathfrak{S}$; we now assign a continuous (zigzagged) path to each realization of the point measure.

Definition 9.4 (Assigning paths for a given $s \in \mathfrak{S}$). Define the map $\Phi_1 = \Phi_{1,s} : \mathcal{M}_T \to C[0, T]$ as follows:

- First, label the (countably many) atoms on $(0, 1]$ from right to left as $a_1, a_2, \ldots$, i.e., the closest one on the left to 1 as a_1, the second closest as a_2, etc., and note that $1 = a_1$ is possible; also label the atoms on $(1, T]$, from the closest to the furthest as $b_1, b_2, \ldots$.
- Assign label "N" to all intervals (the union of which is denoted by $S_1^{(N)}$) $[a_i, a_{i+1})$, which are such that in s, the corresponding letter is N. Here "corresponding" means that c_0 corresponds to $[a_1, b_1)$, and for $i \geq 1$, c_i corresponds to $[b_i, b_{i+1})$, while c_{-i} corresponds to $[a_{i+1}, a_i)$.
- Do the same for S, E and W.

The path we obtain makes steps to the North (up) resp. to the West (left), South (down), East (right) on $S_1^{(N)}$, resp. $S_1^{(W)}, S_1^{(S)}, S_1^{(E)}$.

Let $\mu \in \mathcal{M}_T$. For $0 < r \leq T$, we define the vertical and horizontal components of the path as

$$\Phi_1^{\mathsf{vert}}(\mu)(r) := \mathsf{Leb}((0, r] \cap S_1^{(N)}) - \mathsf{Leb}((0, r] \cap S_1^{(S)}), \text{ with}$$
$$\Phi_1^{\mathsf{vert}}(\mu)(0) := 0,$$

$$\Phi_1^{\mathsf{hori}}(\mu)(r) := \mathsf{Leb}((0, r] \cap S_1^{(E)}) - \mathsf{Leb}((0, r] \cap S_1^{(W)}), \text{ with}$$
$$\Phi_1^{\mathsf{hori}}(\mu)(0) := 0,$$

where Leb is the Lebesgue measure on the real line. Then $\Phi_1^{\mathsf{vert}}(\mu)(\cdot)$ is well defined and continuous on $[0, T]$. Then

$$|\Phi_1^{\mathsf{vert}}(\mu)(r)|, |\Phi_1^{\mathsf{hori}}(\mu)(r)| \leq r, \ 0 < r \leq T.$$

Just like in Chapter 6 discussing the coin-turning walk, one can show that for $s \in \mathfrak{S}, T > 0$ given,

(a) $\Phi_{1,s} : \mathcal{M}_T \to C[0, T]$ is a continuous and uniformly bounded functional, when the former space is equipped with the vague topology and the latter with the supremum norm $|\,.\,| = |\,.\,|_{[0,T]}$,

(b) as $n \to \infty$, $N^{(n)} \to \mathsf{Poiss}(b)$ in law (using the vague topology of measures on $(0, T]$).

To finish the proof, let P^{uni} be a law on $\mathfrak{S}$ obtained by choosing C_i uniformly in $\{N, W, S, E\}$ and doing it independently for all $i \in \mathbb{Z}$, and let

$$Q(\cdot) := P^{\mathrm{uni}}(\cdot \mid \forall i \in \mathbb{Z} : \ C_i \neq C_{i+1}).$$

Furthermore, let $T > 1$ and $F : C([0, T], |\cdot|) \to \mathbb{R}$ be a bounded continuous functional. By (a) above, $F \circ \Phi_{1,s} : (\mathcal{M}_T, \text{vague}) \to \mathbb{R}$ is a bounded continuous functional too. Hence, by (b) above,

$$\lim_n E((F \circ \Phi_{1,s})(N^{(n)})) = E((F \circ \Phi_{1,s})(\mathsf{Poiss}(b))).$$

Finally, bounded convergence yields that

$$\lim_n \int_{\mathfrak{S}} E((F \circ \Phi_{1,s})(N^{(n)}))\, Q(\mathrm{d}s) = \int_{\mathfrak{S}} E((F \circ \Phi_{1,s})(\mathsf{Poiss}(b)))\, Q(\mathrm{d}s).$$

$$(9.50)$$

The right-hand side of (9.50) is the expectation of F applied on the zigzag process with parameter b, while the left-hand side is the limit of those terms where the zigzag process is replaced by $S^{(n)}$ (all processes restricted on $[0, T]$). Since F was an arbitrary bounded continuous functional, (9.50) means that the processes S^n converge weakly on the time interval $[0, T]$. $\qquad\square$

9.7 Some Open Problems

In this section, we formulate some open problems.

Problem 9.4 (Monotonicity). Fix $d \geq 1$. Is it true that if $p'_n \leq p_n$ for $n \geq 1$ and the walk exhibits strong transience for the sequence $\{p_n\}$, then the same holds for the sequence $\{p'_n\}$? (Compare with the last sentence in Example 9.1.)

Problem 9.5 (Critical case). Fix $d \geq 2$. In the critical case ($p_n = \text{const}/n$ for large n), we conjecture that the walk exhibits strong transience, which, of course, would readily follow form monotonicity (cf. the last sentence in Example 9.1) and that this property is inherited to its scaling limit, the zigzag process too.

Problem 9.6 (Transient dimensions). Is it true that the walk always exhibits strong transience whenever $d = 3$? Clearly, monotonicity would imply this, since when $p_n = p = 1$, S is a simple symmetric random walk.

In the periodic case, for example, using the notation of the proof of Theorem 9.2, when $d \geq 3$, it is known (see, e.g., Spitzer (1976)) that $\mathbb{P}(\lim_n |S_n^*| = \infty) = 1$. We know that for all $\tau_n < m < \tau_{n+1}$ $(1/r)|S_m - S_n^*|$ is uniformly bounded by a geometrically distributed variable, with parameter p_{n_0}. However, it is not clear whether this is enough to control S via S^*.

Problem 9.7 (Slow decay). What happens when $d = 2$ and the p_n's decay slowly as $n \to \infty$? We already know that if $p_n = \text{const} \in (0, 1)$, then the walk is recurrent. An interesting question is whether the answer changes if, for example, $p_n \sim 1/\log n$.

9.8 Appendix

Here we state and prove two lemmas that were needed in the proofs. Some versions of these statements are presumably known, but since we could not find a proper reference, we present their proofs here.

Lemma 9.3. *Let $s \in (0, 1/2]$, $s_0 \in \mathbb{R}$, and $M \geq 2$ be an integer such that $Ms \geq 1$. Moreover, let $a_k = ks + s_0 \pmod 1$, for $k = 1, 2, \ldots, M$.*

Then[4]

$$N_{s,M} := |\{k : a_k \in [0, 1/2)\}| \geq \frac{2}{15}M.$$

Proof. First, assume that $s \geq 1/4$. Since the length of the interval $[1/2, 1)$ is $1/2$, and $s \leq 1/2$, for any triple $\{a_i, a_{i+1}, a_{i+2}\}$, $i \geq 1$, at least one element must not lie in $[1/2, 0)$. Therefore, $N_{s,M} \geq \lfloor M/3 \rfloor \geq M/5$.

Next, assume that $s < 1/4$, and define

$$\tau_k := \inf\{i : is \geq k\}, \quad k = 0, 1, 2, \ldots.$$

Let $K \in \mathbb{Z}$ be such that $\tau_K \leq M$, $\tau_{K+1} > M$; since $Ms \geq 1$, we have $K \geq 1$. Then we have

$$0 < a_{\tau_k+1} < a_{\tau_k+2} < \cdots < a_{\tau_{k+1}} \leq 1 \quad \text{for each } k = 0, 1, 2, \ldots.$$

In each such increasing sequence, since $s < 1/4$, we have at least $\lfloor 1/(2s) \rfloor \geq 1/(3s)$ elements in the segment $[0, 1/2)$ of length $1/2$. The total number of elements in this sequence, $\tau_{k+1} - \tau_k \leq \lceil 1/s \rceil \leq 5/(4s)$. Consequently,

$$N_{s,M} \geq K \times \frac{1}{3s} \quad \text{while } M \leq (K+1) \times \frac{5}{4s},$$

so $N_{s,M}/M \geq 2/15$ since $K \geq 1$. $\qquad\square$

Lemma 9.4. *Let* $(q_i)_{i=1}^{\infty}$ *form a probability distribution* ($q_i \geq 0$, $\sum_{i=1}^{\infty} q_i = 1$). *Assume that for some* $a > 0$ *and a positive integer* M *we have*

$$q_j \geq \frac{a}{M} \quad \text{for } j = 1, 2, \ldots, M.$$

Let $h(s) := \sum_{j=1}^{\infty} q_j |\cos(js)|$. *Then*

$$h(s) \leq \max\left(1 - c_1' as^2 M^2, 1 - c_2' a\right), \quad 0 \leq s \leq \frac{\pi}{2},$$

for some absolute constants $c_1', c_2' > 0$.

---

[4]The constant $2/15$ is definitely suboptimal. The true constant is probably $1/3$, with equality achieved for $M = 3, 6, 9, \ldots$.

Proof. First note that

$$|h(s)| \leq \sum_{j=1}^{M} q_j |\cos(js)| + \sum_{j=M+1}^{\infty} q_j = 1 - \sum_{j=1}^{M} q_j (1 - |\cos(js)|)$$

$$\leq 1 - \frac{a}{M} \sum_{j=1}^{M} (1 - |\cos(js)|). \tag{9.51}$$

We use the elementary inequality

$$1 - \cos \alpha \geq \frac{\alpha^2}{4}, \quad 0 \leq \alpha \leq \frac{\pi}{2}, \tag{9.52}$$

which holds since since $\psi(\alpha) := 1 - \cos \alpha - \frac{\alpha^2}{4}$ has the properties that $\psi(0) = 0$ and $\psi'(\alpha) = \sin \alpha - \frac{\alpha}{2} > 0$ for $\alpha \in (0, \pi/2)$.

Case 1: $M \geq 2$. Let $M_1 = \lfloor M/2 \rfloor \in [M/3, M/2]$. If $tM_1 \leq \pi/2$, then by (9.52)

$$\sum_{j=1}^{M} (1 - |\cos(js)|) \geq \sum_{j=1}^{M_1} (1 - \cos(js)) \geq \sum_{j=1}^{M_1} \frac{j^2 s^2}{4}$$

$$= \frac{M_1(M_1+1)(2M_1+1)s^2}{24} \geq \frac{M^3 s^2}{12 \cdot 3^3}.$$

On the other hand, if $sM_1 \geq \pi/2$, then $sM \geq 2sM_1 \geq \pi$. Let $\tilde{s} = \frac{s}{\pi} \leq \frac{1}{2}$, then $\tilde{s}M \geq 1$ and by Lemma 9.3 (setting $j_0 := -1/4$), in the set $\{j\tilde{s} \bmod 1, \ j = 1, 2, \ldots, M\}$ there will be at least $\frac{2M}{15}$ elements which lie in the segment $[1/4, 3/4]$; for the corresponding indices j, this implies that

$$|\cos(sj)| \leq \max_{x \in [\frac{\pi}{4}, \frac{3\pi}{4}]} |\cos x| = \cos \frac{\pi}{4} = \frac{1}{\sqrt{2}}.$$

Hence,

$$\sum_{j=1}^{M} (1 - |\cos(js)|) \geq \frac{2M}{15} \left(1 - \frac{1}{\sqrt{2}}\right),$$

and consequently,

$$\frac{a}{M}\sum_{j=1}^{M}(1 - |\cos(js)|) \geq a \cdot \begin{cases} \frac{M^2s^2}{324}, & \text{if } sM_1 \leq \pi/2; \\ \\ \frac{2-\sqrt{2}}{15}, & \text{otherwise.} \end{cases}$$

Case 2: $M = 1$. Since $s \in [0, \pi/2]$, (9.52) yields that

$$\frac{a}{M}\sum_{j=1}^{M}(1 - |\cos(js)|) = a(1 - \cos s) \geq \frac{as^2}{4} = \frac{aM^2s^2}{4}.$$

This together with (9.51) imply the result.   □

Chapter 10

Urn-Related Random Walk with Drift $\rho x^\alpha / t^\beta$

10.1 Introduction

In this chapter, we study a stochastic processes X_t, which may loosely be described as a random walk on $\mathbb{R}_+$ (or more generally on $\mathbb{R}$) with the asymptotic drift given by

$$\mu_t := \mathbb{E}(X_{t+1} - X_t \mid X_t = x) \sim \rho \frac{|x|^\alpha}{t^\beta},$$

where ρ, α and β are some fixed constants, and the exact meaning of "$\sim$" is made precise later. Our goal is to establish when this process is recurrent or transient, by finding the whole line of phase transitions in terms of (α, β). We also analyze some critical cases, when the value of ρ becomes important as well. Note that because of symmetry, it is sufficient to consider only these processes on $\mathbb{R}_+$, and from now on we assume that $X_t \geq 0$ a.s. for all values of t.

The original motivation comes from an open problem related to Friedman urns, posed in Freedman (1965). In certain regimes of these urns, to the best of our knowledge, it is still unknown whether the number of balls of different colors can overtake each other infinitely many times with a positive probability. We do not describe this problem in more detail here rather we refer the reader directly to Section 10.6.

Incidentally, the class of stochastic processes we are considering covers simultaneously not only the Friedman urn but also the walk with an asymptotically zero drift, first probably studied by Lamperti, see Lamperti (1960) and Lamperti (1963). His one-dimensional walks with drift depending only on the position of the particle naturally arise when proving recurrence of the simple random walk on $\mathbb{Z}^1$ and $\mathbb{Z}^2$ and transience on $\mathbb{Z}^d$, $d \geq 3$. They can be used of course for answering the question of recurrence for a much wider class of models, notably those involving polling systems, for example, see Aspandiiarov *et al.* (1996) and Menshikov and Zuyev (2001). It will not be surprising if the model we are considering also covers some other probabilistic models, of which we are unaware at the moment.

We study the random walk whose drift depends *both on time and the position* of a particle, hence this can viewed as an example of a non-homogeneous Markov chain. Throughout this chapter, we assume that

$$(\alpha, \beta) \in \Upsilon = \{(\alpha, \beta) : \ \beta > \alpha \text{ and } \beta \geq 0\}$$

to avoid the situations when the drift becomes unbounded and the borderline cases (the only exception is $\alpha = \beta = 1$). We show that under some regularity conditions, the walk is transient when (α, β) lie in the following area

$$\mathbf{Trans} = \{(\alpha, \beta) : \ 0 \leq \beta < 1, \ 2\beta - 1 < \alpha < \beta\} \subset \Upsilon$$

and recurrent for (α, β) in

$$\mathbf{Rec} = \Upsilon \setminus \overline{\mathbf{Trans}} = \{(\alpha, \beta) : \ \beta \geq 0, \ \alpha < \min(\beta, 2\beta - 1)\},$$

where $\overline{\mathbf{Trans}}$ denotes the closure of the set $\mathbf{Trans}$. (Please see Figure 10.1.) In the special critical case $\alpha = \beta = 1$, we show that the walk is transient for $\rho > 1/2$ and recurrent for $\rho < 1/2$. An example of such a walk with $\alpha = \beta = 1$ is the process on $\mathbb{Z}_+$ with the following jump distribution:

$$\mathbb{P}(X_{t+1} = n \pm 1 \mid X_t = n) = \frac{1}{2} \pm \frac{\rho n}{2t},$$

where $t = 0, 1, 2, \ldots$. This walk is analyzed in Section 10.6.

Throughout this chapter, we need the following hypotheses. Let X_t be a stochastic process on $\mathbb{R}_+$ with jumps $D_t = X_t - X_{t-1}$ and let $\mathcal{F}_t = \sigma(X_0, X_1, \ldots, X_t)$. Let a be some positive constant.

(H1) Uniform boundedness of jumps: There exists a constant $B_1 > 0$ such that $|D_t| \leq B_1$ for all $t \in \mathbb{Z}_+$ a.s.

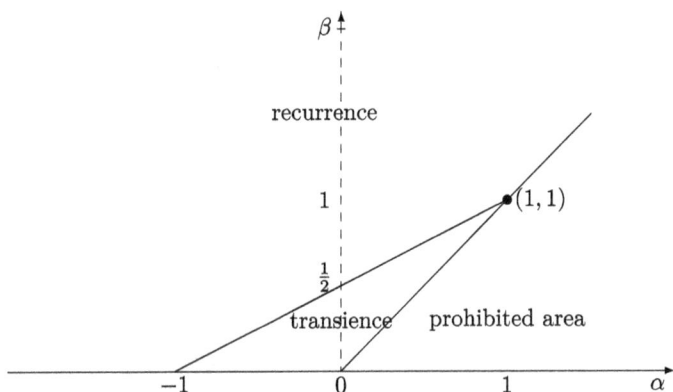

Fig. 10.1. Diagram for (α, β).

(H2) Uniform non-degeneracy on $[a, \infty)$: There exists a constant $B_2 > 0$ such that whenever $X_{t-1} \geq a$, $\mathbb{E}(D_t^2 \mid \mathcal{F}_{t-1}) \geq B_2$ for all $t \in \mathbb{Z}_+$ a.s.

(H3) Uniform boundedness of time to leave $[0,a]$: The number of steps required for X_t to exit the interval $[0, a]$ starting from any point inside this interval is uniformly stochastically bounded above by some independent random variable $W \geq 0$ with a finite mean $\mu = \mathbb{E}W < \infty$, that is, for all $s \in \mathbb{R}_+$, when $X_s \leq a$, one has

$$\forall x \geq 0: \quad \mathbb{P}(\eta(s) \geq x \mid \mathcal{F}_s) \leq \mathbb{P}(W \geq x),$$

where $\eta(s) = \inf\{t \geq s : X_t > a\}$.

The rest of this chapter is organized as follows. In Section 10.2, we prove some technical lemmas. In Section 10.3, we formulate the exact statement about the transience of the process X_t and prove it, while in Section 10.4, we do the same for recurrence. We also study some borderline cases in Section 10.5 and present an open problem in Section 10.5.3. Finally, we apply our results to generalized Pólya and Friedman urns in Section 10.6.

10.2 Some Technical Tools

First, we need the following important law of iterated logarithms for martingales.

Lemma 10.1 (LILM; Proposition (2.7) in Freedman (1975)).
Suppose that $S = \{S_n\}_{n\in\mathbb{N}}$ is a martingale adapted to the filtration $\{\mathcal{F}_n\}_{n\in\mathbb{N}}$ and $\Delta_n = S_n - S_{n-1}$ are its differences. Let $T_n := \sum_{i=1}^{n} \mathbb{E}(\Delta_i^2 \,|\, \mathcal{F}_{i-1})$, $\sigma_b := \inf\{n : T_n > b\}$, and

$$L(b) := \operatorname{ess\,sup}_{\omega} \ \sup_{n \le \sigma_b(\omega)} |\Delta_n(\omega)|.$$

Assume that $L(b) = o(b/\log\log b)^{1/2}$ as $b \to \infty$. Then

$$\limsup_{n\to\infty} \frac{S_n}{\sqrt{2T_n \log\log T_n}} = 1 \quad a.s. \ on \ \{T_n \to \infty\}.$$

Lemma 10.2. *Let X_t, $t = 1, 2, \ldots$ be a sequence of random variables adapted to filtration $\mathcal{F}_t$ with differences $D_t = X_t - X_{t-1}$ satisfying (H1), (H2) and (H3) for some $a > 0$. Assume that on the event $\{X_t \ge a\}$,*

$$\mathbb{E}(D_{t+1} \,|\, \mathcal{F}_t) \ge 0 \ a.s.$$

Then for any $A > 0$,

$$\mathbb{P}(\exists t : \ X_t > A\sqrt{t}) = 1.$$

Note that both the assumptions and conclusion of Lemma 10.2 are weaker than those of Lemma 10.1, but it is Lemma 10.2 what we use further in the text.

Proof. First, we are going to essentially "freeze" the process X_t whenever it enters the interval $[0, a]$, where it is not a submartingale, until the moment when X_t exits from this interval. Define the function $s(t) : \mathbb{Z}_+ \to \mathbb{Z}_+$ such that $s(0) = 0$ and for $t \ge 0$

$$s(t+1) = \begin{cases} t+1, & \text{if } X_t > a \text{ or } X_{t+1} > a \\ s(t), & \text{otherwise.} \end{cases}$$

Let $\tilde{X}_t = X_{s(t)}$. Then $\tilde{X}_n$ is a submartingale satisfying (H1), perhaps with a new constant $\tilde{B}_1 = B_1 + a$. Indeed, when $X_t > a$, $\tilde{X}_t = X_t$ and $\tilde{X}_{t+1} = X_{t+1}$, so $\mathbb{E}(\tilde{X}_{t+1} - \tilde{X}_t \mid \mathcal{F}_t) = \mathbb{E}(D_{t+1} \mid \mathcal{F}_t) \ge 0$. When $X_t < a$ (and so is $\tilde{X}_t < a$), either $X_{t+1} < a$ and then $s(t+1) = s(t)$ implying $\tilde{X}_{t+1} = \tilde{X}_t$, or $X_{t+1} \ge a$ in which case $\tilde{X}_{t+1} = X_{t+1} \ge a > \tilde{X}_t$.

Let

$$\tilde{D}_n = \tilde{X}_n - \tilde{X}_{n-1},$$

$$Z_n = \mathbb{E}(\tilde{D}_n \mid \mathcal{F}_{n-1}) \geq 0,$$

$$S_n = X_n - Z_1 - Z_2 - \cdots - Z_n.$$

Then

$$\mathbb{E}(S_n - S_{n-1} \mid \mathcal{F}_{n-1}) = \mathbb{E}(X_n - X_{n-1} - Z_n \mid \mathcal{F}_{n-1}) = 0$$

whence S_n is a martingale with differences $\Delta_n := S_n - S_{n-1} = \tilde{D}_n - Z_n$. Note that

$$\mathbb{E}(\Delta_n^2 \mid \mathcal{F}_{n-1}) = \mathbb{E}((S_n - S_{n-1} - Z_{n-1})^2 \mid \mathcal{F}_{n-1})$$

$$= \mathbb{E}(\tilde{D}_n^2 \mid \mathcal{F}_{n-1}) - Z_n^2. \tag{10.1}$$

Let $\eta_0 = 0$ and for $k = 1, 2, \ldots,$

$$\zeta_k = \inf\{t \geq \eta_{k-1} : \ X_t \leq a\},$$

$$\eta_k = \inf\{t \geq \zeta_k : \ X_t > a\}$$

be the consecutive times of entry in and exit from $[0, a]$. Then $\tilde{W}_k := \eta_k - \zeta_k$ are stochastically bounded by i.i.d. random variables $W_1, W_2, \ldots$ with the distribution of W. Therefore,

$$\limsup_{m \to \infty} \frac{\sum_{i=1}^m \tilde{W}_i}{m} \leq \mu \ \text{a.s.}$$

and consequently the number

$$I_n = \{t \in \{0, 1, \ldots, n-1\} : t \notin [\zeta_k, \eta_k) \text{ for some } k\}$$

$$= \{t \in \{0, 1, \ldots, n-1\} : X_t > a\}$$

of those times which do not belong to some "frozen" interval $[\zeta_k, \eta_k)$ satisfies a.s.

$$|I_n| \geq \frac{n}{2\mu} \tag{10.2}$$

for n sufficiently large.

Next, since $\tilde{D}_n$'s are bounded, we have $|\Delta_n| \leq |\tilde{D}_n| + |\mathbb{E}(\tilde{D}_n \,|\, \mathcal{F}_{n-1})| \leq 2(B_1 + a)$. Therefore, $L(b) \leq 2(B_1 + a)$ and the conditions of Lemma 10.1 are met. First, suppose that

$$T_n = \sum_{i=1}^{n} \mathbb{E}(\Delta_i^2 \,|\, \mathcal{F}_{i-1}) \to \infty,$$

then

$$\limsup_{n \to \infty} \frac{S_n}{\sqrt{2 T_n \log\log T_n}} = 1 \quad \text{a.s.}$$

Therefore, for infinitely many n's, we would have

$$S_n \geq \sqrt{T_n \log\log T_n}.$$

Using (10.1), this results in

$$X_n = \sum_{i=1}^{n} Z_i + S_n \geq \sum_{i=1}^{n} Z_i + \sqrt{\sum_{i=1}^{n} (\mathbb{E}(\tilde{D}_i^2 \,|\, \mathcal{F}_{i-1}) - Z_i^2) \log\log T_n}$$

$$\geq \sum_{i \in 1 + I_n} Z_i + \sqrt{\sum_{i \in 1 + I_n} (\mathbb{E}(D_i^2 \,|\, \mathcal{F}_{i-1}) - Z_i^2) \log\log T_n} \qquad (10.3)$$

since $i-1 \in I_n$ implies $X_{i-1} > a$ and consequently $\tilde{D}_i = D_i$ (note that each term in the sums above is non-negative). Let $0 \leq k \leq |I_n|$ be the number of those Z_i's, $i \in I_n$, such that $Z_i < \sqrt{B_2/2}$. Then (10.3) together with $\mathbb{E}(D_i^2 \,|\, \mathcal{F}_{i-1}) \geq B_2$ yield

$$X_n \geq (|I_n| - k)\sqrt{\frac{B_2}{2}} + \sqrt{\frac{k B_2 \log\log T_n}{2}} \geq \sqrt{\frac{n B_2 \log\log T_n}{2\mu}}$$

for n large enough, taking into account the fact that $T_n \leq B_1^2 n$ and inequality (10.2). This implies the statement of Lemma 10.2, since we assumed $T_n \to \infty$.

On the other hand, on the complementary event $\sum_{i=1}^{\infty} \mathbb{E}(\Delta_i^2 \,|\, \mathcal{F}_{i-1}) < \infty$, by, e.g., theorem in Chapter 12 in Williams (1991), S_n converges a.s. to a finite quantity S_∞, and we obviously must also have $\mathbb{E}(\tilde{D}_n^2 \,|\, \mathcal{F}_{n-1}) - Z_n^2 \to 0$ yielding

$$\liminf_{i \to \infty, \ i:\ X_{i-1} > a} Z_i \geq \sqrt{B_2}.$$

Combining this with (10.2), we obtain

$$\liminf_{n \to \infty} \frac{X_n}{n} = \liminf_{n \to \infty} \frac{S_n + Z_1 + Z_2 + \cdots + Z_n}{n} \geq \frac{\sqrt{B_2}}{2\mu}$$

which is an even stronger statement than the one needed. □

Lemma 10.3. *Fix $a > 0$, $c > 0$, $\gamma \in (0,1)$, and consider a Markov process X_t, $t = 0, 1, 2, \ldots$ on $\mathbb{R}_+$ with jumps $D_t = X_t - X_{t-1}$, for which the hypotheses (H1) and (H2) hold. Suppose that for some large $n > 0$ the process starts at $X_0 \in (a, \gamma n]$ and that on the event $\{a \leq X_t \leq n\}$*

$$\mathbb{E}(D_t \mid \mathcal{F}_{t-1}) \leq \frac{c}{n}.$$

Let

$$\tau = \inf\{t : X_t < a \text{ or } X_t > n\}$$

be the time to exit $[a, n]$. Then

(i) *$\tau < \infty$ a.s.,*
(ii) *$\mathbb{P}(X_\tau < a) \geq \nu = \nu(\gamma, c, B_2) > 0$ uniformly in n.*

Proof. First, let us show that the process X_t must exit $[a, n]$ in a finite time. Since $|D_t| \leq B_1$, by Markov inequality for non-negative random variables for any $\varepsilon > 0$, we have

$$\mathbb{P}(B_1^2 - D_t^2 \geq (1 - \varepsilon^2)B_1^2 \mid \mathcal{F}_{t-1}) \leq \frac{\mathbb{E}(B_1^2 - D_t^2 \mid \mathcal{F}_{t-1})}{(1 - \varepsilon^2)B_1^2} \qquad (10.4)$$

$$\leq (1 - \varepsilon^2)^{-1} \left(1 - B_2 B_1^2\right).$$

Hence, for a sufficiently small $\varepsilon > 0$, the RHS of (10.4) can be made smaller than 1, whence there is a $\delta > 0$ such that

$$\mathbb{P}(D_t^2 \geq (\varepsilon B_1)^2 \mid \mathcal{F}_{t-1}) > 2\delta.$$

In turn, this implies that at least one of the probabilities $\mathbb{P}(D_t \geq \varepsilon B_1 \mid \mathcal{F}_{t-1})$ or $\mathbb{P}(D_t \leq -\varepsilon B_1 \mid \mathcal{F}_{t-1})$ is larger than δ. Hence, from any starting point, the walk can exit $[a, n]$ in at most $n/(\varepsilon B_1)$ steps with probability at least $\delta^{n/(\varepsilon B_1)}$, yielding that $\varepsilon B_1 \tau / n$ is stochastically bounded by a geometric random variable with parameter $\delta^{n/(\varepsilon B_1)}$, which is not only finite but also has all finite moments.

To prove the second claim of the lemma, we first establish the following elementary inequality. Fix a $k \geq 1$ and consider the function $g(x) = (1-x)^k - 1 + kx - k(k-1)x^2/4$. Since $g(0) = 0$, $g'(0) = 0$ and $g''(x) = k(k-1)((1-x)^{k-2} - 1/2) \geq 0$ for $|x| \leq 1/(2k)$, we have $g(x) \geq 0$ on this interval. Consequently,

$$(1-x)^k - 1 \geq -kx + \frac{k(k-1)x^2}{4} \quad \text{for } x \in \left[-\frac{1}{2k}, \frac{1}{2k}\right]. \qquad (10.5)$$

Now, let $Z_t = 2n - X_t$ and $Y_t = Z_t^k$ for some $k \geq 1$ to be chosen later. Suppose that $n > 2kB_1$. Then, on the event $\{X_t \in [a, n]\}$, we have $Z_t \in [n, 2n]$ yielding $|D_{t+1}/Z_t| \leq B_1/n \leq 1/(2k)$ and thus by (10.5) we have

$$\mathbb{E}(Y_{t+1} - Y_t \mid \mathcal{F}_t) = Y_t \mathbb{E}\left(\left(1 - \frac{D_{t+1}}{Z_t}\right)^k - 1 \mid \mathcal{F}_t\right)$$

$$\geq kY_t \left[-\frac{\mathbb{E}(D_{t+1} \mid \mathcal{F}_t)}{Z_t} + \frac{(k-1)\mathbb{E}(D_{t+1}^2 \mid \mathcal{F}_t)}{4Z_t^2}\right]$$

$$\geq kY_t \left[-\frac{c}{nZ_t} + \frac{(k-1)B_2}{4Z_t^2}\right]$$

$$\geq kY_t \left[-\frac{c}{n^2} + \frac{(k-1)B_2}{16n^2}\right] > 0,$$

once $k > 1 + 16c/B_2$.

Hence, $Y_{t \wedge \tau}$ is a non-negative submartingale. By the optional stopping theorem,

$$\mathbb{E}(Y_\tau) \geq Y_0 \geq [(2-\gamma)n]^k.$$

On the other hand,

$$\mathbb{E}(Y_\tau) = \mathbb{E}(Y_\tau; \ X_\tau < a) + \mathbb{E}(Y_\tau; \ X_\tau > n)$$

$$\leq (2n)^k \mathbb{P}(X_\tau < a) + n^k(1 - \mathbb{P}(X_\tau < a)),$$

yielding

$$\mathbb{P}(X_\tau < a) \geq \frac{(2-\gamma)^k - 1}{2^k - 1} =: \nu > 0. \qquad \square$$

Lemma 10.4. *Assume that* X_t, $t = 0, 1, \ldots$ *is a submartingale satisfying* (H1). *Then for any* $x > 0$,

$$\mathbb{P}\left(\inf_{0 \le t \le hx^2} X_t < X_0 - bx\right) \le c(h, b, B_1) = \frac{4hB_1^2}{b^2}.$$

Proof. Let $Z_n = \mathbb{E}(X_{n+1} - X_n \mid \mathcal{F}_n) \ge 0$. Then

$$S_t = X_0 - (X_t - Z_1 - Z_2 - \cdots - Z_t)$$
$$= (X_0 - X_t) + Z_1 + \cdots + Z_t \ge X_0 - X_t$$

is a square-integrable martingale with $S_0 = 0$ since $|S_n| \le |X_0| + 2nB_1$. Moreover, since

$$\mathbb{E}\left((S_t - S_{t-1})^2 \mid \mathcal{F}_{t-1}\right) = \mathbb{E}\left((X_t - X_{t-1} - Z_t)^2 \mid \mathcal{F}_{t-1}\right)$$
$$= \mathbb{E}\left((X_t - X_{t-1})^2 \mid \mathcal{F}_{t-1}\right) - Z_t^2 \le B_1^2,$$

we have

$$A_n := \sum_{t=1}^{n} \mathbb{E}\left((S_t - S_{t-1})^2 \mid \mathcal{F}_{t-1}\right) \le nB_1^2.$$

By Doob's maximum $\mathbb{L}^2$ inequality (see Durrett, 1996, pp. 254–255),

$$\mathbb{E}\left(\sup_{0 \le m \le n} |S_m|^2\right) \le 4\mathbb{E}S_n^2 = 4A_n \le 4nB_1^2.$$

Consequently, by Chebyshev's inequality,

$$\mathbb{P}\left(\inf_{0 \le t \le hx^2} X_t < X_0 - bx\right) = \mathbb{P}\left(\sup_{0 \le t \le hx^2} X_0 - X_t > bx\right)$$
$$\le \mathbb{P}\left(\sup_{0 \le t \le hx^2} |S_t| > bx\right)$$
$$< \frac{4(hx^2)B_1^2}{b^2 x^2} = \frac{4hB_1^2}{b^2}. \qquad \square$$

10.3   Transience

We now establish conditions for the transience of the process.

Theorem 10.1. *Consider a Markov process X_t, $t = 0, 1, 2, \ldots$, on $\mathbb{R}_+$ with increments $D_t = X_t - X_{t-1}$, satisfying* (H1), (H2) *and* (H3) *for some $a > 0$. Suppose that for t sufficiently large, on the event $\{X_t \geq a\}$, we have either*

(i) *for some $\rho > 1/2$*

$$\mathbb{E}(D_{t+1} \mid \mathcal{F}_t) \geq \frac{\rho X_t}{t}$$

or

(ii) *for some $\rho > 0$ and $(\alpha, \beta) \in$ **Trans**,*

$$\mathbb{E}(D_{t+1} \mid \mathcal{F}_t) \geq \frac{\rho X_t^{\alpha}}{t^{\beta}}.$$

Then X_t is transient in the sense that for any starting point $X_0 = x$ we have

$$\mathbb{P}\left(\lim_{t \to \infty} X_t = \infty\right) = 1.$$

Proof. Consider $Y_t = t/X_t^2$. Then

$$Y_{t+1} - Y_t = \frac{t+1}{(X_t + D_{t+1})^2} - \frac{t}{X_t^2}$$

$$= \frac{t+1}{X_t^2}\left[\frac{1}{(1 + D_{t+1}/X_t)^2} - \frac{1}{1 + 1/t}\right]$$

$$\leq \frac{t+1}{X_t^2}\left[\frac{1}{t} - 2\frac{D_{t+1}}{X_t} + 3\frac{D_{t+1}^2}{X_t^2} + O\left(\frac{D_{t+1}}{X_t}\right)^3\right]$$

yielding

$$\mathbb{E}(Y_{t+1} - Y_t \mid \mathcal{F}_t) \leq \frac{1 + 1/t}{X_t^2}Q_t, \qquad (10.6)$$

where

$$Q_t = 1 - 2\rho\frac{t^{1-\beta}}{X_t^{1-\alpha}} + 3B_1^2\frac{t}{X_t^2} + O(X_t^{-3}).$$

Consider two cases:

(i) $\alpha = \beta = 1$, then $Q_t = 1 - 2\rho + 3B_1^2 \frac{t}{X_t^2} + O(X_t^{-3})$.

(ii) $(\alpha, \beta) \in \mathbf{Trans}$.

In the first case, Q_t and hence (10.6) are negative as long as $Y_t = t/X_t^2 \leq r$ for some positive constant $r < (2\rho - 1)/3B_1^2$. (Note that this would imply $X_t \geq \sqrt{t/r} \geq a$ for large enough t). Fix an arbitrary small $\varepsilon > 0$ and suppose that for some time s we have $Y_s = s/X_s^2 \leq \varepsilon r$. Let

$$\tau = \tau(s) = \inf\{t > s : Y_t \geq r\}.$$

Then $Y_{t \wedge \tau}$ is a non-negative supermartingale, hence it a.s. converges to some random limit $Y_\infty = \lim_{t \to \infty} Y_t$. By Fatou's lemma, $\mathbb{E}Y_\infty \leq Y_s \leq \varepsilon r$. On the other hand,

$$\mathbb{E}Y_\infty = \mathbb{E}(Y_\infty; \ \tau < \infty) + \mathbb{E}(Y_\infty; \ \tau = \infty) \geq r\mathbb{P}(\tau < \infty),$$

hence $\mathbb{P}(\tau < \infty) \leq \varepsilon$.

Finally, to show that for any $\varepsilon > 0$ with probability 1 there is an s such that $s/X_s^2 \leq \varepsilon r$, we apply Lemma 10.2. Consequently, $\mathbb{P}(\tau(s) = \infty$ for some $s) = 1$ yielding that $\limsup_{t \to \infty} t/X_t^2 \leq r$ a.s., and thus $\mathbb{P}(X_t \to \infty) = 1$.

Now, consider case (ii) and observe that $0 \leq \beta < 1$ and $1 - \alpha > 0$. Suppose that

$$X_t^{2-2\delta} \leq t \leq X_t^2 \quad \text{for some } \delta \in \left(0, \frac{1 + \alpha - 2\beta}{2(1 - \beta)}\right).$$

Then

$$Q_t = 1 - 2\rho \frac{t^{1-\beta}}{X_t^{1-\alpha}} \left(1 - \frac{3B_1^2}{2\rho} \frac{t^\beta}{X_t^{1+\alpha}}\right) + O(X_t^{-3}).$$

Since $t \leq X_t^2$, and $2\beta < \alpha + 1$,

$$\frac{t^\beta}{X_t^{1+\alpha}} \leq \frac{X_t^{2\beta}}{X_t^{1+\alpha}} = \frac{1}{X_t^{1+\alpha-2\beta}} = o(1),$$

therefore

$$Q(t) \le 1 - 2\rho \frac{X^{2(1-\beta)(1-\delta)}}{X_t^{1-\alpha}} (1 - o(1)) + O(X_t^{-3})$$

$$= 1 - 2\rho X^{2(1-\beta)(1-\delta)-(1-\alpha)} (1 - o(1)) < 0$$

since $2(1-\beta)(1-\delta)-(1-\alpha) > 0$ due to the choice of δ. Therefore, on the event $\{X_t^{2-2\delta} \le t \le X_t^2\}$, Y_t is a supermartingale by inequality (10.6).

Define the following sets:

$$M := \{s, x \ge 0 : x^{2-\delta} > s\},$$

$$R := \{s, x \ge 0 : x^{2-2\delta} > s\},$$

$$L := \{s, x \ge 0 : x^2 < s\}.$$

By Lemma 10.2, there will be infinitely many times s for which $s \le X_s^2/2$, so that $(s, X_s) \notin L$. Fix such an s and let

$$\tau = \tau(s) = \inf\{t > s : (t, X_t) \in L \cup M\}.$$

Then $Y^{(*)} := Y_{t \wedge \tau(s)}$ is a bounded supermartingale which a.s. converges to $Y_\infty^{(*)}$; we have $\mathbb{E} Y_\infty^{(*)} \le 1/2$ and as before obtain that on the event $\{\tau < \infty\}$, $\mathbb{P}(Y_\tau \in L) \le 1/2$ independently of s. Therefore, either $\tau(s) = \infty$ for some s which implies transience immediately or by Borel–Cantelli lemma there will be infinitely many times s for which $(s, X_s) \in M$. From now assume that the latter is the case.

Consider the sequence of stopping times when (t, X_t) crosses the curve $t = X_t^{2-\delta}$ then reaches either area L or area R before crossing this curve again. Rigorously, suppose that for some $t = \sigma_0$ we have $(t, X_t) \in M$ and it has just entered area M. Set

$$\eta_0 = \inf\{t > \sigma_0 : (t, X_t) \in L \cup R\}.$$

Then for $k \ge 0$ let

$$\sigma_{k+1} = \begin{cases} \inf\{t > \eta_k : (t, Y_t) \in M\}, & \text{if } (\eta_k, X_{\eta_k}) \in L \\ \inf\{t > \eta_k : (t, Y_t) \notin M\}, & \text{if } (\eta_k, X_{\eta_k}) \in R, \end{cases}$$

$$\eta_{k+1} = \inf\{t > \sigma_{k+1} : (t, Y_t) \in L \cup R\}.$$

We thus have

$$\sigma_0 < \eta_0 < \sigma_1 < \eta_1 < \sigma_2 < \eta_2 \ldots.$$

It could, of course, happen that one of these stopping times is infinite, hence all the remaining ones too; however this would imply that $(t, X_t) \notin L$ for all large t, which would in turn imply transience (recall though that we assumed that we visit the area M infinitely often). Let us therefore assume from now on that all η_k's and σ_k's are finite.

For $t \geq \sigma_k$, $k \geq 0$, consider a supermartingale $Y^{(k)} = Y_{t \wedge \eta_k}$. Since the jumps of X_t are bounded, $X_{\sigma_k} = \sigma_k^{1/(2-\delta)} + O(1)$ and $\mathbb{E} Y_{\eta_k} \leq Y_{\sigma_k} = X_{\sigma_k}^{-\delta}(1 + o(1))$ and as before, we obtain that

$$\mathbb{P}\left((\eta_k, X_{\eta_k}) \in L \mid \mathcal{F}_{\sigma_k}\right) \leq X_{\sigma_k}^{-\delta}(1 + o(1)) = \frac{1 + o(1)}{\sigma_k^{\delta/(2-\delta)}}. \quad (10.7)$$

On the other hand, starting at (σ_k, X_{σ_k}) it takes a lot of time for (t, X_t) to reach L, and also if $(\eta_k, X_{\eta_k}) \in R$, it takes a lot of time to exit M since the walk has to go against the drift. More precisely,

$$\sigma_{k+1} - \sigma_k \geq (\eta_k - \sigma_k) 1_{(\eta_k, X_{\eta_k}) \in L} + (\sigma_{k+1} - \eta_k) 1_{(\eta_k, X_{\eta_k}) \in R}. \quad (10.8)$$

Set $x := X_{\sigma_k}$,

$$h = \frac{1}{2(2B_1 + 1)^2} < \frac{1}{8B_1^2},$$

and observe that since $2hx^2 - x^{2-\delta} > hx^2$

$$\left\{ \inf_{0 \leq i \leq hx^2} X_{\sigma_k+i} \geq x\sqrt{2h} \right\} \subseteq \{(\sigma_k + i, X_{\sigma_k+i}) \notin L \; \forall i \in [0, hx^2]\}. \quad (10.9)$$

By Lemma 10.4, the probability of the LHS of (10.9) is larger than

$$1 - \frac{4hB_1^2}{(1 - \sqrt{2h})^2} = \frac{1}{2}.$$

Similarly, when $(\eta_k, X_{\eta_k}) \in R$, set $y := X_{\eta_k} > x^{\frac{2-\delta}{2-2\delta}}$. Since

$$(y-x)^{2-\delta} - x^{2-\delta} > x^{\frac{(2-\delta)^2}{2-2\delta}}(1+o(1)) - x^{2-\delta} = x^{2 + \frac{\delta^2}{2-2\delta}}(1+o(1)) \gg x^2,$$

we have

$$\left\{ \inf_{0 \le i \le \frac{x^2}{8B_1^2}} X_{\eta_k+i} \ge y - x \right\} \subseteq \left\{ (\eta_k + i, X_{\eta_k+i}) \in M \;\; \forall i \in \left[0, \frac{x^2}{8B_1^2} \right] \right\}.$$

(10.10)

By Lemma 10.4, the probability of the LHS of (10.10) is also less than $1/2$. Therefore, since $a < 1/(8B_1^2)$, from (10.8) we obtain

$$\mathbb{P} \left(\sigma_{k+1} - \sigma_k \ge a X_{\sigma_k}^2 \right) > \frac{1}{2}.$$

On the other hand, provided σ_k is large enough,

$$a X_{\sigma_k}^2 = a \sigma_k^{\frac{2}{2-\delta}} (1 + o(1)) > 3\sigma_k,$$

so with probability strictly larger than $1/2$, independently of the past, we have $\sigma_{k+1} \ge 4\sigma_k$ (and in any case, $\sigma_{k+1} \ge \sigma_k$). By the strong law of large numbers, $\operatorname{card}\{k \in \{1, 2, \ldots, K\} : \; \sigma_k \ge 4\sigma_{k-1}\} \ge K/2$ for all large K, yielding that for some $C_1 > 0$

$$\sigma_k \ge C_1 4^{k/2} = C_1 2^k \quad \text{for all large } k.$$

Consequently, the probability in (10.7) is bounded by

$$\frac{1 + o(1)}{(C_1 2^k)^{\frac{\delta}{2-\delta}}},$$

which is summable over k. By the Borel–Cantelli lemma, only finitely many events $\{(\eta_k, X_{\eta_k}) \in L\}$ occur, or, equivalently, for large times $(t, X_t) \notin L$. This yields transience. $\qquad \square$

10.4   Recurrence

We now turn our attention to the recurrence of the process.

Theorem 10.2. *Consider a Markov process X_t, $t = 0, 1, 2, \ldots$, on $\mathbb{R}_+$ with increments $D_t = X_t - X_{t-1}$, satisfying (H1) and (H2) for some $a > 0$. Suppose that on the event $\{X_t \ge a\}$ either*

(i) *for some $\rho < 1/2$*

$$\mathbb{E}(D_{t+1} \mid \mathcal{F}_t) \le \frac{\rho X_t}{t}$$

or

(ii) *for some $\rho > 0$ and $(\alpha, \beta) \in$* **Rec**

$$\mathbb{E}(D_{t+1} \mid \mathcal{F}_t) \le \frac{\rho X_t^\alpha}{t^\beta}.$$

Then X_t is "recurrent" in the sense that for any starting point $X_0 = x$ we have

$$\mathbb{P}(\exists\, t \ge 0 \ such\ that\ X_t < a) = 1.$$

Hence, also $\mathbb{P}(X_t < a$ infinitely often$) = 1$.

Proof. Let $Y_t = X_t^2 / t \ge 0$ and assume $X_t \ge a$. Then

$$Y_{t+1} - Y_t = \frac{(X_t + D_{t+1})^2}{t+1} - \frac{X_t^2}{t} = \frac{2t X_t D_{t+1} - X_t^2 + t D_{t+1}^2}{t(t+1)}$$

whence

$$(t+1)\mathbb{E}(Y_{t+1} - Y_t \mid \mathcal{F}_t) = \mathbb{E}(D_{t+1}^2 \mid \mathcal{F}_t) + [2 X_t \mathbb{E}(D_{t+1} \mid \mathcal{F}_t) - X_t^2/t]$$

$$\le B_1^2 + (2\rho \kappa_t - 1)\, X_t^2/t$$

$$\le B_1^2 - (1 - 2\rho \kappa_t)\, Y_t, \tag{10.11}$$

where

$$\kappa_t = \frac{X_t^{\alpha-1}}{t^{\beta-1}}.$$

Consider the following three cases:

(i) $\alpha = \beta = 1$, then $\kappa_t = 1$.

(ii-a) $\beta > \alpha \ge 1$, then since $X_t \le B_1 t$, we have $\kappa_t \le B_1^\alpha / t^{\beta-\alpha} \to 0$ as $t \to \infty$.

(ii-b) $\alpha < 1$, $\beta > (\alpha+1)/2$, then whenever $Y_t \ge 1$ we have $X_t^{2(\alpha-1)} \le t^{\alpha-1}$ and

$$\kappa_t \le \frac{t^{(\alpha-1)/2}}{t^{\beta-1}} = \frac{1}{t^{\beta-(\alpha+1)/2}} \to 0 \quad as\ t \to \infty.$$

(Note also that (ii-a) and (ii-b) together cover the set **Rec**.)

In all the cases, we can find $t_0 > 0$ such that $2\rho\kappa_t < 1$ for all $t \geq t_0$ and thus by (10.11)

$$\mathbb{E}(Y_{t+1} - Y_t \,|\, \mathcal{F}_t) \leq 0 \tag{10.12}$$

(under the assumption that $X_t^2 \geq t$ in case (ii-b)). Let $s \geq t_0$, and set $\tau(s) = \inf\{t \geq s : Y_t \leq r\}$. Equation (10.12) yields that for each s, $\{Y_{t \wedge \tau(s)}, \ t \geq s\}$ is a supermartingale, therefore a.s. there is a limit $Y_\infty = \lim_{t \to \infty} Y_{t \wedge \tau(s)}$.

Suppose that for some $s \geq t_0$ the event $\{\tau(s) = \infty\}$ occurs. Then $X_t^2 / t \to Y_\infty$ a.s., and hence $X_t^2 < (1 + Y_\infty)t$ for all sufficiently large t. On the other hand, if $\tau(s) < \infty$ for *all* $s \geq t_0$, there will be an infinite sequence of times $t_1, t_2, \ldots$ such that $X_{t_k}^2 \leq t_k$. We conclude that in both cases, there is a possibly random value Z such that $X_t^2 \leq Zt$ for infinitely many times $t_k \in \mathbb{Z}_+$, $k = 1, 2, \ldots$.

First, suppose that $\alpha \geq 0$. Then for a fixed t_k, define a process $X_t' = X_{t+t_k}$. Set $n = 2X_{t_k}$ and $\gamma = 1/2$, and observe that the process X_t' satisfies the conditions of Lemma 10.3 with some $c = c(2\beta - \alpha, \rho, Z) > 0$. Indeed, when $X_t \leq n$, the drift of X_t is at most of order $n^\alpha / t^\beta \sim 1/n^{2\beta - \alpha} \leq 1/n$ since $2\beta - \alpha \geq 1$ and $\alpha \geq 0$. Hence, there is a constant $\nu > 0$, independent of t_k, such that

$$\mathbb{P}(X_t, \ t \geq t_k, \ \text{reaches } [0, a] \text{ before } [n, \infty) \,|\, \mathcal{F}_{t_k}) \geq \nu.$$

Therefore, by the second Borel–Cantelli lemma (Durrett, 1996, p. 240) $\{X_t \leq a\}$ for infinitely many t's.

Suppose now that $\alpha < 0$. Consider $W_t = X_t^{1-\nu}$ for some $0 < \nu < 1$. Then

$$\mathbb{E}(W_{t+1} - W_t \,|\, \mathcal{F}_t) = X_t^{1-\nu} \mathbb{E}\left((1 + D_{t+1}/X_t)^{1-\nu} - 1 \,|\, \mathcal{F}_t\right)$$

$$= (1 - \nu) X_t^{1-\nu} \mathbb{E}\left(\frac{D_{t+1}}{X_t} - \frac{\nu}{2}\frac{D_{t+1}^2}{X_t^2} + O(X_t^{-3}) \,|\, \mathcal{F}_t\right)$$

$$\leq (1 - \nu) X_t^{-1-\nu}\left(\frac{X_t^{1+\alpha}}{t^\beta} - \frac{\nu B_2}{2} + O(X_t^{-3})\right). \tag{10.13}$$

Let $n = n_k = \sqrt{Zt_k}$ so that $X_{t_k} \leq n$. Since $2\beta > 1 + \alpha$, we can fix an $\zeta > 1$ such that

$$2\beta > \zeta(1 + \alpha).$$

Consider the process W_t for $t \in [t_k, \eta]$, where

$$\eta = \eta_k := \inf\{t \geq t_k : X_t \leq a \text{ or } X_t \geq n^\varsigma\}.$$

Then $\eta < \infty$ a.s. from the same argument as in part (i) of Lemma 10.3.

Moreover, for $t \in [t_k, \eta]$,

$$\frac{X_t^{1+\alpha}}{t^\beta} \leq \frac{(n^\varsigma)^{1+\alpha}}{(n^2/Z)^\beta} = \frac{Z^\beta}{n^{2\beta - \varsigma(1+\alpha)}} \to 0 \text{ as } k \to \infty$$

since $t_k \to \infty$ and hence $n_k \to \infty$. Therefore, the right-hand side of (10.13) is negative and thus $W_{t \wedge \eta}$ is a supermartingale. Consequently, by the optional stopping theorem,

$$n^{1-\nu} \geq X_{t_k}^{1-\nu} = W_{t_k} \geq \mathbb{E}(W_\eta \,|\, \mathcal{F}_{t_k}) \geq (1 - p)(n^\varsigma)^{1-\nu},$$

where $p = p_k = \mathbb{P}(X_\eta \leq a \,|\, \mathcal{F}_{t_k})$. This implies

$$p_k \geq 1 - \frac{1}{n_k^{(\varsigma-1)(1-\nu)}} \to 1 \text{ as } k \to \infty$$

finishing the proof of the theorem. $\square$

10.5 Some Particular Cases

We are now going to discuss some particular choices of the parameters α, β.

10.5.1 Case $\alpha = \beta \geq 0$

Since we can always rescale the process X_t by a positive constant, in this section, we assume that $B_1 = 1$. Then, in turn, it is also reasonable to restrict our attention only to the case $\rho \leq 1$, since if the jumps of X_t can be indeed close to 1 with a positive probability, we might have $X \approx t$, and the drift of order $\rho(X/t)^\beta$ with $\rho > 1$ would imply that the drift is in fact larger than $1 = B_1$ leading to a contradiction, so the model would not be properly defined.

Theorem 10.3 ($\alpha = \beta < 1$). *Consider a Markov process X_t, $t = 0, 1, 2, \ldots$, on $\mathbb{R}_+$ with increments $D_t = X_t - X_{t-1}$, satisfying (H1),*

(H2) *and* (H3) *for some* $a > 0$. *Assume that for some* $\beta < 1$ *and* $\rho \in (0, 1]$ *on the event* $\{X_t \geq a\}$

$$\mathbb{E}(D_{t+1} \mid \mathcal{F}_t) \geq \rho \left(\frac{X_t}{t} \right)^{\beta}.$$

Then X_t *is transient.*

Proof. The proof is identical to the proof of Theorem 10.1, case (ii).
□

Theorem 10.4 ($\alpha = \beta > 1$). *Consider a Markov process* X_t, $t = 0, 1, 2, \ldots$, *on* $\mathbb{R}_+$ *with increments* $D_t = X_t - X_{t-1}$, *satisfying* (H1) *and* (H2) *for some* $a > 0$. *Assume that for some* $\beta > 1$ *and* $\rho < 1$, *on the event* $\{X_t \geq a\}$, *the process* X *satisfies*

$$\mathbb{E}(D_{t+1} \mid \mathcal{F}_t) \leq \rho \left(\frac{X_t}{t} \right)^{\beta}.$$

Then X_t *is recurrent.*

Proof. Fix $\zeta \in (\rho, 1)$ and consider $Y_t = X_t / t^{\zeta}$. Then, calculating as before, we obtain

$$\mathbb{E}(Y_{t+1} - Y_t \mid \mathcal{F}_t) \leq \frac{X_t}{t(t+1)^{\zeta}} \left(\rho(X_t/t)^{\beta-1} - \zeta \right).$$

Since $\limsup X_t/t \leq B_1 = 1$ and $\rho < \zeta$, for large t this is negative and hence Y_t is a non-negative supermartingale converging almost sure. On the other hand, $\zeta < 1$, thus implying

$$\lim_{t \to \infty} \frac{X_t}{t} = \lim_{t \to \infty} \frac{Y_t}{t^{1-\zeta}} = 0 \text{ a.s.}$$

and consequently since $\beta > 1$ for some sufficiently large t we have $(X_t/t)^{\beta-1} < 1/4$. Therefore, for large t,

$$\mathbb{E}(D_{t+1} \mid \mathcal{F}_t) \leq \rho \left(\frac{X_t}{t} \right)^{\beta-1} \times \frac{X_t}{t} \leq \frac{1}{4} \frac{X_t}{t},$$

and hence X_t is recurrent by Theorem 10.2.
□

The following statement immediately follows from Theorems 10.1 and 10.2.

Corollary 10.1 ($\alpha = \beta = 1$). *Assume that X_t is a process satisfying* (H1), (H2) *and* (H3) *for some $a > 0$:*

(i) *If for some $\rho < 1/2$,*

$$\mathbb{E}(D_{t+1} \,|\, \mathcal{F}_t) \le \frac{\rho X_t}{t} \quad \text{when } X_t \ge a,$$

then X_t is recurrent.

(ii) *If for some $\rho > 1/2$,*

$$\mathbb{E}(D_{t+1} \,|\, \mathcal{F}_t) \ge \frac{\rho X_t}{t} \quad \text{when } X_t \ge a,$$

then X_t is transient.

10.5.2 Case $\alpha \le 0$, $\beta = 0$

In this case, the drift is of order ρ/X_t^ν, where $\nu = -\alpha \ge 0$. This is the situation resolved by Lamperti (1960, 1963).

Theorem 10.5 ($\alpha = -1$, $\beta = 0$). *Suppose that X_t is a process satisfying* (H1) *and* (H2) *for some $a > 0$. Then, when $X_t \ge a$,*

(i) *if for some $\rho \le 1/2$,*

$$\mathbb{E}(D_{t+1} \,|\, \mathcal{F}_t) \le \rho \, \frac{\mathbb{E}(D_{t+1}^2 \,|\, \mathcal{F}_t)}{X_t}$$

holds, then X_t is recurrent,

(ii) *if for some $\rho > 1/2$,*

$$\mathbb{E}(D_{t+1} \,|\, \mathcal{F}_t) \ge \rho \, \frac{\mathbb{E}(D_{t+1}^2 \,|\, \mathcal{F}_t)}{X_t}$$

holds, then X_t is transient.

Corollary 10.2 ($\alpha \in (-\infty, -1) \cup (-1, 0)$, $\beta = 0$). *Assume that X_t is a process satisfying* (H1) *and* (H2) *for some $a > 0$. Then, when $X_t \ge a$,*

(i) *if for some $\nu > 1$,*

$$\mathbb{E}(D_{t+1} \,|\, \mathcal{F}_t) \le \frac{\rho}{X_t^\nu}$$

holds, then X_t is recurrent,

(ii) *if for some $\nu < 1$,*

$$\mathbb{E}(D_{t+1} \mid \mathcal{F}_t) \geq \frac{\rho}{X_t^\nu}$$

holds, then X_t is transient.

10.5.3 Case $2\beta - \alpha = 1$, $-1 \leq \alpha \leq 1$: open problem

Two cases, namely, $\alpha = \beta = 1$ and $\alpha = -1$, $\beta = 0$, are already covered. It is also straightforward to see that when $\alpha = 0$, $\beta = 1/2$, by the law of iterated logarithm, the process is recurrent for any ρ.

Cases $-1 < \alpha < 0$, $\beta = \frac{1}{2}(\alpha + 1)$ and $0 < \alpha < 1$, $\beta = \frac{1}{2}(\alpha + 1)$: unfortunately, we cannot find a general sensible criteria to separate recurrence and transience here and leave this as an **open problem**.

10.6 Application to Urn Models

We close the chapter by discussing the application of the above to certain urn models.

Fix a constant $\sigma > 0$. Consider a Friedman-type urn process (W_n, B_n), $n \geq 1$, with the following properties. We choose a white ball with probability $W_n/(W_n + B_n)$ and a black ball with a complementary probability; whenever we draw a white (black resp.) ball, we add a random number A of white (black resp.) balls and $\sigma - A$ black (white resp.) balls. For simplicity, suppose $0 \leq A \leq \sigma$ a.s. A special case when A is not random is considered in Freedman (1965). Following his notations, let $\alpha = \mathbb{E}A$, $\beta = \sigma - \alpha$ and $\rho = (\alpha - \beta)/\sigma = (\alpha - \beta)/(\alpha + \beta)$. In addition, assume that $\alpha > \beta > 0$.

It turns out that this urn can be coupled with a random walk described above. Indeed, for $t = 0, 1, 2, \ldots$, set $X_t = |W_t - B_t|/(\beta - \alpha) \in \mathbb{Z}_+ \subset \mathbb{R}_+$. Without much loss of generality, assume that the process starts at time $(W_0 + B_0)/\sigma \in \mathbb{Z}$, then $t = (W_t + B_t)/\sigma \in \mathbb{Z}$. Consequently, once $X_t \neq 0$ holds, we obtain that

$$\mathbb{E}(X_{t+1} - X_t \mid \mathcal{F}_t) = \frac{1}{2}\left(1 + \frac{(\beta - \alpha)X_t}{\sigma t}\right)(+1)$$

$$+ \frac{1}{2}\left(1 - \frac{(\beta - \alpha)X_t}{\sigma t}\right)(-1) = \frac{\rho X_t}{t}.$$

Corollary 3.3 in Freedman (1965) states that when $\rho > 1/2$, $W_n - B_n = W_0 - B_0$ (equivalently, $X_n = 0$) occurs for finitely many n with a *positive probability*, and after the corollary, the author says that he does not know whether this event occurs a.s. On the other hand, our Theorem 10.2 answers this question in the affirmative — indeed, a.s. there will be finitely many times when the difference between the number of white and black balls in the urn equals a particular constant.

In conclusion, we mention that the interested reader will find more material on urn models in Janson (2004) as well as in Pemantle (2007).

Bibliography

Abramowitz, M. and Stegun, I.A. (1964). *Handbook of mathematical functions with formulas, graphs, and mathematical tables, National Bureau of Standards Applied Mathematics Series*, Vol. 55 (For sale by the Superintendent of Documents, U.S. Government Printing Office, Washington, DC).

Arratia, R., Barbour, A.D., and Tavaré, S. (1999). The Poisson-Dirichlet distribution and the scale-invariant Poisson process, *Combin. Probab. Comput.* **8**, 5, pp. 407–416.

Aspandiiarov, S., Iasnogorodski, R., and Menshikov, M. (1996). Passage-time moments for nonnegative stochastic processes and an application to reflected random walks in a quadrant, *Ann. Probab.* **24**, 2, pp. 932–960.

Benaïm, M., Bouguet, F., and Cloez, B. (2017). Ergodicity of inhomogeneous Markov chains through asymptotic pseudotrajectories, *Ann. Appl. Probab.* **27**, 5, pp. 3004–3049.

Billingsley, P. (1999). *Convergence of probability measures*, 2nd edn., Wiley Series in Probability and Statistics: Probability and Statistics (John Wiley & Sons, Inc., New York), doi:10.1002/9780470316962, https://doi.org/10.1002/9780470316962, a Wiley-Interscience Publication.

Bouguet, F. and Cloez, B. (2018). Fluctuations of the empirical measure of freezing Markov chains, *Electron. J. Probab.* **23**, pp. Paper No. 2, 31.

Bowman, F. (1958). *Introduction to Bessel functions* (Dover Publications Inc., New York).

Breiman, L. (1992). *Probability, Classics in Applied Mathematics*, Vol. 7 (Society for Industrial and Applied Mathematics (SIAM), Philadelphia, PA), corrected reprint of the 1968 original.

Cénac, P., de Loynes, B., Offret, Y., and Rousselle, A. (2020). Recurrence of multidimensional persistent random walks. Fourier and series criteria, *Bernoulli* **26**, 2, pp. 858–892.

Cénac, P., Le Ny, A., de Loynes, B., and Offret, Y. (2018). Persistent random walks. I. Recurrence versus transience, *J. Theoret. Probab.* **31**, 1, pp. 232–243.

Chen, A.Y. and Renshaw, E. (1994). The general correlated random walk, *J. Appl. Probab.* **31**, 4, pp. 869–884.

Davydov, J.A. (1970). The invariance principle for stationary processes, *Teor. Verojatnost. i Primenen.* **15**, pp. 498–509.

Davydov, J.A. (1973). Mixing conditions for Markov chains, *Teor. Verojatnost. i Primenen.* **18**, pp. 321–338.

Di Crescenzo, A. (2002). Exact transient analysis of a planar random motion with three directions, *Stoch. Stoch. Rep.* **72**, 3–4, pp. 175–189.

Dietz, Z. and Sethuraman, S. (2007). Occupation laws for some time-nonhomogeneous markov chains. *Electron. J. Probab.* **12**, 23, pp. 661–683.

Dobrušin, R.L. (1956). Central limit theorem for nonstationary Markov chains. II, *Teor. Veroyatnost. i Primenen.* **1**, pp. 365–425.

Dolgopyat, D. and Sarig, O. M. (2023). *Local limit theorems for inhomogeneous Markov chains, Lecture Notes in Mathematics*, Vol. 2331 (Springer, Cham).

Doyle, P.G. and Snell, J.L. (1984). *Random walks and electric networks, Carus Mathematical Monographs*, Vol. 22 (Mathematical Association of America, Washington, DC).

Drogin, R. (1972). An invariance principle for martingales, *Ann. Math. Statist.* **43**, pp. 602–620.

Durrett, R. (1996). *Probability: theory and examples*, 2nd edn. (Duxbury Press, Belmont, CA).

Dynkin, E.B. and Yushkevich, A.A. (1969). *Markov processes: Theorems and problems* (Plenum Press, New York), translated from the Russian by James S. Wood.

Engländer, J. and Volkov, S. (2018). Turning a coin over instead of tossing it, *J. Theoret. Probab.* **31**, 2, pp. 1097–1118.

Engländer, J. and Volkov, S. (2022). Conservative random walk, *Electron. J. Probab.* **27**, pp. Paper No. 138, 29.

Engländer, J., Volkov, S., and Wang, Z. (2020). The coin-turning walk and its scaling limit, *Electron. J. Probab.* **25**, pp. Paper No. 3, 38.

Erdős, P. (1965). Extremal problems in number theory, in *Proc. Sympos. Pure Math., Vol. VIII* (Amer. Math. Soc., Providence, RI), pp. 181–189.

Fayolle, G., Malyshev, V.A., and Menshikov, M.V. (1995). *Topics in the constructive theory of countable Markov chains* (Cambridge University Press, Cambridge).

Freedman, D.A. (1965). Bernard Friedman's urn, *Ann. Math. Statist.* **36**, pp. 956–970.

Freedman, D.A. (1975). On tail probabilities for martingales, *Ann. Probability* **3**, pp. 100–118.

Fristedt, B. and Gray, L. (1997). *A modern approach to probability theory*, Probability and its Applications (Birkhäuser Boston, Inc., Boston, MA).

Gantert, N. (1990). Laws of large numbers for the annealing algorithm, *Stochastic Process. Appl.* **35**, 2, pp. 309–313.

Goldstein, S. (1951). On diffusion by discontinuous movements, and on the telegraph equation, *Quart. J. Mech. Appl. Math.* **4**, pp. 129–156.

Goodman, G.S. (1974). The existence of intensities of countable state nonstationary Markov transition functions, *Ann. of Math. (2)* **99**, pp. 545–552.

Grimmett, G.R., Menshikov, M.V., and Volkov, S.E. (1996). Random walks in random labyrinths, *Markov Process. Related Fields* **2**, 1, pp. 69–86, disordered systems and statistical physics: Rigorous results (Budapest, 1995).

Gut, A. (2005). *Probability: a graduate course*, Springer Texts in Statistics (Springer, New York).

Janson, S. (2004). Functional limit theorems for multitype branching processes and generalized Pólya urns, *Stochastic Process. Appl.* **110**, 2, pp. 177–245.

Jones, G.L. (2004). On the Markov chain central limit theorem, *Probab. Surv.* **1**, pp. 299–320.

Kac, M. (1974). A stochastic model related to the telegrapher's equation, *Rocky Mountain J. Math.* **4**, pp. 497–509, reprinting of an article published in 1956.

Kallenberg, O. (2017). *Random measures, theory and applications*, Probability Theory and Stochastic Modelling, Vol. 77 (Springer, Cham).

Karatzas, I. and Shreve, S.E. (1991). *Brownian motion and stochastic calculus*, Graduate Texts in Mathematics, Vol. 113, 2nd edn. (Springer-Verlag, New York).

Keller, N. and Klein, O. (2022). Proof of Tomaszewski's conjecture on randomly signed sums, *Adv. Math.* **407**, pp. Paper No. 108558, 39.

Kesten, H. (1969). A sharper form of the Doeblin-Lévy-Kolmogorov-Rogozin inequality for concentration functions, *Math. Scand.* **25**, pp. 133–144.

Kolesnik, A.D. (2021). *Markov random flights*, Monographs and Research Notes in Mathematics (CRC Press, Boca Raton, FL).

Lamperti, J. (1960). Criteria for the recurrence or transience of stochastic process. I, *J. Math. Anal. Appl.* **1**, pp. 314–330.

Lamperti, J. (1963). Criteria for stochastic processes. II. Passage-time moments, *J. Math. Anal. Appl.* **7**, pp. 127–145.

Lawler, G.F. and Limic, V. (2010). *Random walk: a modern introduction*, Cambridge Studies in Advanced Mathematics, Vol. 123 (Cambridge University Press, Cambridge).

Levin, D.A. and Peres, Y. (2017). *Markov chains and mixing times* (American Mathematical Society, Providence, RI), second edition of [MR2466937], With contributions by Elizabeth L. Wilmer, With a chapter on "Coupling from the past" by James G. Propp and David B. Wilson.

Lyons, R. (1988). Strong laws of large numbers for weakly correlated random variables, *Michigan Math. J.* **35**, 3, pp. 353–359.

Mahmoud, H.M. (2009). *Pólya urn models*, Texts in Statistical Science Series (CRC Press, Boca Raton, FL).

Mazliak, L. and Shafer, G. (eds.) (2022). *The splendors and miseries of martingales—their history from the casino to mathematics*, Trends in the History of Science (Birkhäuser/Springer, Cham).

Menshikov, M., Popov, S., and Wade, A. (2017). *Non-homogeneous random walks*, Cambridge Tracts in Mathematics, Vol. 209 (Cambridge University Press, Cambridge).

Menshikov, M. and Zuyev, S. (2001). Polling systems in the critical regime, *Stochastic Process. Appl.* **92**, 2, pp. 201–218.

Meyer, C.D. (2023). *Matrix analysis and applied linear algebra*, 2nd edn. (Society for Industrial and Applied Mathematics (SIAM), Philadelphia, PA).

Montgomery-Smith, S.J. (1990). The distribution of Rademacher sums, *Proc. Amer. Math. Soc.* **109**, 2, pp. 517–522.

Nguyen, H.H. (2012). A new approach to an old problem of Erdos and Moser, *J. Combin. Theory Ser. A* **119**, 5, pp. 977–993.

Orsingher, E. and Ratanov, N. (2002). Planar random motions with drift, *J. Appl. Math. Stochastic Anal.* **15**, 3, pp. 205–221.

Peligrad, M. (2012). Central limit theorem for triangular arrays of non-homogeneous Markov chains, *Probab. Theory Related Fields* **154**, 3–4, pp. 409–428.

Pemantle, R. (2007). A survey of random processes with reinforcement, *Probab. Surv.* **4**, pp. 1–79.

Pólya, G. and Szegő, G. (1998). *Problems and theorems in analysis. I*, Classics in Mathematics (Springer-Verlag, Berlin), series, integral calculus, theory of functions, Translated from the German by Dorothee Aeppli, Reprint of the 1978 English translation.

Proctor, R.A. (1982). Solution of two difficult combinatorial problems with linear algebra, *Amer. Math. Monthly* **89**, 10, pp. 721–734.

Ratanov, N. and Kolesnik, A.D. (2022). *Telegraph processes and option pricing*, 2nd edn. (Springer-Verlag, Berlin).

Rényi, A. (1970). *Foundations of probability* (Holden-Day, Inc., San Francisco, Calif.-London-Amsterdam).

Rogozin, B.A. (1961). On the increase of dispersion of sums of independent random variables, *Teor. Verojatnost. i Primenen.* **6**, pp. 106–108.

Saloff-Coste, L. and Zúñiga, J. (2007). Convergence of some time inhomogeneous Markov chains via spectral techniques, *Stochastic Process. Appl.* **117**, 8, pp. 961–979.

Saloff-Coste, L. and Zúñiga, J. (2009). Merging for time inhomogeneous finite Markov chains. I. Singular values and stability, *Electron. J. Probab.* **14**, pp. 1456–1494.

Saloff-Coste, L. and Zúñiga, J. (2011). Merging and stability for time inhomogeneous finite Markov chains, in *Surveys in stochastic processes*, EMS Ser. Congr. Rep. (Eur. Math. Soc., Zürich), pp. 127–151.

Sárközi, A. and Szemerédi, E. (1965). Uber ein Problem von Erdos und Moser, *Acta Arith.* **11**, pp. 205–208.

Seneta, E. (2006). *Non-negative matrices and Markov chains*, Springer Series in Statistics (Springer, New York), revised reprint of the second (1981) edition.

Seneta, E. (2014). Inhomogeneous Markov chains and ergodicity coefficients: John Hajnal (1924–2008), *Comm. Statist. Theory Methods* **43**, 7, pp. 1296–1308.

Sethuraman, S. and Varadhan, S.R.S. (2005). A martingale proof of Dobrushin's theorem for non-homogeneous Markov chains, *Electron. J. Probab.* **10**, 36, pp. 1221–1235.

Shohat, J.A. and Tamarkin, J.D. (1943). *The Problem of Moments*, American Mathematical Society Mathematical Surveys, Vol. I (American Mathematical Society, New York).

Sinaĭ, Y.G. (1982). The limit behavior of a one-dimensional random walk in a random environment, *Teor. Veroyatnost. i Primenen.* **27**, 2, pp. 247–258.

Solomon, F. (1975). Random walks in a random environment, *Ann. Probability* **3**, pp. 1–31.

Spitzer, F. (1976). *Principles of random walk*, 2nd edn. (Springer-Verlag, New York, Heidelberg).

Stanley, R.P. (1980). Weyl groups, the hard Lefschetz theorem, and the Sperner property, *SIAM J. Algebraic Discrete Methods* **1**, 2, pp. 168–184.

Stroock, D.W. (2011). *Probability theory*, 2nd edn. (Cambridge University Press, Cambridge), an analytic view.

Sullivan, B.D. (2013). On a conjecture of Andrica and Tomescu, *J. Integer Seq.* **16**, 3, Article 13.3.1, p. 6.

Szász, D. and Tóth, B. (1984). Persistent random walks in a one-dimensional random environment, *J. Statist. Phys.* **37**, 1–2, pp. 27–38.

Tang, W. and Tang, F. (2023). The Poisson binomial distribution—old & new, *Statist. Sci.* **38**, 1, pp. 108–119.

Tao, T. and Vu, V.H. (2009). Inverse Littlewood-Offord theorems and the condition number of random discrete matrices, *Ann. of Math. (2)* **169**, 2, pp. 595–632.

Taylor, H.M. and Karlin, S. (1998). *An introduction to stochastic modeling*, 3rd edn. (Academic Press, Inc., San Diego, CA).

Vasdekis, G. and Roberts, G.O. (2023). Speed up zig-zag, *Ann. Appl. Probab.* **33**, 6A, pp. 4693–4746.

Zúñiga, J.V. (2008). *Merging of some time homogeneous and inhomogeneous Markov chains* (ProQuest LLC, Ann Arbor, MI), thesis (Ph.D.)–Cornell University.

Index

www.ingramcontent.com/pod-product-compliance
Lightning Source LLC
Chambersburg PA
CBHW050544190326
41458CB00007B/1905

More from Kogan Page

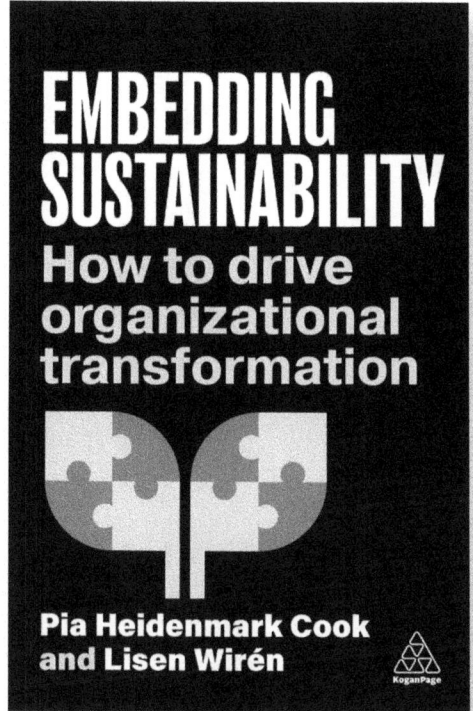

www.koganpage.com

KoganPage

From 4 December 2025 the EU Responsible Person (GPSR) is:
eucomply oÜ, Pärnu mnt. 139b – 14, 11317 Tallinn, Estonia
www.eucompliancepartner.com

9 781398 618527